SOLID WASTE MANAGEMENT

Technology Assessment

SOLID WASTE MANAGEMENT
Technology Assessment

by General Electric Company

Van Nostrand Reinhold-General Electric Series

VNR **VAN NOSTRAND REINHOLD COMPANY**
NEW YORK   CINCINNATI   ATLANTA   DALLAS   SAN FRANCISCO
LONDON   TORONTO   MELBOURNE

Van Nostrand Reinhold Company Regional Offices:
New York Cincinnati Atlanta Dallas San Francisco

Van Nostrand Reinhold Company International Offices:
London Toronto Melbourne

Copyright © 1975 by the General Electric Company

Library of Congress Catalog Card Number: 76-45815
ISBN: 0-442-22648-9

Manufactured in the United States of America

Published by Van Nostrand Reinhold Company
450 West 33rd Street, New York, N.Y. 10001

Published simultaneously in Canada by Van Nostrand Reinhold Ltd.

15 14 13 12 11 10 9 8 7 6 5 4 3 2

Library of Congress Cataloging in Publication Data

General Electric Company.
    Solid waste management.

    "Van Nostrand Reinhold-General Electric series."
    Bibliography: p.
    Includes index.
        1. Refuse and refuse disposal. I. Title.
TD791.G46   1976   628'.3   76-45815
ISBN 0-442-22648-9

FOREWORD

This book contains a thorough analysis of all of the methods for processing and disposing of solid waste currently in use or being considered. It includes not only the conventional disposal processes, such as landfill, composting, and incineration, but also the newer resource recovery technologies, such as pyrolysis, which are now emerging from the laboratory. In addition to analysis of complete processes, discussions pertaining to the merits of key pieces of equipment, such as shredders, air separators, and conveyors, are also included.

The analysis includes a description of the technical features of each process and how it works, a review of operating history and experience to date, and estimates of both capital and operating costs, as well as scale-up considerations. Numerous tables are provided that allow easy comparison of competing processes in terms of net energy recovered, effluents, and the weight of materials to be landfilled.

This book should be of interest to anyone having responsibility for solid waste disposal as well as those considering the installation of new solid waste processing equipment. It should be particularly helpful to solid waste officials who are responsible for the planning and procurement of solid waste facilities but who too often have neither the time nor the resources to conduct their own comparative study to arrive at the best, environmentally sound, solid waste system.

This title, published in 1973, was updated in November 1974 to include information pertaining to processes most promising to solid waste management. It reports an objective, broad based evaluation of existing and emerging technology. It was tailored by the General Electric Company to support the parallel Connecticut Solid Waste Management Study conducted in 1972.

The material has been directed at the known solid waste processes or disposal techniques, and has been organized as follows:

    Comparison of Resource Recovery Strategies
    Refuse Fuel Recovery Processes
    Pyrolysis Processes
    Composting Processes
    Hybrid System Design
    Sanitary Landfills
    Baling and Balefills
    Facilities and Components

TABLE OF CONTENTS

TABLE OF CONTENTS (Cont'd)

TABLE OF CONTENTS (Cont'd)

TABLE OF CONTENTS (Cont'd)

TABLE OF CONTENTS (Cont'd)

LIST OF ILLUSTRATIONS

LIST OF ILLUSTRATIONS (Cont'd)

LIST OF ILLUSTRATIONS (Cont'd)

LIST OF TABLES

LIST OF TABLES (Cont'd)

LIST OF TABLES (Cont'd)

LIST OF TABLES (Cont'd)

SOLID WASTE
MANAGEMENT
Technology Assessment

Section 1

SUMMARY

The Technology Assessment study of the potential processes applicable to the Connecticut Solid Waste Management Program has indicated that energy or fuel recovery with some form of material recovery can be more than competitive compared to modern, low-pollution incinerators, and may even offer advantages over sanitary landfill, where hauling distances are appreciable. Cost comparisons in this report are strictly tentative, however, and are not the primary purpose of the report.

Where comparisons are made between capital and operating costs for various processes, no allowances are made for future inflation, and the comparisons are made without regard to distances between waste sources, facilities, and markets for materials. The financial comparisons shown in this report are therefore useful primarily for ranking the different processes for further evaluation. The various resource recovery processes are also compared on the basis of energy yields, material yields, and the development status of the processes.

These rankings have been used to narrow the choices to: 1) the dry or wet fuel process used in conjunction with an electric utility fossil fuel boiler and 2) the pyrolysis processes used to manufacture oil or gas. These processes offer the most promising and innovative technology identified. They solve the problem of disposal with minimum volume, offer resource recovery of a critical commodity (fuel), and have the potential for reasonable cost to the State of Connecticut, causing a minimum amount of pollution.

The technology assessment study has also indicated that the many name-brand processes actually are very similar in their details. Most use a shredder, an air classifier, and a magnetic separator, and some use aluminum- and glass-extraction equipment. The pyrolysis processes add the extra processing step to produce oil or gas; the composting process is a similar additional step that may be useful in closing down old landfill sites. The technology available for solid waste processing is therefore actually limited to material separation into combustible and noncombustible streams, pyrolysis, and composting. Separation plants can be designed to suit the user's need and are therefore not dependent on a proprietary process.

The major weakness in the emerging technologies appears to be that few have accumulated much operating experience. Pyrolysis has only been demonstrated on a pilot plant scale, and air classification of solid wastes has not been reduced to a science. Fortunately, the Environmental Protection Agency has made several grants that will create full-scale separation and pyrolysis plants during the next two years. In addition, earlier grants have created plants to evaluate the preparation and use of dry fuel for electric utility boilers. Another grant sponsors a plant demonstrating many of the steps in preparation of utility boiler fuel by wet pulping.

Section 2

COMPARISON OF RESOURCE RECOVERY STRATEGIES

In reviewing the alternative strategies for resource recovery, not only the costs and advantages of each strategy must be assessed, but also the technical feasibility, the environmental impacts, the anticipated schedule of availability of the technology, and effects of changing factors in the technical-economic climate. This section of the report deals with these factors, starting first with a discussion of the strategy options. In later sections, the specific processes for implementing the strategies are dealt with in more detail.

ALTERNATIVE STRATEGIES

In everyday practice, the current strategy in the United States is to either use landfill methods or incinerate the refuse. Both approaches are disposal strategies only, with the incineration providing volume reduction at the expense of air pollution and higher capital and operating costs. In a few specific situations, when properly executed, landfill is truly a land reclamation process.

The alternative strategies involving resource recovery can be categorized in four general areas:

- Material recovery
- Energy recovery
- Land recovery
- Combinations of the above

MATERIAL RECOVERY

Material recovery can be performed by wet or dry front-end processing of the refuse as received from collection trucks. Examples of such systems include the dry separation system being developed by the U.S. Bureau of Mines in College Park, Maryland and the wet pulping system being demonstrated at the Black-Clawson plant in Franklin, Ohio, under an Environmental Protection Agency demonstration grant. In the first system, the valuable products are recovered by a system of dry shredding and air, magnetic, and electrostatic separation techniques.

A back-end material recovery system has been developed by the U.S. Bureau of Mines in College Park, Maryland, where incinerator residue is processed for recovery of metals using mineral beneficiation techniques common to the mining industry.

3

ENERGY RECOVERY

Energy recovery of refuse can be accomplished by a variety of techniques. The most straightforward in concept is to use a water-wall incinerator as a boiler for steam recovery. This approach is characterized by high capital and operating costs and long construction times and requires a nearby user of the steam. It is a well developed technology, however, and is used widely in Europe, where, in some cases, the steam is used in on-site turbines to generate electrical power. The latter application is even more capital-intensive and is not efficient compared to modern, large-scale powerplants.

In the United States, no power generating incinerators are known to exist, but several steam generating incinerators are now operating, the largest being the Chicago Northwest incinerator, which went on-line in 1971. The City of Nashville is planning an installation where a steam generating incinerator will provide steam heat in the winter and chilled water in the summer for air conditioning of a nearby office complex.

Another energy recovery approach is to convert the refuse into a fuel to be consumed elsewhere. The average pound of untreated municipal refuse has an approximate heating value of 4700 Btu (see "Composition and Heat Value of Average Municipal Refuse" in Section 3). When properly processed, it can serve as a usable low-sulfur fuel.

One approach is to shred or wet-pulp the refuse and then burn it as a supplemental fuel in an electrical utility boiler. When used as an auxiliary fuel, the corrosive effects and particulate emissions from refuse burning are diluted by the primary fuel to more acceptable levels than possible when burning refuse alone. The dry shredding approach is now being demonstrated in the City of St. Louis, where shredded refuse is being burned in a Union Electric Company boiler which burns pulverized coal as the primary fuel supplemented by up to 20-percent refuse. (Percentages are expressed in terms of heat value.)

A. M. Kinney, Inc. has proposed a process of wet-pulping the refuse using Black-Clawson equipment. At Menlo Park, California, the Combustion Power Corporation is developing a refuse fuel preparation process where combustible materials are separated from the noncombustibles by a system of shredding and air separation. This technique was developed as part of the CPU-400 system, where the resulting fuel powers a gas-turbine-driven generator. The fuel is equally suitable for use as a supplementary fuel in utility boilers.

The City of St. Louis, A. M. Kinney, and Menlo Park processes are described in more detail in Section 3, "Refuse Fuel Recovery Processes," of this report. All of these systems incorporate some degree of material recovery from the noncombustible stream. All processes produce some residue for landfill, the most obvious one being bottom ash from the combustion process. Any utility boiler burning the refuse fuel must be equipped with bottom-ash handling capability. This requirement limits the utility boiler application to existing units originally designed to burn coal. Many such units are in service in the United States, and new units now in the design stage could be modified to handle bottom ash.

An alternative to preparing fuel by shredding or wet pulping can be found in various pyrolysis processes. Pyrolysis can generally be classified as the thermal breakdown of material in the absence or near absence of oxygen, such as can be found in coke and charcoal ovens. Recent developments of pyrolysis of solid waste include processes to produce an oil (the Garrett process) and to produce various forms of low-Btu fuel gases (Union Carbide Corporation and the Urban Research and Development Corporation).

Other pyrolysis processes are intended to serve as high-volume-reduction, low-pollution incinerators with steam heat recovery (Monsanto-Landgard and Torrax) but can also be used to generate low-Btu fuel gas. (Pyrolysis systems are described in Section 4, "Pyrolysis Processes.")

COMPOSTING AND BIOCHEMICAL PROCESSES

Composting is the biochemical degradation of organic material into a humuslike material that is useful as a soil conditioner or as a feedstock for making fertilizer and other higher value products such as wallboard. It can also be considered as a possible fuel and has been burned experimentally in a plant of the Del Marva Power and Light Corporation.

The art of composting has been applied to agriculture for centuries and has had considerable application to solid waste overseas. Most composting processes in the United States have proven uneconomic, but three processes involving high-rate digesters and one using windrowing do show promise. One, the Fairfield-Hardy plant in Altoona, Pennsylvania has processed over 100,000 tons of municipal solid waste to date and has established a continuing market for its product. This and the three other processes (Ecology, Inc., Conservation International, and Eweson/Geochemical) are described in Section 5, "Composting Processes."

It is felt that compost processes may not only be economically viable processes for producing salable products, but may also provide a practical solution to closing down existing landfills that do not meet standards. The use of compost as a cover material for existing noncompliant dumps and the way in which the Eweson/Geochemical process, in particular, offers a relatively portable process for producing a cover or fill for difficult landfill areas are discussed under "Uses of Compost as a Soil Improvement and Cover Material," in Section 5.

Biochemical processes involving anaerobic digestion of solid wastes are also being developed to produce methane gas of pipeline quality (Ref. 1). In general, the refuse is mixed with water to form a slurry and is put into digesters to achieve bacterial decomposition under anaerobic conditions. Processes under development are still proprietary, and little data are available. It is estimated that these gasification processes are eight to ten years away from commercial application.

COMPARISON OF PROCESSES

The usefulness of the various processes to meet the State of Connecticut's needs is a function of the process economics, the type of input and output streams, and the expected timetable of development of the process to the point where it is technically feasible for large-scale operation. The following paragraphs compare the various processes by these criteria.

PROCESS CAPITAL AND OPERATING COSTS

In a study performed for the President's Council on Environmental Quality, the Midwest Research Institute has performed an economic comparison of the various resource recovery processes (Ref. 2). The study used as its basis of comparison a plant receiving 1000 tons per day of packer truck waste, operating 300 days per year and having a 20-year economic life.

Studies were made for specific proprietary processes and were then lumped into average values for the various generic categories of processes. Results for this study are to be considered more comparative than absolute, because cost figures are computed on a national average basis and reflect costs at an earlier point in time than is the case in the State of Connecticut system.

Table 1 summarizes the results of the study, showing comparative investment and operating costs. Included are costs for nearby landfill and for remote landfill involving a rail haul distance of 100 miles. No resource value is recovered from the landfill operations or from the straight incinerator operation, whereas some value is assigned to the resources recovered from the other processes.

Table 1 shows that the only system that approaches nearby landfill in net cost per ton is the fuel recovery process. As described in the following paragraphs, the generic fuel recovery process consists of the preparation of shredded or pulped refuse fuel from which ferrous metals are extracted and sold separately. Material recovery and pyrolysis are much higher in net cost than fuel recovery but are less than remote landfill costs.

Composting is nearly the same as remote landfill in net cost. When comparing initial capital investment, the resource recovery costs fall in about the same ranking as does the net cost per input ton. However, judging initial capital costs of the recovery processes versus landfill, from Table 1, is very

6

Table 1

SUMMARY OF SYSTEM ECONOMICS*

System Concept	Investment	Total Annual Cost	Resource Value	Net Annual Cost	Net Cost/Input Ton
Incineration only	$ 9,299	$2,303	$ 0	$2,303	$7.68
Incineration and residue recovery	10,676	2,689	535	2,154	7.18
Incineration and steam recovery	11,607	3,116	1,000	2,116	7.05
Incineration plus steam and residue recovery	12,784	3,508	1,535	1,973	6.57
Incineration and electrical energy recovery	17,717	3,892	1,200	2,692	8.97
Pyrolysis	12,334	3,287	1,661	1,626	5.42
Composting (mechanical)	17,100	2,987	1,103	1,884	6.28
Materials recovery	11,568	2,759	1,328	1,431	4.77
Fuel recovery	7,577	1,731	920	811	2.70
Sanitary landfill (close in)	2,472	770	0	770	2.57
Sanitary landfill (remote)	2,817	1,781	0	1,781	5.94

*Based on municipally owned, 1000-ton/day plant with 20-year economic life, operating 300 days/year.

Source: Midwest Research Institute.

risky because of the potential divergence of specific land values from the national averages used by the Midwest Research Institute. In addition, certain specific processes in each generic group may deviate significantly from the group average. For example, the Conservation International composting process described under "Varro (Ecology, Inc.) Process" in Section 5 is significantly lower, and the Ecology, Inc. process also described is significantly higher than the compost process average.

Two important factors affecting costs in Table 1 are the economies of scale and the value of the recovered resources. Figure 1 shows the Midwest Research Institute prediction of the effect of plant capacity on net operating cost. Note that the nearby and remote landfill operation costs are relatively insensitive to design capacity, whereas all resource recovery processes show decided advantage in changing to large-scale facilities. For very large-scale facilities (2000-ton/day capacity), fuel recovery is shown to be cheaper than the nearby landfill approach, and material recovery, pyrolysis, and composting lie well below remote landfill costs. These cost predictions by the Midwest Research Institute were based on established scale factors for the unit processes involved and on scaling information furnished by the process developers.

The results of Table 1 and Figure 1 are based on specific values for each recovered resource. At best, these values can be considered approximate es-

7

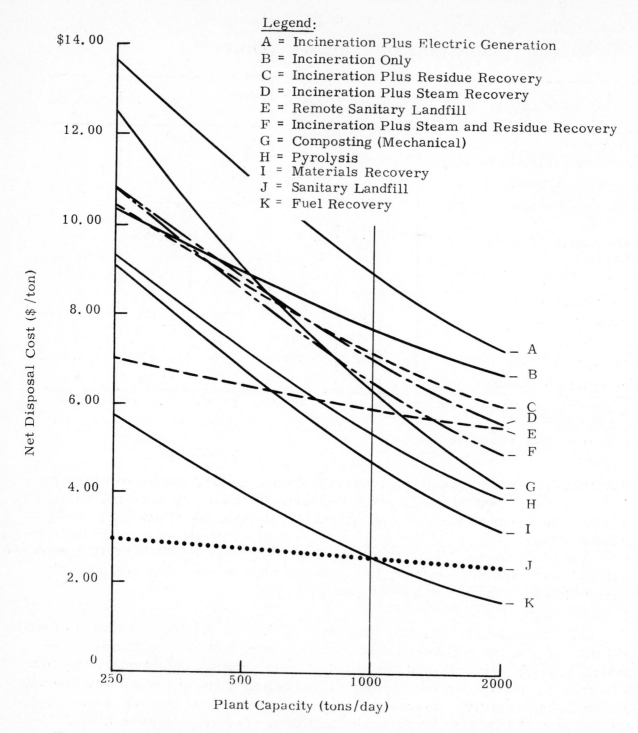

Legend:
A = Incineration Plus Electric Generation
B = Incineration Only
C = Incineration Plus Residue Recovery
D = Incineration Plus Steam Recovery
E = Remote Sanitary Landfill
F = Incineration Plus Steam and Residue Recovery
G = Composting (Mechanical)
H = Pyrolysis
I = Materials Recovery
J = Sanitary Landfill
K = Fuel Recovery

Figure 1. Net Disposal Costs Associated with Municipally Owned
Resource Recovery Systems at Various Plant Capacities
(20-Year Economic Life, 300 Day/Year Operation)
(Source: Midwest Research Institute)

8

timates and are subject to varying market conditions. The results assume that all of the recovered resources can be sold.

Table 2 lists the net operating costs of a 1000 ton-per-day plant for the various processes when the assumed value of the recovered resource is varied

Table 2

EFFECT OF RESOURCE VALUE VARIATION
ON RECOVERY SYSTEM ECONOMICS

System Concept	Net Operating Cost ($/ton) Resource Value as Percent of Estimated Value			
	150%	100%	50%	0%
Incineration and residue recovery	6.29	7.18	8.08	8.96
Incineration and steam recovery	5.39	7.05	8.72	10.38
Incineration plus steam and residue recovery	4.02	6.57	9.13	11.69
Incineration and energy recovery	6.98	8.97	10.98	12.98
Pyrolysis -- oil recovery	2.65	5.42	8.18	10.96
Composting	4.44	6.28	8.12	9.95
Materials recovery	2.56	4.77	6.98	9.20
Fuel recovery	1.17	2.70	4.24	5.77

Source: Midwest Research Institute

from 0 to 150 percent of the values used in Table 1 and Figure 1. These variations are compared graphically, in Figure 2, with costs of a nearby sanitary landfill, a remote sanitary landfill, and a modern incinerator. Fuel recovery is shown to be the resource recovery process that is least sensitive to value variations of the product and will be less expensive than remote land-fill, even when the fuel is given away.

The above figures all point to fuel recovery as the most economically attractive resource recovery process, with material recovery, pyrolysis, and composting being more expensive but viable alternatives.

Comparable figures for the specific Connecticut situation are available in the final report of the Connecticut Solid Waste Management Plan. Capital cost estimates have been made by General Electric on the basis of the estimating procedure outlined under "Hybrid System" in Section 6, where various front-end systems are compared. These cost figures, which exclude land and site preparation costs are:

Process	Cost
Garrett pyrolysis	
Front end only	$ 15,300,000
Complete system	$ 25,400,000
Combustion Power Corporation (CPU-400 System), front end only	$ 11,600,000
Modified St. Louis process (air classifier added)	$ 11,700,000
Hybrid system	$ 13,800,000

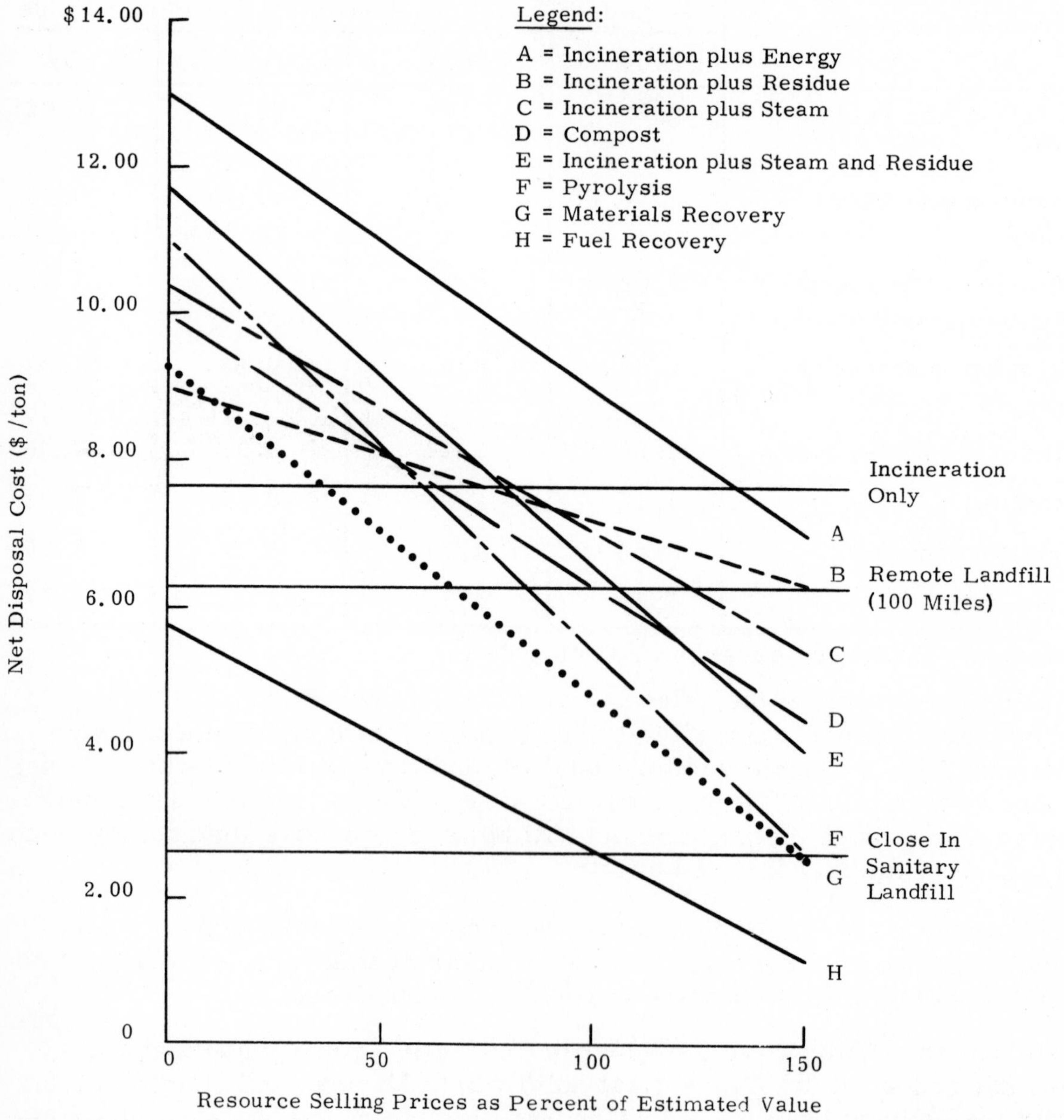

Legend:

A = Incineration plus Energy
B = Incineration plus Residue
C = Incineration plus Steam
D = Compost
E = Incineration plus Steam and Residue
F = Pyrolysis
G = Materials Recovery
H = Fuel Recovery

Figure 2. Sensitivity of System Economics to Market Value of Recovered Resources (1000 tons/day) (This figure assumes that all the resources recovered can be sold.) (Source: Midwest Research Institute)

10

The hybrid system is a dry shreddding, fuel preparation system with material recovery. The hybrid system evolved in this study as an attempt to optimize a system for the Connecticut situation.

An additional capital cost estimate of interest concerns the A. M. Kinney/ Black-Clawson wet pulping process for preparing fuel. This process is described under "A. M. Kinney/Black Clawson Process" in Section 3, where a capital cost figure of $13,700,000 is deduced by scaling down cost figures for the 2400-ton-per-day plant that has been proposed for Hempstead, Long Island.

Data on incinerator costs for the Connecticut area are found in recent studies performed for the Greater Bridgeport Region (Ref. 3). These studies included water-wall incinerators and refractory wall incinerators of various capacities, as well as a prefabricated incinerator. All incinerators were to be equipped with either scrubbers or electrostatic precipitators, to meet air emission standards.

Figure 3 shows a plot of the capital costs per ton per day that were derived from the General Electric studies.

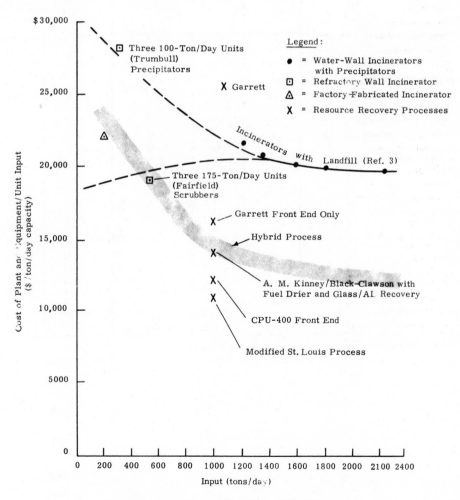

Figure 3. Capital Costs

11

Table 3

COMPARISON OF OUTPUT STREAMS BASED
ON 1000 TON/DAY INPUT OF PACKER REFUSE

Generic Type	Solid Fuel Preparation Process							
	St. Louis		A. M. Kinney/Black-Clawson					
Process	As Is	With Air Classifier	As Is	with Fuel Drier and Glass/Al Recovery	CPU-400	Hybrid	Union Carbide	URI
Types of refuse accepted	Packer	Packer	Packer	Packer	Packer, Bulky	Packer, Bulky	Packer, Bulky	Packer,
Sewage in (tons/day)	None	None	None	None	Up to 1100	None	None	Non
Outputs (tons/day)								
Ferrous metal	76*	72*	65*	70	50*	72	--	--
Aluminum	--	--	--	5	5*	5	--	--
Other nonferrous metals	--	--	--	--	2*	--	--	--
Glass	--	--	--	60	30*	30	--	--
Carbon char	--	--	--	--	--	--	--	--
Humus or compost	--	--	--	--	--	--	--	--
Fiber pulp	--	--	--	--	--	--	--	--
Slag or aggregate	--	--	--	--	--	--	220	250
Solid fuel (tons/day)	924*	632*	1100	643	519	632	--	--
Percent moisture	--	24	50	25	0	24	--	--
Fuel oil (tons/day)	--	--	--	--	--	--	--	--
Fuel gas (scf/day)	--	--	--	--	--	--	25×10^6	$33 \times$
Total useful product to ship (tons/day)	1000?	704	1165	778	606	739	220	250
Waste, junk (tons/day)	0	296	135*	142	75	261	--	--
Ash in fuel (tons/day)	200*	59	55*	48	75	59	--	Unknown
Total to dispose (tons/day)	200	355	190*	190	150	320	0	Unknown
Energy Outputs								
Steam flow (ppm)	--	--	--	--	--	--	--	--
Steam pressure (psi)	--	--	--	--	--	--	--	--
Gross energy in steam Btu/day	--	--	--	--	--	--	--	--
Gross energy in fuel Btu/day	9.4×10^9	7.4×10^9	7.37×10^9*	6.46×10^9	6.74×10^9	7.4×10^9	7.05×10^9*	6×10
Energy Inputs								
Oil or gas used (Btu/day)	--	--	--	--	--	--	--	--
Electrical (Btu/day) (34% efficient)	0.5×10^9	0.75×10^9	0.36×10^9	0.36×10^9	0.36×10^9	0.80×10^9	1.17×10^9	$0.17 \times$
Net energy output (Btu/day)	8.9×10^9	6.65×10^9	7.01×10^9	6.10×10^9	6.38×10^9	6.60×10^9	5.88×10^9	$5.38 \times$
Energy Conversion Efficiency	0.947	0.707	0.746	0.649	0.649	0.702	0.626	0.62

*Actual number or scale-up of number from the process developer.

Pyrolysis Process			High Temperature Incinerator	Compost Process				Mixed Processes
Monsanto	Torrax	Garrett	American Thermogen	Ecology	Geochemical/ Eweson	Fairfield Hardy	Conservation International	Hercules
Packer, Bulky	Packer, Bulky	Packer	Packer, Bulky	Packer	Packer	Packer	Packer	Packer, Bulky
None	None	None	None	None	330*	250*	None	308*
70*	--	66.5*	--	50*	50	50*	--	54*
--	--	--	--	--	--	--	--	7*
--	--	--	--	--	--	--	--	--
--	--	56*	--	--	--	--	--	85*
80*	--	51*	--	--	--	--	--	48*
--	--	--	--	500*	600	300*	600*	182*
--	--	--	--	--	--	--	--	240*
170*	250	--	250	--	--	--	--	--
--	--	--	--	--	--	--	--	--
--	--	228*	--	--	--	--	--	--
--	--	--	--	--	--	--	--	--
320		401.5	250	550	650	350*	600*	616*
--	--	70	--	--	80	175	--	Unknown
--	--	4.8	--	--	--	--	--	--
0	0	74.8	--	0	80	175	0	Unknown
200,000	250,000	--	191,700	--	--	--	--	--
250	150	--	300	--	--	--	--	--
5.51×10^9	6.86×10^9	--	5.35×10^9	--	--	--	--	--
--	--	4.8×10^9	--	4.2×10^9	4.0×10^9	4.2×10^9	1.6×10^9	Unknown
1.0×10^9	2.0×10^9		Unknown (proprietary)	Unknown	Unknown	Unknown	Unknown	Unknown
0.18×10^9	0.27×10^9	1.3×10^9	--	Unknown	Unknown	Unknown	Unknown	Unknown
4.33×10^9	4.59×10^9	3.5×10^9	Unknown	Unknown	Unknown	Unknown	Unknown	Unknown
0.461	0.488	0.372	<0.569	<0.447	<0.426	<0.447	<0.170	Unknown

13

As expected, the incinerator cost figures are higher than the costs for resource recovery processes. Some divergence of points occurs for the incinerators in the 200 to 1200 ton-per-day range, so the cost curves for the two extremes are shown as dotted lines in this range. Comparing the preliminary capital cost figures from Figure 3 to the investment figures in Table 1, it is evident that the General Electric estimates for the Connecticut case run just about twice the Midwest Research Institute estimates for the national average case. The important point here is that the relative capital costs of the processes considered fall in the same ranking order by either method.

Operating cost figures have not yet been generated by the General Electric Company, so net cost figures per input ton are not yet available for comparison. However, on the basis of the figures available, the initial conclusion that fuel recovery is the most economically attractive resource recovery process appears to remain valid.

INPUT AND OUTPUT STREAMS

Another way of comparing the processes is to compare input and output streams. Comparison has been made on the basis of a 1000-ton-per-day input capacity plant. In all cases, the output stream composition was based on the assumption that the entire 1000 tons of input per day were packer truck refuse of the type specified in Table 14, but the normal acceptance of bulky waste by the process was duly noted.

Table 3 summarizes the results of this comparison for the various processes that received concentrated study. Note that several processes have the ability to accept sewage or sewage sludge. In the case of the two composting processes accepting sewage, the sewage is a necessary input to the process.

The output material stream of each process is also noted in Table 3. In all cases, but one, the tonnage of solid fuel produced is given with the moisture included. The one exception is the CPU-400 system case, where the information was provided to give the fuel output on a dry weight basis only. The CPU-400 system fuel will of course have some moisture content.

The total useful product to ship includes all of the solid or liquid fuel and useful material recovered in the process. Questionable items that may or may not be marketable, such as carbon char and slag or aggregate, are included as useful products to be shipped. Also, note that the A. M. Kinney process, as is, produces 1165 tons of product for 1000 tons of refuse. The difference, of course, is due to water in the 50-percent moist fuel.

Many of the processes produce waste or junk, at the process site, which must be disposed of. In addition, all of the solid and liquid fuels have an ash content, and this ash must be disposed of after burning the fuel at the utility.

14

When considering energy in Table 3, gross energy is first shown. This energy is the heat content of the steam produced or the energy due to the higher heating value of the fuel produced. Energy inputs can be in the form of electricity or in the form of oil or gas fuel that must be purchased for use in the process. To put everything on a common basis, the electrical energy used is converted to the equivalent heat value of fuel required to generate the electricity at an average plant efficiency of 34 percent. The net energy output is the gross energy output minus the energy inputs. Energy conversion efficiency is the net energy output divided by the heat content of the solid waste input.

Table 3 shows that, in terms of both gross and net energy delivered per day for a 1000-ton-per-day plant, the St. Louis process, as is, leads with a gross output of 9.4×10^9 Btu per day. However, Section 3 of this report, under "St. Louis Fuel Process," shows that air classification must be added to make it a viable process. With the air classification, the St. Louis process energy outputs are expected to drop to a 7.4×10^9-Btu-per-day gross and a 6.65×10^9-Btu-per-day net.

Similarly, the A.M. Kinney/Black-Clawson process, as is, produces a high energy output, but the moisture content is too high to be acceptable in a utility boiler. When a drier is added to produce a 25-percent moist fuel, the A.M. Kinney/Black-Clawson process produces a gross energy output of 6.46×10^9 Btu per day and a net value of 6.10×10^9 Btu per day. In scanning the net energy yields of the solid fuel recovery processes, Table 3 shows that the viable processes produce a net energy output ranging from 6.1×10^9 to 6.65×10^9 Btu per day, the variation falling well within the accuracy of these calculations. The hybrid process, CPU-400 process, and modified A.M. Kinney/Black-Clawson process all offer reasonable material recovery yields on top of their energy recovery. (The hybrid approach is believed to offer the greater reliability potential.)

In reviewing the pyrolysis systems, a greater spread in yields is seen. The highest net energy yield is obtained from the Union Carbide process, which produces a fuel gas. Net energy yield is of the same order of magnitude as that of the solid fuel preparation processes, while the only material recovery is an innocuous frit suitable as a construction aggregate or for direct landfilling.

The Urban Research and Development Corporation produces a lower heating value gas in larger quantities, so net heat produced is only seven percent less than that of the Union Carbide process. Residue is basically similar in quantity and quality. Data obtained from the Urban Research and Development Corporation was more generalized than that which was supplied by Union Carbide, however, and the comparison between the two process yields must be considered to be only approximate.

Monsanto and Torrax recover energy in the form of steam having roughly 75 percent of the heat value of the gas producing pyrolysis processes. Both require supplementary fossil fuels, in quantity, to operate their processes. Torrax residue is very similar to that of Union Carbide and the Urban Research and Development Corporation and may be marketable. Monsanto also produces a similar slag aggregate, plus ferrous metals and a carbon char. If a market cannot be found for the char, it will present a disposal problem, due to its pH of 12.

The most sophisticated range of products from the pyrolysis systems is provided by the Garrett Corporation. There is significant material recovery of ferrous metals and pure, noncolor-sorted glass, plus a pyrolytic oil. This oil, called "Garboil," is similar to No. 6 fuel oil, but does require stainless steel storage and handling facilities. The net energy yield of the oil product is less than half that of the solid fuel preparation and gas pyrolysis processes. However, the fuel oil product is storable and easily transportable. Because the front end of the Garrett process does a thorough job of separating the inorganic from the organic stream, the process offers real opportunities for even more material separation from the inorganic stream in the future.

At the start of the study, there was some hope that compost products might be useful boiler fuels, because of their small particle size and low sulfur content. However, Table 3 shows that of the three compost products tested, the gross energy yield for the 1000-ton-per-day plant was only 4.0 to 4.2 x 10^6 Btu per day. This, plus a high ash content in the samples tested, caused concentration on the pyrolysis and solid fuel processes in the energy strategy.

The Hercules process, shown in the last column of Table 3, uses a mixture of technologies to arrive at an assortment of material outputs.

The highest net energy yields, then, are produced by the solid fuel preparation and the slagging pyrolysis processes that produce fuel gas. Although the extent of material recovery varies from one process to another, virtually all processes except the slagging pyrolysis processes can be extended to include additional material recovery at very little sacrifice in net energy recovery. Limited front-end separation may even be possible on the slagging pyrolysis systems, but feasibility and desirability of this step have not yet been established.

These comments do not address the effects of process tolerances and variabilities in feed on the various material output specifications. To market the products, these data must be known. A study of these effects is reported in Reference 4.

DATES OF AVAILABILITY OF PROCESSES

The various processes being considered are in varying degrees of readiness for application. Table 4 lists the status of the various processes as of 1973. The processes having sizable throughputs to date include the City of St. Louis

Table 4

MAJOR RESOURCE RECOVERY PROCESSES

Process	Largest Capacity Unit (tons/day)	Current Process Rate (tons/day)	Operating Since	Total Tons to Date
Fuel/Material Recovery				
St. Louis	600	200	April 1972	15,000
CPU-400	75	Intermittent	1971	1000
A.M. Kinney	None	--	--	None
Material Recovery				
Franklin, Ohio	150	50	1971	Unknown
Bureau of Mines	750	Under Construction	--	--
Pyrolysis				
Monsanto	35	Closed	2 yr	Unknown
Garrett	4	Intermittent	2 yr	Unknown
Union Carbide	5	Intermittent		Unknown
URDC	120	Intermittent	Sept 1972	Unknown
Torrax	75	Intermittent		Unknown
Hercules	None	--	--	None
Composting				
Ecology, Inc.	150	40	1971	4000
Geochemical/Eweson	30	15	May 1972	1000+
Fairfield Hardy	25	25	1964	100,000
Conserservation International	350	350	1957	100,000
Steam Recovery				
Chicago N.W.	1600	1600	1971	Unknown

process, the Ecology, Inc. process, the Fairfield-Hardy process, and the Conservation International process. The latter two processes have operated for years and have accumulated a very large backlog in processing compost. All other processes are in earlier stages of development and are either in the pilot plant stage or are just now emerging into larger scale operation.

Table 5 lists a projected timetable of availability of the processes, based on known grants issued by the Environmental Protection Agency or on other firm plans. The processes marked with an asterisk do not yet have firm implementation plans, and dates are assumed on the basis of a firm go-ahead by June 1973. All dates assumed no delays due to the approval process or to strikes.

Table 5

ESTIMATED TIMETABLE OF AVAILABILITY

Process	Earliest Full-Scale Demonstration Plant in U.S.					Earliest Full-Scale Operating Plant			Earliest 1000-Ton/Day Plant Operating
	Design Capacity, Input Solid Waste (tons/day)	Startup	Location	Process Running Normally	User Feasibility Established	Startup	Probable. Capacity (tons/day)	Process and User Acceptance	Startup
Steam recovery									
Chicago N.W.	1600	1971	Chicago	1973	In Europe	1976	--	January 1977	January 1976
Fuel/Material Recovery									
St. Louis (coal)	650	1972	St. Louis	1974	1975	--	--	--	NA
Hybrid (oil)	≈ 500	1975*	Devon	1976	1977	1975	500	1977	1978
CPU-400, front end	75	1971	Menlo Park	1973	1974	1976	600	1978	1979
	680	1976*	Montgomery County	1977	1978				
A. M. Kinney/Black-Clawson		1975*	Hempstead	June 1976	1977	1979	1000	1980	1979
Pyrolysis									
Monsanto (steam)	1000	1975	Baltimore	1975	1977	1979	500	1981	1979
Garrett (oil)	200	1975	San Diego	1975	1977	1979	500	1981	1982
Union Carbide (gas)	200	1974	West Virginia	1975	1977	1979		1978	1982
URDC (gas)	120	1972	East Granby	1975*	--	1976	240	1978	1980
Torrax (steam)	75	1971	Tonawanda	TBD**	TBD	TBD	TBD	TBD	TBD
Hercules	1500	1975	Wilmington	1975	1976	1975			1979
Composting									
Ecology, Inc.	150	1971	Brooklyn	1973	--	1975	150	1976	--
Eweson/Geochemical	30	1972	Big Sandy	1973	--	1975	120	1976	--
Fairfield Hardy	25	1964	Altoona	Established	--	1975	300	197.	--
Fairfield Hardy	600	19.	San Juan	1974	--	1975			--
Conservation International	350	19.	Jamaica	Established		1975	350	1976	--
Material recovery									
Franklin, Ohio	150	1971	Franklin	1974	--	--	--	--	--
Bureau of Mines (dry)	--	TBD	College Park	TBD	--	--	--	--	--
Bureau of Mines incinerator residue	750	1974	Lowell	1975	--	--	--	--	--

*Plans to implement are not yet firm. Dates are based on a firm go-ahead by June 1973.

**TBD = To be determined.

18

The fuel and material recovery processes and the pyrolysis processes producing utility fuels are assumed to require two years, from startup, before determining feasibility by the utility user. It is believed that this length of time, running on a new fuel, is required by the utility to establish long-term corrosion and pollution effects. Where processes are being implemented elsewhere, the schedule is based on a wait-and-see approach, implementation by a user being delayed until feasibility at a reasonable throughput level (200 tons/day or more) is proven elsewhere. The resulting schedule of availability of the processes is also shown in Table 5.

BUILDING BLOCKS OF ENERGY STRATEGY -- A CONNECTICUT EXAMPLE

Tables 4 and 5 show that a fuel and material recovery process could conceivably be installed in Connecticut by 1975. If installed, the process would require two years of operational testing before its feasibility for use in a utility boiler would be fully established. During this two-year period, operation on a 500-ton-per-day basis is anticipated. If, after 1 to 1-1/2 years of operation on this basis, the results appear encouraging, it would be reasonable to consider expanding the fuel processing plant to consider servicing, by this process, more utility boilers with auxiliary fuel, and a January 1978 date for achieving startup of the expanded 1000-ton-per-day plant is feasible on this basis.

The use of refuse-derived supplementary solid fuel in utility boilers, however, will be limited by the expected load factor on the boiler. This situation is demonstrated in Figure 4, which shows the expected refuse consumption of

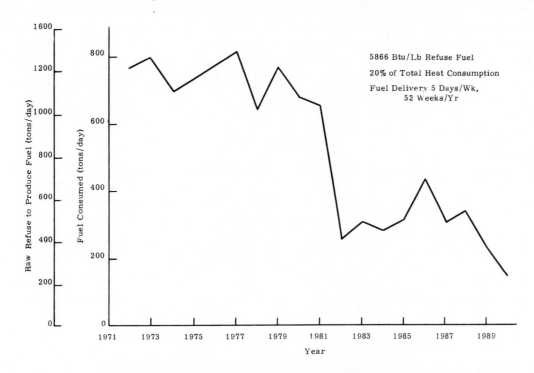

Figure 4. Expected Refuse Consumption of Norwalk Harbor Units 1 and 2 (One Operated at a Time)

19

Norwalk Harbor Units 1 and 2 of the Northeast Utilities system (Ref. 5). The plot in Figure 4 is based on the assumption, that refuse will be burned as an auxiliary fuel in only one boiler at a time and that burning will be at a rate of 20 percent of the total heat input to that boiler. The particular boiler burning refuse is expected to be the boiler running at base load at that time.

The startup of a large nuclear plant in 1981 is expected to drastically reduce the economic dispatch of the fossil-fueled utility boilers at that time, and no longer will it be possible to count on at least one boiler operating near 100-percent capacity. As a result, in 1981 the capacity for burning refuse will reduce proportionately. Most fossil fuel burning units will have the same drop in capacity as that shown for the Norwalk Harbor units (Figure 4). This drop in capacity can be partially offset by increasing the rate of refuse burning from 20 percent (by Btu content) to higher values, although this change will involve considerable risk of added corrosion and atmospheric emissions. The capacity drop can be further offset by using more than one boiler at a time in each plant. If this is done, some bypassing arrangement will be needed to handle the refuse during normal outages of the boilers for planned maintenance.

The answer to this dilemma is believed to be to use pyrolysis products as a fuel for the utility boilers and for other industrial customers. Pyrolysis processes permit cleaning of the oil or fuel gas being produced, thus reducing the air pollution and boiler corrosion problems that limit solid refuse fuel to use as a supplementary fuel. Table 5 shows that pyrolysis processes generally can be started in Connecticut by 1979, with completion of the two-year acceptance tests by 1981. Assuming encouraging results from at least one pyrolysis process during the acceptance tests, it is possible to foresee a large-scale pyrolysis plant starting in 1982, to provide either gas fuel, oil fuel, or steam derived from refuse, to supply all or part of the requirements of one or more fossil-fueled utility boilers. The schedule could be accelerated at the price of added risk.

Section 3

SOLID REFUSE FUEL RECOVERY PROCESSES

Raw, untreated, municipal refuse has been shown to have a heating value in the range of 4000 to 5000 Btu/pound. This heat content has been utilized widely in European incinerators to generate steam, and some steam recovery incinerators are now in use in the United States. The capital cost of steam-generating incinerators is high, however, and markets for steam are not always situated nearby (steam is a perishable commodity).

Considerable activity has recently been generated for using properly prepared refuse as a supplementary solid fuel for utility boilers. In this section, mechanically prepared refuse is discussed. Section 4, "Pyrolysis Processes," deals with transformation of refuse into liquid and gaseous fuels by pyrolysis.

Although it cannot be stored indefinitely, prepared refuse is a much less perishable commodity than steam. In addition, when considering capital costs, the major capital expense of the boiler has already been borne by the utility user. The rising costs of fuel as the utilities switch to lower sulfur fuels make the consideration of refuse as a fuel more attractive. Having an average sulfur content ranging between 0.1 and 0.2 percent and a reasonable heating value, refuse is now seriously being studied as a supplementary fuel for utility boilers. Adoption of its use can be a major step in alleviating both the solid waste crisis and the energy crisis.

To prepare refuse as a solid fuel, it is necessary to size-reduce it, in order to make it more uniform and to increase the surface area for proper burning. Because most modern utility boilers use some form of suspension burning and no grate, refuse fuel particles must be relatively small (2 in. or less). Composting is one method used for producing fuel of small size; this method is discussed separately in Section 5, "Composting Processes."

This section of the report deals with mechanical methods of refuse fuel preparation. Three processes are described: two processes accomplish size reduction by shredding; one process uses wet pulping. Provision is made for various degrees of recovery of valuable materials from each process.

The characteristics of municipal refuse are discussed under "Composition and Heat Value of Average Municipal Refuse," in this section, both as a fuel source and as a source of materials. Seasonal variations and future trends in the refuse composition are also described.

The important considerations to be made in the application of refuse as a supplementary solid fuel in State of Connecticut electric utility plants is discussed below under "Application Considerations."

ST. LOUIS FUEL RECOVERY PROCESS

By implication, the first record of a logical system is contained in a U.S. patent No. 3,387,574, awarded to J.F. Mullen of the Combustion Engineering Company in Windsor, Connecticut and issued June 11, 1968 (filed Nov. 14, 1966). The first two cover pages of the patent are shown in Figure 5. This patent describes a shredded, high-moisture-content fuel being delivered by a conveyor system to a set of rotating fuel feeders and by air locks to a pressurized, pneumatic conveyor.

The pneumatic conveyor blows the shredded fuel into a tangentially fired, rectangular furnace. This system is conceptually similar to the one being utilized at the City of St. Louis demonstration plant.

The City of St. Louis demonstration plant fuel process is a straightforward process to convert packer truck refuse to a product to be used as an auxiliary fuel in a pulverized coal-burning utility boiler. The fuel consists of refuse shredded to 1-1/2-inch nominal size, with ferrous metals removed. Since startup in April 1972, the demonstration plant in St. Louis had disposed of 15,000 tons of refuse by March 1, 1973, achieving normal firing rates as high as 15 percent of the rated heat value of the boiler. Operating at a partial load, firing rates of 40-percent refuse (by heating value) have been maintained for one five-hour test period (Ref. 6).

The St. Louis process is among the lowest capital cost processes under consideration and produces a product (fuel) that will be of increasing value as the nation's energy problems develop further. However, questions of effect on stack emissions, long-term corrosion of boiler components, and compatibility with oil-fired boilers remain to be answered.

PROCESS DESCRIPTION

Fuel Preparation

Figure 6 shows a simplified diagram of the process used to prepare the fuel as originally installed and Figure 7 shows the present process after the addition of air classification.

Initial operations demonstrated that when milled solid waste was burned in suspension and when only the magnetic metals were removed prior to firing, several undesirable effects resulted. These effects included severe internal abrasion of pneumatic pipe lines, stoppage of pneumatic feeders due to solid particles becoming lodged between the rotors and the housing, and the inclusion in the bottom ash of not only inert materials such as glass, nonferrous metals, and ceramics, but also particles of wood, dense plastics, rubber, and leather, which although normally combustible, had sufficient mass to prevent their being burned in suspension. The significance of these effects was great

22

June 11, 1968   J. F. MULLEN   3,387,574
SYSTEM FOR PNEUMATICALLY TRANSPORTING HIGH-MOISTURE FUELS
SUCH AS BAGASSE AND BARK AND AN INCLUDED FURNACE
FOR DRYING AND BURNING THOSE FUELS IN SUSPENSION
UNDER HIGH TURBULENCE

Filed Nov. 14 1966   6 Sheets—Sheet 1

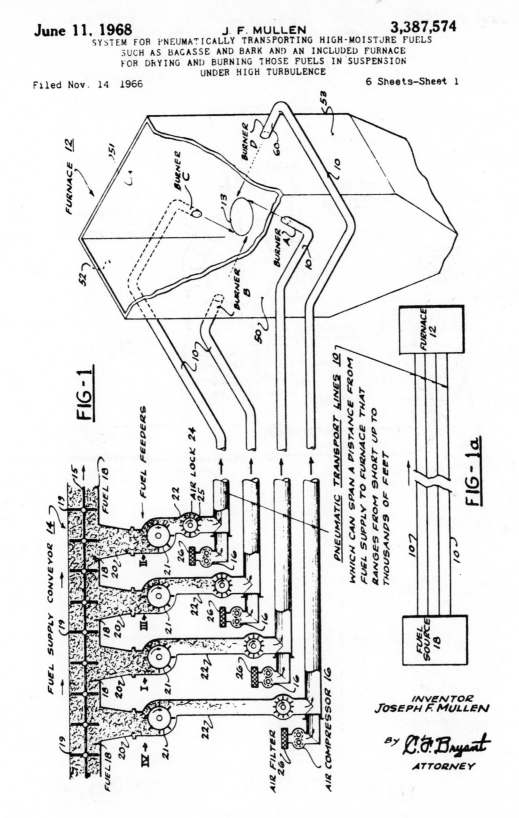

Figure 5.   Cover Pages of U.S. Patent for Pneumatically Transporting
High-Moisture Fuels (Sheet 1 of 2)

June 11, 1968   J. F. MULLEN   3,387,574

SYSTEM FOR PNEUMATICALLY TRANSPORTING HIGH-MOISTURE FUELS
SUCH AS BAGASSE AND BARK AND AN INCLUDED FURNACE
FOR DRYING AND BURNING THOSE FUELS IN SUSPENSION
UNDER HIGH TURBULENCE

Filed Nov. 14, 1966   6 Sheets-Sheet 2

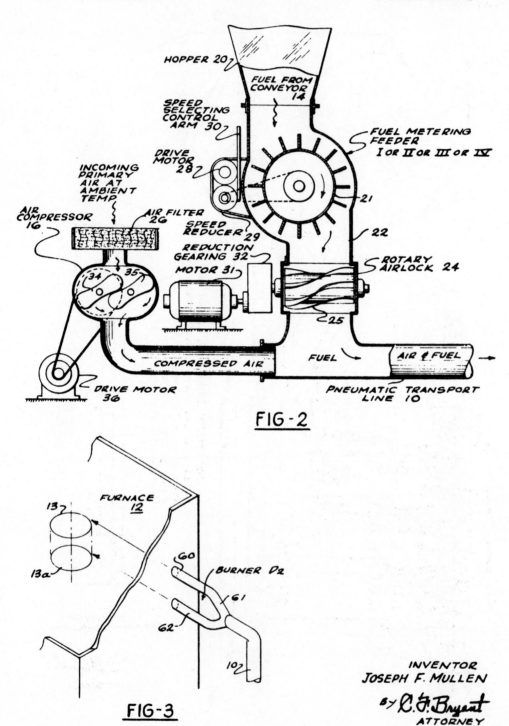

FIG-2

FIG-3

INVENTOR
JOSEPH F. MULLEN

BY C. F. Bryant
ATTORNEY

Figure 5.   Cover Pages of U. S. Patent for Pneumatically Transporting
High-Moisture Fuels (Sheet 2 of 2)

24

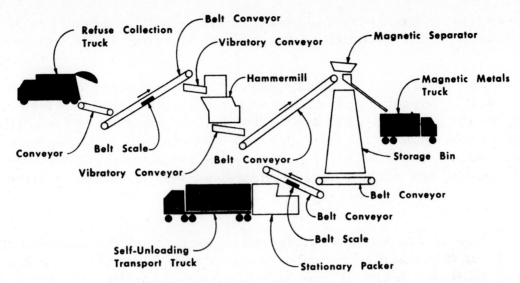

Figure 6. Components of St. Louis Refuse Processing Facilities
(Source: Ref. 7)

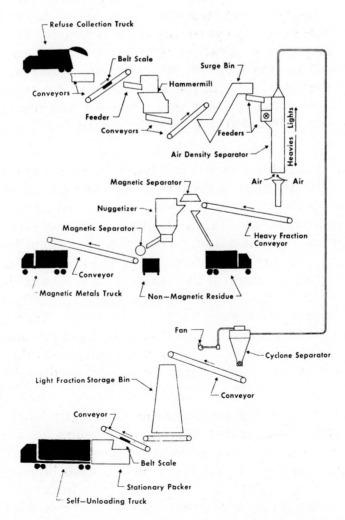

Figure 7. Solid Waste Processing Facilities

enough to result in the decision to install an air density separator, in order to eliminate the heavier particles from the material fired to the boilers.

In the fuel preparation process, packer refuse is dumped on the floor of the 100- by 125-foot receiving building and is pushed onto the belt feed conveyor by means of a front-end loader. The belt conveyor feeds a vibrating conveyor, which smooths the input flow to a 1250-hp, 45-ton-per-hour Gruendler shredder, which, in one operation, reduces the refuse to 1-1/2-inch nominal size. (Actually, 96% of the refuse is under 1 in. and 50% is under 3/8 in. [Ref. 7].)

Two-stage milling would be preferred on permanent installations, but the Gruendler shredder used at the 650-ton-per-day St. Louis plant is a 900-rpm, horizontal shaft, hammermill type, using 200-pound hammers mounted on an 80-inch-long rotor having a hammer circle diameter of 60 inches.

From the hammermill, in the process as first installed, the milled material is discharged to another vibrating conveyor, which feeds an inclined belt conveyor leading to a 1200-cubic-yard-capacity (approximately 150 tons) Miller-Hofft storage bin. Magnetic separation of ferrous metals was effected at the head pulley of this belt conveyor, with the magnetic materials discharged directly to trucks for disposal. From the storage bin, the supplementary fuel is conveyed by an auger conveyor to a belt conveyor and then to a stationary packer for loading onto self-unloading transfer trailers.

The air density separator, shown in Figure 8, following the philosophy of the original installation, also is a very simple device consisting of a chamber into which the milled solid waste is introduced, with an upward current of air of sufficient velocity to separate the light, fluffy, readily combustible fraction of the waste from the heavy particles. The device as installed permits control of both the separating velocity and the carrying velocity by means of automatically activated vanes on the inlet of the fan pulling the carrying air through the separator. Further control of the separating velocity is provided by a manual adjustment of the throat area of the separator. Additional flexibility of the device is provided by hinged panels on both the back and the front of the separation chamber. This latter feature permits both the direction of the airflow and the direction of the solids flow to be modified to a limited degree.

The heavy fraction from the air classifier is magnetically scalped, with the nonmagnetic residue discharging directly to a truck for disposal. The magnetic material passes through a second grinder or nuggetizer, which reduces the particle size of the ferrous fraction and also increases the density by compaction of the fragments. A second magnetic separator on the output of the nuggetizer strips off any nonferrous particles broken off from the ferrous material.

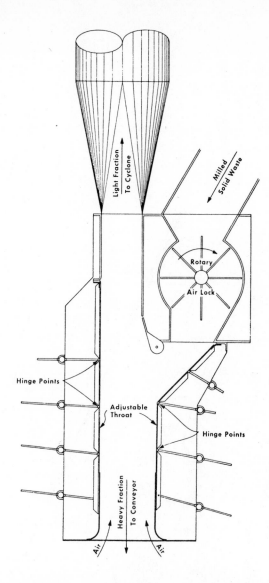

Figure 8. Air Density Separator

Figure 9 shows an overall view of the fuel processing plant. The processing plant at St. Louis has a total of 2000 hp installed (Ref. 8) and occupies approximately two acres. Design is based on 650-ton-per-day refuse input on a two-shift operating basis (the third shift is for maintenance). The design output is 600 tons per day of fuel and 50 tons per day of ferrous metal scrap.

Powerplant Processing

At St. Louis, the transfer trailers are hauled 18 miles (two hours round trip) to the Merramec powerplant of the Union Electric Company, where the receiving and fuel handling system is as shown in Figure 10. The transfer trailer unloads into a city-owned receiving building containing a receiving bin, a belt conveyor, and the feeder for a pneumatic conveyor. The 12-inch-diameter pneumatic conveyor feeds a 75-ton Atlas storage bin, which provides enough surge capacity to assure continuity of burning in case of minor refuse

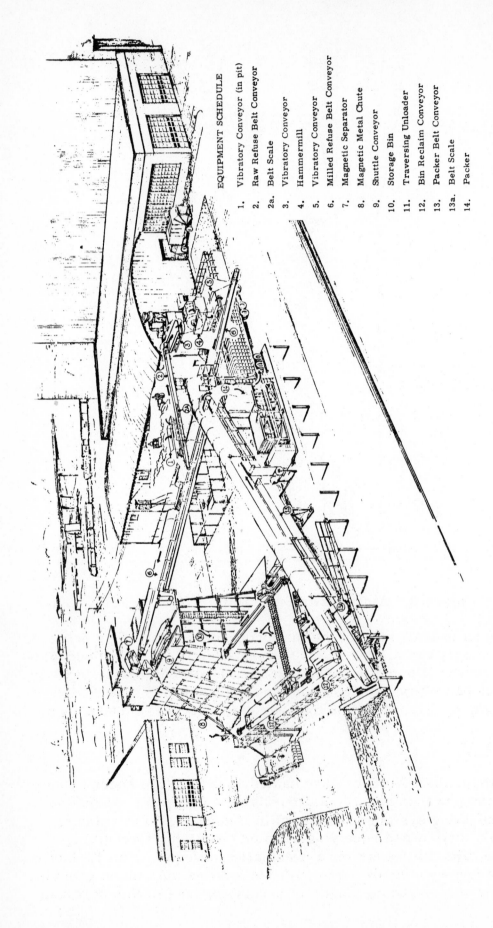

EQUIPMENT SCHEDULE

1.   Vibratory Conveyor (in pit)
2.   Raw Refuse Belt Conveyor
2a. Belt Scale
3.   Vibratory Conveyor
4.   Hammermill
5.   Vibratory Conveyor
6.   Milled Refuse Belt Conveyor
7.   Magnetic Separator
8.   Magnetic Metal Chute
9.   Shuttle Conveyor
10. Storage Bin
11. Traversing Unloader
12. Bin Reclaim Conveyor
13. Packer Belt Conveyor
13a. Belt Scale
14. Packer

Figure 9. Diagrammatic View of St. Louis Supplementary Fuel Processing Plant
(Source: Horner & Shifrin)

28

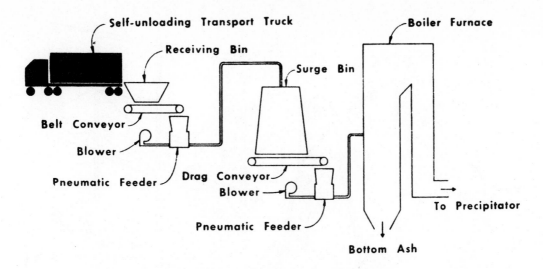

Figure 10. Schematic Diagram of Receiving and Fuel Handling System
(Source: Ref. 9)

delivery interruptions. The bin also serves as a refuse metering device and divides the single stream of refuse into four streams for feeding to the pneumatic transport systems that deliver the refuse to each corner of the boiler furnace.

The Atlas bin (Figure 11) has a capacity of 8600 cubic feet. Its reverse-sloping walls are intended to minimize bridging. The 40-foot-diameter bin floor is a concrete slab with four drag-chain conveyors imbedded in it. Four sweep-bucket trains (made of heavy metal buckets open at the bottom) ride on the bin floor. One end of each train is free and the other is pinned to a circumferential ring, which is rotated slowly around the bin floor by a hydraulic motor. As the ring rotates, the free end of the bucket train tends to go to the center of the bin floor; in the process, the buckets fill with refuse and the refuse falls into the drag-chain conveyor trenches. This procedure results in a fairly even split of flow to each drag-chain conveyor. A crude metering of the total refuse flow is obtained by varying the speed of the bucket train drive ring and by varying the speed of the drag chain conveyor.

The refuse from each drag chain conveyor is feed to a rotary air lock for introduction to the pneumatic feed to the boiler. This air lock is similar to a paddle wheel rotating in a cylindrical housing with tight clearances. Pressure in the pneumatic feed line varies from 1 psig at no load to 2.5 psig at a refuse transport rate of 5 tons per hour per feed line. Each of the four pneumatic feed lines serves one corner of the boiler. A total airflow of 9000 scfm is used in the feed lines. The feed lines are each 700 feet long and are made up of 8- and 10-inch Schedule 40 pipe.

The refuse is fired in a 125-megawatt-rated, tangentially fired, Combustion Engineering boiler having a maximum gross output of 142 megawatts and steam conditions of 1250 psig, 950°F superheater output, and 950°F reheater

29

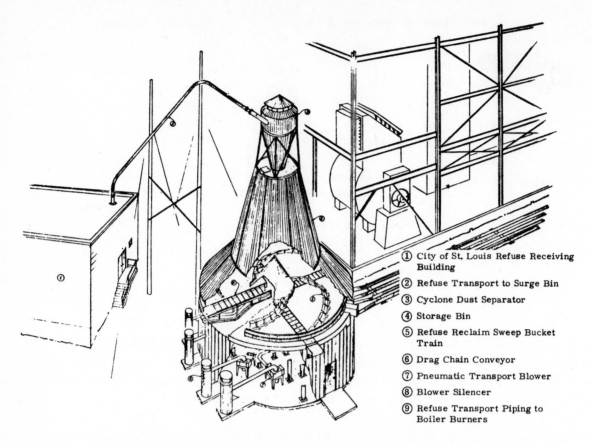

① City of St. Louis Refuse Receiving
   Building
② Refuse Transport to Surge Bin
③ Cyclone Dust Separator
④ Storage Bin
⑤ Refuse Reclaim Sweep Bucket
   Train
⑥ Drag Chain Conveyor
⑦ Pneumatic Transport Blower
⑧ Blower Silencer
⑨ Refuse Transport Piping to
   Boiler Burners

Figure 11. Refuse Firing Prototype (Source: Ref. 9)

output. Although not a modern unit (20 years old), the general configuration
is basically the same as that for other units now going into service for 2400
psig, 1000°F superheater output, and 1000°F reheater output operation at the
Union Electric Company (Ref. 9).

At each corner of the boiler there is a series of four coal burners de-
ployed in a vertical array. Each burner injects pulverized coal in a horizontal
direction tangential to the furnace wall. This configuration produces a vortex-
like fireball in the furnace and provides good suspension burning with sufficient
residence time for good combustion of the coal. At each corner, between each
level of coal burners, is a pair of gas burners, one over the other. These gas
burners currently are not being used.

By removing one gas nozzle in each corner of the furnace, space was
provided for a refuse injection nozzle that is tiltable ±15 degrees from horizontal by means of a handwheel. These 10-inch-diameter nozzles are located
at the midpoint of the vertical array, and most operation of the nozzles has
been carried out with them tilted in the fully up position. The feed system is
designed to operate with all four corners feeding refuse or with just two corners feeding refuse.

Although mixing of the coal and refuse prior to injection into the boiler
was considered, this approach was not used because:

30

- Spontaneous combustion in the coal pulverizers was a concern.

- Union Electric did not want to jeopardize the proper operation of the boiler combustion control system.

As used, the Union Electric boiler operates at a relatively constant refuse fuel flow rate and adjusts for load swings by varying the coal flow rate. The only direct connection between the boiler control and the refuse firing system is through the boiler trip circuitry. A boiler trip automatically trips the refuse feed system.

Input and Output Streams

The input to the St. Louis demonstration plant is strictly packer truck waste; no bulky refuse is handled. Design values are 650-ton-per-day input, on a two-shift operating basis. Design output is 600 tons per day of fuel and 50 tons per day of ferrous metal scrap.

As part of the St. Louis demonstration plant, a continuing monitoring of the refuse fuel is being conducted. A statistical analysis is being conducted to determine the average value and statistical spread of critical parameters of the refuse. Results for the first four months of operation (Ref. 10) are given in Tables 6 and 7 and are compared to the coal fired in the boiler. Assay values for the refuse fuel are on an as-received basis at the powerplant and are given as a percent by weight.

Additional discussion of refuse as a fuel appears below under "Composition and Heat Value of Average Municipal Refuse."

The data in Table 6 and in the preceding text can be linearly extrapolated to a 1000-ton-per-day plant, which will produce:

- 76 tons per day of ferrous metals

Table 6

COAL VERSUS REFUSE FUEL

Assay	St. Louis Refuse Fuel	Bituminous High-Volatile Pennsylvania Coal
Moisture (%)	27.8	2.6
Sulfur (%)	0.15	1.3
Chloride (%)	0.34	Not available
Ash (%)	21.6	9.1
Btu/lb	4631	13,610

- 924 tons per day of fuel having a gross heating value of 9.4×10^9 Btu

- 200 tons per day of ash from the powerplant

Table 7

COAL VERSUS REFUSE FUEL
(Comparison on Heat Value Basis)

Assay	St. Louis Refuse Fuel	Bituminous High-Volatile Pennsylvania Coal	Refuse/Coal Ratio
Btu of sample	13,610	13,610	1
Sample weight	2.939 lb	1 lb	2.939
Moisture weight	0.817 lb	0.026 lb	31.42
Sulfur weight	0.0044 lb	0.013 lb	0.339
Chloride weight	0.0110 lb	Not available	--
Ash weight	0.634 lb	0.091 lb	6.976

The heating value of the fuel has been adjusted upward slightly to coincide with year-around average values reported below under "Composition and Heat Value of Average Municipal Refuse."

General Electric calculations have been made to determine the output stream for the case of perfect separation of combustibles from noncombustibles and for the case where realistic separation efficiencies (described under "Composition and Heat Value of Average Municipal Refuse") are applied.

Results for a 1000-ton-per-day input plant are summarized in Table 8, using the suggested standard refuse composition listed in Table 14.

Table 8

OUTPUT STREAM OF ST. LOUIS PROCESS
WITH AIR SEPARATION ADDED

Output	With 100% Separation Efficiencies	With Realistic Separation Efficiencies
Ferrous metals (tons/day)	76	72
Fuel (tons/day)	794	632
Process waste to landfill (tons/day)	130	296
Ash from powerplant (tons/day)	37.3	58.7
HHV of fuel (Btu/lb)	5919	5866
Gross total daily heat value of fuel (Btu/day)	9.40×10^9	7.4×10^9
Estimated heat equivalent of power used at 34% efficiency (Btu/day)	0.75×10^9	0.75×10^9
Net total daily heat value of fuel (Btu/day)	8.65×10^9	6.65×10^9

Physical Requirements and Manpower

The 650-ton-per-day fuel processing facility at St. Louis occupies approximately two acres of land and has a total of 2000 hp, installed (Ref. 8). Manpower requirements are, for the present one-shift operation:

- Two front-end loader operators
- One control room operator
- One yard man (unloads bin and loads trucks)
- Two truck drivers
- One supervisor

Maintenance on an as-needed basis is performed by one welder and two helpers, for hammer retipping.

Equipment includes:

- Two front-end loaders
- Two 75-yard transfer trailers plus tractors
- Two dump trucks for ferrous metal disposal

At the powerplant, the facilities are as shown in Figure 10 and occupy less than one acre. One Atlas bin operator is required for each shift at the powerplant.

OPERATING HISTORY AND EXPERIENCE

The St. Louis demonstration plant was designed solely to demonstrate, at minimum cost, the feasibility of burning processed refuse as a supplementary fuel in coal-fired utility boilers. Minimum cost was emphasized in the design, and the fuel processing plant was installed next to the South St. Louis incinerator, which serves both as a backup disposal facility and as a source of manpower for some maintenance services. No redundancy in process equipment is provided, and processes are designed to minimize capital expense of equipment.

The St. Louis demonstration plant was first operated on April 4, 1972. From that time, through November 30, 1972, the plant has delivered 11,000 tons of fuel (Ref. 11) to the Merramec powerplant. Operation at the fuel processing plant has been on a one-shift basis, with a maintenance crew retipping half of the hammers of the shredder each night. The fuel delivery rate during the spring of 1973 was 2200 to 2400 tons per month, representing about 35 percent of the available capacity, on a one-shift basis. The prime reason quoted is that erosion problems in the fuel transport system at the powerplant required a cutback in throughput. By early 1974, after installation of the air classifier, a run of over 3000 tons of air classified material through the pneumatic system confirmed that air classification would eliminate jamming and erosion in the feed system.

Fuel Plant Operating Experience

Major operating problems at the fuel processing plant include feed problems in the output storage bin, shredder maintenance, and input conveyor problems.

Originally, a vibratory conveyor was installed to receive refuse from the front-end loaders and to provide metering of the refuse onto the belt conveyor feeding the shredder. An acoustic coupling between this conveyor and the building roof resulted in large amplitude oscillations of the roof, and the conveyor was deactivated. Now the front-end loader discharges into a chute feeding the belt conveyor, and a leveling bar on the belt conveyor tends to level its load.

Shredder maintenance is not unusual for this service. Hammers are re-tipped after every 600 tons of throughput. A welder can retip 16 of the 32 hammers in an eight-hour shift. This particular shredder is required to perform very strenuous service in that it must reduce the trash to 1-1/2-inch particles in one stage of shredding. If funding had permitted, the City of St. Louis would have preferred to operate with a coarse shredder, followed by magnetic separation and air separation prior to final shredding to the small size. In their view, this procedure would significantly reduce maintenance (Ref. 7).

Problems with the output storage bin are related to the unique physical characteristics of shredded refuse. The bin is approximately 60 feet high and holds about 150 tons of shredded refuse when full (Ref. 8). Anticipated problems with fires caused by spontaneous combustion in the bin did not materialize. Bulk temperatures of 120°F were the highest recorded during the summer of 1972.

Problems with packing of the fuel in the bin have occurred. Overnight settling of a 10-foot depth to 6 or 8 feet is typical (Ref. 8). The resulting packing produces auger starting torques that are 350 percent of predicted values. The City of St. Louis is solving this problem by only partially filling the bin, piling the fuel at one end. The auger conveyor is run to the empty end when not in use.

Another problem in the bin is bridging, but this problem is minimized by the reverse slope of the bin walls. Other minor problems of blowing pieces of refuse have now been solved by proper shielding and sealing at critical points in the conveyor system.

In summary, the fuel processing plant experience has demonstrated the relatively high shredder maintenance to be expected from single-stage shedding of unsorted refuse down to 1-1/2-inch nominal size. Other problems center in the material handling area, and suitable work-arounds have been made.

Powerplant Operating Experience
================================

Between April 4, 1972 and November 30, 1972, the Merramec No. 1 unit had burned 400,000 tons of coal and 11,000 tons of shredded refuse, with the monthly refuse consumption averaging between 2200 and 2500 tons for the last four months of that period. (The unit was shut down in December for periodic inspection and maintenance.) Because refuse is fired at a rate between 10 and 15 tons per hour for periods of 4 to 16 hours per day on a five-day-week basis, fuel rates of 12 to 13 percent are typical values when firing refuse and running at full load. Rates up to 40 percent (by Btu value) have been achieved when the boiler is running at partial load.

Experience burning wet refuse has been excellent. In October 1972, the moisture content reached 35 percent many times, and the boiler operators actually preferred to maintain a high refuse/coal ratio under those conditions, because the wet refuse burned better than wet coal. The last week before the planned November shutdown, the unit burned 40 percent refuse by weight to 60 percent coal during the 12 to 16 hours per day that the refuse feed system was manned at the powerplant. This situation corresponds, on a heat value basis, to 22-percent refuse firing. No discernible effect on boiler plant efficiency was observed during the test period.

Major operational problems at the powerplant during that time were traceable to the presence of glass and nonferrous metals in the refuse fuel. These materials cause erosion at elbows of the pneumatic feeder lines and jamming of the rotary air locks in the pneumatic feeders. The air locks (described above under "Powerplant Processing") have 0.003-inch clearances that are jammed as often as 30 times a day. This situation required continual attendance by an operator, to clear jams, although a jam can usually be cleared in a minute. This operator is a major cost item to the utility. Because of this high cost, three shifts were not run.

Erosion problems do not show up in straight pipe runs, but do show up as small holes at elbows. All elbows in the pneumatic feeder are relatively large-diameter sweeps of Schedule 40 steel pipe, but were not provided with wear plates in the initial installation because of cost considerations.

The glass, nonferrous metals, and larger organic particles in the fuel had presented another problem, to the Union Electric Company, in that they do not burn out in the boiler fire. The result was contamination of the bottom ash. Whereas the Union Electric Company had sold bottom ash for sanding roads, the presence of these contaminants made it unmarketable. Ash handling problems in the furnace, however, were not aggravated by these materials. Burners were tilted up the maximum 15 degrees to increase residence time, but the problem persisted.

After installation of the air classifier, the supplementary fuel has not contained the larger particles of abrasive materials, although fine particles of glass and ceramic materials are still embedded in the paper and plastic particles and therefore are still present in the fuel delivered to the powerplant. It is anticipated that provisions will still be necessary to accommodate a certain amount of pipeline abrasion at those points where changes of direction are necessary. Present indications are that elbows with replaceable sections on the outside radius will accommodate this requirement.

The initial installations of the St. Louis prototype facilities provided for a replacement of 10 percent of the normal pulverized coal with solid waste. The firing facilities were designed to accommodate at least a 20-percent fuel replacement. Initial operations demonstrated that a 15-percent fuel replacement could be maintained easily without any apparent adverse effects upon the boiler and that substantially higher rates of firing of the refuse fuel appeared to be practical.

Corrosion Potential. As part of the testing program of the St. Louis project, probes have been inserted at strategic locations in the boiler, to determine whether corrosion potential is any greater when solid waste is fired in combination with coal than when coal is fired alone. The results of this phase of the testing program are not yet available.

One of the materials in normal solid waste that is the subject of some concern from the corrosion standpoint is chlorine. The analytical work being performed in connection with the St. Louis prototype project includes chlorine determinations. In 210 samples of the solid waste delivered to the powerplant, from which only magnetic metals had been removed, it was found that the chlorine content varied from 0.13 to 0.95 percent, with an average of 0.29 percent. Indications are that less than 50 percent of the chlorine in normal municipal solid waste would be in the form of chlorinated plastics, with the probability that the remainder would be in the form of common salt.

The only type of plastic in major use that will produce a chlorine compound when burned is polyvinyl chloride, which will produce hydrochloric acid. Polyvinyl chloride comprises about 15 percent of the total plastics commonly found in domestic solid waste, and the total plastic materials of all types in domestic solid waste normally comprise only one to 2 percent of the total waste. Because each pound of polyvinyl chloride contains the equivalent of about 0.6 pounds of hydrochloric acid, the total hydrochloric acid in normal solid waste would be on the order of only 0.09 to 0.18 percent of the total solid waste. This would appear to generally confirm the analytical findings in St. Louis, which were derived from materials containing all of the plastics normally found in municipal solid waste.

Another element sometimes associated with corrosion in powerplant boilers is sulfur. Normal municipal solid waste is essentially sulfur-free. Analysis indicates that the sulfur content of the solid waste delivered to the Union Electric Company facilities ranges from 0.01 to 0.40 percent, with an average of only 0.11 percent.

Future solid waste facilities of the St. Louis type will remove magnetic and nonmagnetic metals, along with other nonburnable materials, from the supplementary fuel before delivery to the powerplants. At least in theory the removal of the metals, together with those elements bonded to them, should have the tendency of reducing the corrosion potential that otherwise might be greater. There is some evidence in support of this premise in European observations.

As part of the investigations to be conducted by the Union Electric Company in St. Louis, special test specimens of metals commonly used in high-temperature areas in boilers will be subjected to the conditions prevailing when solid waste is fired. Two specimens of prestressed austenitic stainless steels, type 347 and type 321, will be tested along with two specimens of ferritic alloys, Type T-9 and Type T-22. It is anticipated that these tests will provide at least partial guidance in evaluating potential high-temperature corrosion.

Slagging. There has been no indication to date in the St. Louis project that solid waste has any greater tendency to form slag than Illinois bituminous coal. Utility personnel have voiced the opinion that the prototype furnace appears to be cleaner when solid waste is fired in combination with coal than when coal is fired alone. The ash fusion temperatures of solid waste, from limited investigations, appear to be of the same order of magnitude as those of Illinois bituminous coal.

Carryover of Unburned Particles. There has been no evidence to date in the St. Louis project of any unburned materials being carried back into the back passes of the boiler by the gas stream. Observations made during periods when the solid waste was being fired showed no visual indication of other than normal quantities of fly ash being carried into the back passes. A complete internal inspection of the boiler after solid waste was fired also disclosed no evidence of carryover.

Stability of Operation. During the prototype operations, the solid waste has been fired consistently at rates of 12 to 15 percent of the total heat requirements of the boiler. Although the solid waste was fired intermittently, no instability in boiler operation was evident. One short-term test was made to attempt to determine the maximum possible burning rate of the solid waste. With one of the units set for operation at 100 megawatts, solid waste was fired to the unit at a rate equivalent to 25 percent of the total fuel requirement. During this test no instability of boiler operation was noted. A survey of the bottom ash indicated no increase in unburned material due to this higher rate of firing. Further testing is planned to determine the maximum percentage of fuel that may be supplied by solid waste without creating unstable conditions in boiler operation.

37

Manpower Requirements. Experience in St. Louis has shown that no additional labor would be required for boiler operation when firing solid waste in combination with pulverized coal. Even with the larger quantities of bottom ash generated when firing solid waste that has not been air classified, with the effect of requiring more frequent sluicing of the bottom ash, no additional labor has been required for ash handling.

The receiving facilities at the Union Electric plant were designed to be unattended, and they are being operated as they were designed. The firing facilities also were designed to be unattended, but the pneumatic feeders were found to be subject to occasional stoppages, because of particles of nonferrous metal becoming jammed between the rotors and the housings of the feeders. This jamming has made it necessary, in the initial operations of the facilities, to have a man in attendance whenever solid waste is being fired. The installation of the air classifier has essentially eliminated the jamming problem, and it is expected that this additional operating labor will no longer be necessary.

Ash Handling and Quality. The total ash content of the solid waste delivered to the Union Electric facilities averaged 15.6 percent after air classification. This average was based upon only 99 analyses, almost all of which were on abnormally wet solid waste. It is anticipated that the average ash content over a longer period of time will be somewhat less, because the waste can be expected to be dryer and particles of inert materials should not cling to the dryer particles of paper.

It has not yet been possible, in the St. Louis prototype, to determine the relative quantities of refuse ash carried by the gas stream to the electrostatic precipitators and that dropping to the bottom ash hopper. It is believed, however, that a substantial portion of the refuse ash will fall to the bottom ash hopper, because the particle sizes of the solid waste ash will be far larger than those of pulverized coal and should have sufficient mass to avoid being carried upward by the relatively low vertical gas velocities in the boiler furnaces. Based upon observations, it is believed that 50 to 60 percent of the refuse ash will fall to the bottom ash hopper.

The sodium content of the total refuse ash is substantially higher than considered acceptable for normal coal ash. However, most of the sodium apparently is included in the inorganic fraction of the solid waste, and most of this fraction falls to the bottom ash hopper of the boiler. This has been borne out of analyses of the fly ash, which indicate the sodium in the mixture of solid waste and refuse ash is about 1.0 percent, essentially the same as for normal coal ash alone.

Analyses of fly ash consisting of a mixture of refuse ash and coal ash show no significant differences from the fly ash from coal alone in the St. Louis installation. The fly ash is being sold, as before, to Portland cement manufacturers.

The bottom ash from classified material is free of most of the larger particles of wood, leather, rubber, and nonmagnetic metals that otherwise are present in the bottom ash resulting from the burning of unclassified solid waste in suspension.

Combustion Control. The St. Louis prototype was designed to feed the solid waste to the boilers at an approximate constant rate, and completely independent of the firing of the normal fuel. Only a rough manual adjustment is provided to feed the solid waste in four approximately equal streams. This premise was adopted to effect simplicity, in order to minimize any additional boiler operating concerns. Control of combustion is effected by the existing boiler combustion controls, which merely increase or decrease the rate of firing of the normal fuel to accommodate the varying heat requirements of the boilers. This system has proven to be satisfactory, and no instability of boiler operation due to the firing of solid waste has been noted.

The only intimate connection between the solid waste firing system and the boiler control system is through the boiler trip circuitry. Should the boiler trip, the refuse feed system is automatically stopped. The pneumatic system, however, continues to operate for a short time, in order to clear the pneumatic pipelines of any solid waste remaining in them.

Erosion. A thorough inspection of the heat transfer surfaces in the prototype boilers at St. Louis had indicated no evidence of erosion of these surfaces due to the firing of solid waste up to this time.

Bridging in Ash Hoppers. During the initial operations, some excess accumulation of material has been noted in the ash hoppers of the prototype boilers. Much of this accumulation has been in the form of burnable particles of wood, plastics, leather, and rubber, which have sufficient mass to prevent their burning in suspension. Some of this material, after falling to the bottom ash hopper, is ignited and continues to burn on the surface of the water in the ash hopper, and as additional material falls on top of it, gradually accumulates up to the throat of the ash hopper. There has been no bridging or other adverse effect noted as a result of this accumulation, and there apparently has been no difficulty in sluicing the ash to the bottom ash pond.

The installation of the air classifier has essentially eliminated the heavier particles of burnable material from the supplementary fuel delivered to the powerplant. The bottom ash, therefore, consists mostly of small particles of inert materials, which would not tend to accumulate in the manner thus far experienced in the prototype.

Ash Lagoons. All of the fly ash produced and collected from the prototype boilers is sold to Portland cement manufacturers. The only ash that is discharged to the ash lagoons, therefore, is bottom ash. As previously stated, the initial burning of solid waste as supplementary fuel was with waste that had not been air classified, and the resulting bottom ash contained particles of unburned wood, plastics, leather, and rubber, as well as an occasional particle of citrus fruit rind or bone. When air classified material is burned, most of the burnable materials are eliminated from the bottom ash, although some are still present.

It has been pointed out that the quantity of bottom ash will increase significantly when solid waste is used as supplementary fuel. Depending upon the rate at which the solid waste is fired, the ash ponds will therefore fill at a much faster rate than when the normal fuel is burned. The quantity of the bottom ash requiring ultimate disposal will consequently be increased.

Storage. No adverse effects have resulted from storing the milled solid waste for several weeks at a time. It would appear that in case of an outage of either the solid waste firing system or the boiler, there normally would be no real necessity of emptying the storage bin or the surge bunkers in the firing systems. It would be easy to make provisions for emptying such facilities if it appeared desirable to do so.

Cold Weather Operation. The St. Louis project has operated during cold weather when the ambient temperature has approached 0°F. Union Electric personnel have indicated that the handling and pneumatic firing of the refuse-fuel has not been affected by cold weather. The only concession to cold weather operation which was made was to start the operation of the pneumatic feeders in sequence (instead of starting them together) in order to provide a smooth transition from fossil fuel firing to combination firing. Preheating of the carrying air in the pneumatic systems has not been found necessary.

The Union Electric Company is quite encouraged by results to date and now feels the process may be applicable to front-fired boilers. It is also felt that firing from only one corner of a tangential boiler is entirely feasible. Some firing at two diagonally opposite corners had been done successfully at the Merramec plant. The opinion of D. L. Klumb is that a 20-percent supplemental fuel rate (by Btu) will easily be reached with air-separated trash and that 30 percent of rated capacity may be possible (Ref. 12). Operation at 20 percent of rated Btu could be feasible at boiler output rates as low as 70 to 80 percent of the rated load, and refuse could be burned as a supplementary fuel down to 50-percent load (Ref. 9).

The major operational obstacle to running at high supplemental fuel rates at reduced load lies in the ability of the primary fuel control in the boiler to compensate for variations of the refuse feed system. Tests have been run, at the Union Electric Company, where satisfactory reduced load operation was observed during five hours of supplemental fuel firing, where the refuse fuel/coal firing ratio was 40 percent on a Btu basis (Ref. 6).

The above results must be considered as preliminary, and many more hours of running with high fuel supplement rates will be needed to establish long-term corrosion effects in the Merramec powerplant boiler. A discussion of corrosion factors, in more detail, appears below under "Application Considerations."

AIR AND WATER POLLUTION CONSIDERATIONS

According to Connecticut regulations, altering an existing powerplant boiler to burn refuse as an auxiliary fuel will reclassify the boiler into the "modified source" category and will reduce the allowable particulate emissions from 0.2 pound per 10^6 Btu down to 0.1 pound per 10^6 Btu. At this point, it is not possible to say whether the St. Louis process will meet either requirement.

Visual observation of the Merramec powerplant stack shows virtually no visible emissions from the refuse-firing boiler; however, no particulate loading tests have been run to date, and it is doubtful that, when they are run, they will be directly applicable to oil-burning plants. A more detailed discussion of air pollution factors in power boilers, using processed refuse as an auxiliary fuel, appears below under "Application Considerations."

Water effluents from the process will be the result of sluicing of the bottom ash into lagoons. It is not believed that this is a pollution problem. Table 7 shows that ash content of the present St. Louis process refuse fuel is about seven times that of coal, when compared on an equal Btu basis. Therefore, the bottom ash disposal problem of an oil-fired utility boiler supplemented at a 15-percent rate by unseparated refuse fuel would be about the same as that of the same boiler when operating using only coal, assuming the same fly-ash/bottom-ash split.

It should be pointed out, however, that air classification of the refuse will reduce the ash content to a level approximately 2.4 times that of coal. Therefore, when running an oil-fired boiler at a 15-percent supplemental firing rate with separated fuel, the total bottom ash would be about 35 percent of that encountered when burning 100-percent coal at the same power output.

SCALING CONSIDERATIONS

The St. Louis process is already being run at scales typical of those expected in State of Connecticut applications. Additional fuel preparation capacity can be obtained by parallel processing lines, and this is the recommended approach, because it will also add reliability to the process system by furnishing a level of redundancy. A 1000-ton-per-day processing system could therefore be accomplished using three parallel processing lines.

The primary concern in scaling up the process lies in the scaling of storage facilities. The tendency of shredded refuse to pack can be aggravated by making storage bins too high. In fact, the Atlas bin at the Merramec powerplant did not work well when refuse was piled to depths over 30 feet, and similar results showed up in the Miller-Hofft bin at the fuel processing plant. It is recommended that vertical scaling be avoided and that any increase in storage capacity be accomplished by increasing horizontal dimensions or by using multiple bins.

COST FACTORS

Capital Costs

Capital costs of the 650-ton-per-day demonstration plant at St. Louis are $2,250,000. These costs include $1,700,000 at the fuel preparation plant and $550,000 at the powerplant. These figures are a barebones demonstration system built on already-prepared sites adjacent to existing major facilities that supply many of the support functions, such as offices and washrooms. No redundancy of processing equipment is provided, and no air classifying equipment is included. The added air classifier and ferrous metal grinder cost about $600,000.

A cost estimate (Ref. 13) for a production-type fuel preparation unit capable of 1000-ton-per-day input on a sustained basis shows a capital cost of $11,700,000 (which includes $3,390,000 for contingencies and $200,000 in sales tax). The estimate includes no costs for land or for site improvement and is expressed in 1973 dollars.

Such a system would include three parallel processing lines, each having an air classifying unit. Provision is also made for grinding of the ferrous metal scrap. Each process line utilizes only a single-stage shredder. A second stage of shredding will increase capital cost by as much as 30 percent, but will result in much lower maintenance and much less downtime.

Operating Costs

No operating cost data have yet been released by the St. Louis demonstration. The only cost factors available are the employment roster and the power requirements listed in this section under "Physical Requirements and Manpower."

Estimates prepared by the Midwest Research Institute (Ref. 2) show an expected operating cost of $798,000 per year for a 1000-ton-per-day plant, with no provision included for sale of products or for amortization. This estimate resolves to a predicted $2.66 per ton of waste input, excluding insurance and administrative charges. This estimate applies to the original process and does not include costs of air classifying.

CONCLUSIONS

The St. Louis process has been identified as the lowest capital-cost resource recovery system (Ref. 2); it compares well with all alternatives, from an operating cost standpoint. The process has now had enough operating time to identify the need for air classification of the shredded refuse.

In addition to air classification, future resource recovery techniques can be added to the process to recover valuable materials from the noncombustible stream. The one disadvantage of dry shredding, however, is that glass recovery is limited, because about 80 percent of the glass is pulverized to a size too small for color sorting. No limitation is presented to recovery of other materials.

The process has significant potential for disposal of solid waste, while conserving fossil fuels on a significant scale; however, it requires considerably more experimentation before it can be accepted without reservation. The questions of long-term corrosion and of air pollution in the form of particulate emissions must be answered. Although there is considerable encouragement from the St. Louis results, there is a real need for experimentation in oil-fired boilers, to determine whether there are any synergistic effects that will occur between refuse fuel and oil. At present, only two units are known to exist that burn oil and refuse in a steam boiler. One unit is in Stuttgart, Germany; the other is in Rochester, New York. A discussion of operating experience at these units (in this section under "Application Considerations") shows that these units are not directly comparable to utility boilers. However, it is felt that both the Stuttgart and Rochester units will give some insight into operating problems with oil and refuse firing.

The process developer, Horner and Shiffrin, Inc., has recognized the desirability of two-stage milling and other processing to reduce the load on the grinders and also to make the noncombustible material available for further processing. Figure 11 shows a process flowsheet recently developed by them based upon the St. Louis experience and proposed for the Capital District region of New York State.

It is believed necessary to run a full-scale experimental test in an oil-fired utility boiler to definitely establish the feasibility of applying the St. Louis demonstration process to an oil-fired situation. In summary, the advantages and disadvantages of the St. Louis process are:

- Advantages

    Capital and operating costs are low.
    Consumption of fossil fuels is significantly reduced.
    Fuel processing technology currently exists.
    Considerable operating experience is available.
    Material recovery processes can be added without greatly
        affecting fuel value.

- Disadvantages

    Long-term boiler effects are not yet established.
    Air pollution effects are not yet established. The powerplant
        may also be reclassified.
    No comparable experience with oil-fired utility boilers is available.
    Use is limited to boilers with bottom ash capability. Ash disposal
        is required in significant amounts.

43

The present single-stage shredding approach, without air classification, results in high shredder maintenance and operational problems at the powerplant. Addition of a second shredding stage is desirable.

Glass recovery is limited by the dry-shredding operation.

A.M. KINNEY/BLACK-CLAWSON FUEL RECOVERY PROCESS

This process basically involves the separation of pulped fuel from other constituents of municipal waste by wet pulping and separation techniques used in the paper industry. The resulting pulped fuel is of very fine consistency, with about a 50-percent moisture content. No experience with extensive use of the fuel in utility boilers exists; however, all elements of the fuel preparation process have been in operation at the material recovery demonstration plant in Franklin, Ohio. This plant is operated by the Black-Clawson Company, with the architect engineer being A. M. Kinney, Inc. Because the two companies are disputing ownership of the process, it has been called "A. M. Kinney/Black-Clawson" in this discussion.

PROCESS DESCRIPTION

Process Flow

The process flow is shown in Figure 12 for a 1000-ton-per-day input plant that uses three Disposall®-like Hydrapulpers to process the input stream. Figure 13 diagrammatically shows the Black-Clawson material recovery process. This process has many components similar to those used in the A. M. Kinney/Black-Clawson fuel recovery process. Much of the technology applied in this process reflects the Black-Clawson Company's experience in designing machinery to make paper from pulp; Figure 13 is included to give an impression of the type of machinery involved.

In either process, unsorted municipal solid waste is dumped by the packer truck into a receiving hopper. After sorting out wood palletts and other bulky items, the process is as described below (for a 1000-ton-per-day plant) (Ref. 14).

Primary Treatment. The mixed refuse is deposited in the receiving hopper sections of the apron-type feed conveyors, which move at a rate controlled by the consistency, or solids content, of the slurry leading the Hydrapulpers. Normal operation requires only two of the three 500-ton-per-day pulpers provided. The third pulper is included for peak process periods and as standby capacity when one of the other units is out of service.

Unpulpable material such as tin cans and stones, most of which is inorganic, is ejected by centrifugal force from the pulper through the junk chute

®Registered trademark of the General Electric Company

44

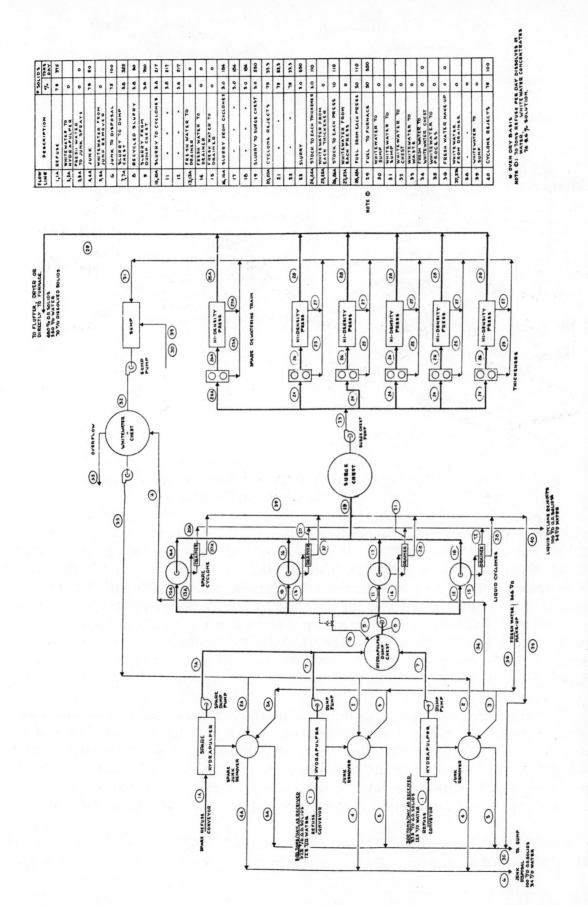

FLOW LINE	DESCRIPTION	% SOLIDS	*TONS DAY
1,1A	REFUSE	7.8	375
1,1A	WHITEWATER TO HYDRAPULPER	0	0
3,3A	FRESH WATER TO JUNK SPRAYS	0	0
4,4A	JUNK	7.8	50
5,5A	WHITEWATER FROM JUNK REMOVER	0	0
6	SLURRY TO DUMP	7.5	100
7,7A	JUNK TO DISPOSAL	3.6	325
8	RECYCLED SLURRY	3.6	50
9	SLURRY FROM DUMP CHEST	3.6	700
10,10A	SLURRY TO CYCLONES	3.6	217
11	"	3.6	217
12	"	3.6	217
13,13A	FRESH WATER TO DRAINER	0	0
14	FRESH WATER FROM DRAINER	0	0
15	FRESH WATER TO DRAINER	0	0
16,16A	SLURRY FROM CYCLONES	3.0	184
17	"	3.0	184
18	"	3.0	184
19	SLURRY TO SURGE CHEST	3.0	550
20,20A	CYCLONE REJECTS	7.5	33.3
21	"	7.8	83.3
22	"	7.8	33.3
23	SLURRY	3.0	550
24,24A	STOCK TO EACH THICKENER	3.0	110
25,25A	WHITEWATER FROM EACH THICKENER	0	0
26,26A	STOCK TO EACH PRESS	10	110
27,27A	WHITEWATER FROM EACH PRESS	0	0
28,28A	FUEL FROM EACH PRESS	50	110
NOTE ①		50	550
30	WHITEWATER TO SUMP	0	0
31	WHITEWATER TO SUMP	0	0
31	WHITEWATER TO WASTE	0	0
33	FRESH WATER TO WHITEWATER CHEST	0	0
33	WHITEWATER TO PROCESS	0	0
36	FRESH WATER MAKE-UP	0	0
37,37A	WHITEWATER FROM DRAINER	0	0
38	WHITEWATER TO SUMP	0	0
39	WHITEWATER TO SUMP	0	0
40	CYCLONE REJECTS	7.8	100

* OVEN DRY BASIS
NOTE ①: 50 TONS REFUSE PER DAY DISSOLVES IN WATER. WHITEWATER CONCENTRATES TO 8.6 % SOLUTION.

Figure 12. A. M. Kinney/Black-Clawson Fuel Recovery Process Flow (Source: A. M. Kinney, Inc.)

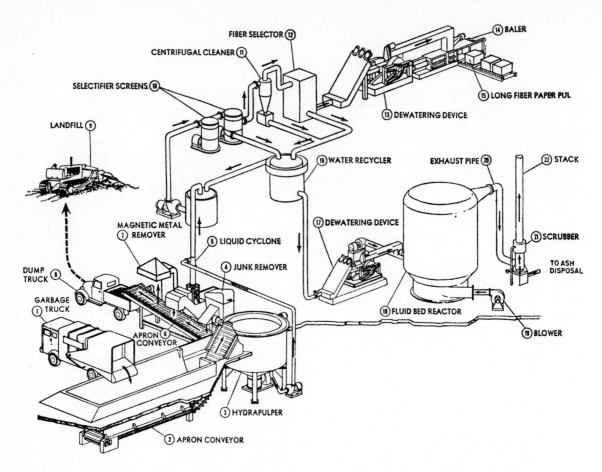

Figure 13. Black-Clawson Material Recovery Process
(Source: Black-Clawson Company)

opening in the side of the Hydrapulper tub. As the ejected material is elevated by the bucket conveyor of the junk remover, it is washed by the system makeup water to remove putrescible material and adherent films. The washed junk is transported by a belt conveyor to a storage bin, from which it is taken to burial. It is anticipated that an average of 100 tons of relatively clean metal is removed each day. Magnetic separation can be installed in the junk remover system to recover an average of 65 tons per day of ferrous metal, which is usually classified as No. 3 bundled scrap and can be sold for about $10 per ton.

The remaining materials in the Hydrapulper are reduced to an aqueous slurry, containing approximately 3 percent solids, by the cutting action of the unit's high-speed rotor. The slurry, after passing through a perforated plate located beneath the rotor, is continuously pumped to the Hydrapulper dump chest, where it is slowly agitated and blended. The dump chest also provides surge volume and a convenient means of emptying the pulper for maintenance.

From the dump chest, the slurry is pumped to any three of the four liquid cyclones that remove glass, metal, and other miscellaneous materials having

46

high terminal settling velocity. This reject material is usually less than 1/2 inch in size. An average of 100 tons per day of this material is thus removed and conveyed to a storage bin to await burial. At present, this material is useful only as a fine aggregate for asphaltic concrete, although sorting methods are expected to be demonstrated that will provide a means of recovering glass and aluminum chips for recycling.

The remaining slurry, at a solids concentration of 3 to 5 percent, is pumped to a surge chest that provides additional blending, surge volume, and capacity for storage of the slurry during maintenance periods. From the surge chest, the slurry is pumped to the dewatering building.

Dewatering. In the dewatering building, the stock is fed directly to five of six dewatering trains. The sixth train is provided as spare and surge capacity. Each train consists of one two-barrel inclined screw thickener, which discharges stock at approximately a 10-percent concentration. This stock is transferred by screw conveyor to a V-cone press, which dewaters it to approximately 50-percent solids. Approximately 1100 tons of material per day are discharged by the presses, of which 480 tons are suspended solids, 70 tons are dissolved solids, and 550 tons are water.

White water from both stages of dewatering is pumped back to the refuse preparation building for reuse in the pulper. Table 9 shows the need for fresh-water makeup to the system. A. M. Kinney, Inc. claims that because

Table 9

SYSTEM WATER BALANCE

Flow	Tons/Day
Output	
With junk removed	34
With cyclone rejects	34
With fuel to furnace	550
Total water removed	618
Input	
In refuse	250
Additional water required	368

of the continuous need for makeup water, there is no contaminated waste water to be discharged to sewers or streams.

Fuel Preparation. The dewatered pulp is conveyed to a fluffer, or light-duty hammermill, which disintegrates the cake produced by the second stage of dewatering, so that the material can be conveyed pneumatically to hoppers located at each boiler feed point. Some drying of the pulp by the transport air is expected. The processed refuse is estimated to have an as-received heating value of 3350 Btu per pound.

Fuel Characteristics and Feed Methods

Studies of solid wastes from various metropolitan areas show that while the refuse composition varies with location and the seasons, the ranges of variation for major components are small. Further, the day-to-day fluctuations of these components are also small. From these results, A. M. Kinney, Inc. concludes that municipal refuse should be convertible to a fuel having nearly uniform day-to-day firing characteristics.

Proximate analyses of pulped refuse indicate that there is uniformity of composition and that the refuse fuel will be similar to that shown in Figure 14. The fuel will also have relative fineness and uniformity of fuel particle size, further enhancing its usefulness as a fuel.

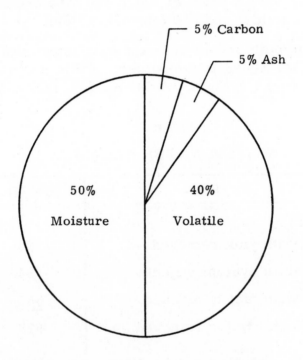

Figure 14. Refuse Fuel Analysis

Input and Output Streams

As estimated by A. M. Kinney, Inc. , the input and output streams for the current 100-ton-per-day process (which does not include glass and aluminum recovery) will be, for a 1000 ton-per-day plant:

- Input

    Packer truck waste   1000 tons/day

48

- Output

Ferrous metal	65 tons/day
Solid fuel (50% moisture)	1100 tons/day
Total product	1165 tons/day
Reject metals from junk remover	35 tons/day
Liquid cyclone rejects (containing aluminum and glass)	100 tons/day
Ash in fuel	55 tons/day
Total for disposal	190 tons/day
HHV of fuel (wet basis)	3350 Btu/lb
Gross total energy/day in fuel	7.37×10^9 Btu/day
Estimated power requirements based on 2000 hp at 34% generation efficiency	0.36×10^9 Btu/day
Net total energy/day	7.01×10^9 Btu/day

Physical Requirements

The physical requirements and other data for a 1000-ton-per-day plant were estimated by A. M. Kinney, Inc. as follows (Ref. 15):

- Land requirement: 10 acres
- Economic life of plant: 15 years
- Operating schedule: 24 hours per day, 365 days per year
- Manpower required:

    One plant manager

    Four supervisors

    Twelve operators

    Seven laborers

- Power requirement: 2000 hp, connected (estimated)

OPERATING HISTORY AND EXPERIENCE

A plant using the A. M. Kinney/Black-Clawson fuel recovery process as such, does not exist. However, the process is basically a modification of the

49

Black-Clawson Fibreclaim Process, which is now in service in Franklin, Ohio.

Plans for the demonstration plant were formulated at the City of Franklin, Ohio in March 1969. This 150-ton-per-day plant handles all of the current municipal waste generation of Franklin and the surrounding communities: Carlisle, Springboro, and Franklin Township. The plant is designed to process the area's projected tonnage through 1990. Currently it is operating on a one-shift, 50-ton-per-day basis.

The plant was constructed under a two-thirds grant from the Bureau of Solid Waste Management (now the office of Solid Waste Management Programs, Environmental Protection Agency) to demonstrate a new solid waste recycling system developed by the Black-Clawson Company in Middletown, Ohio. The project was under the direction of B. F. Eicholz, City Manager of Franklin. A. M. Kinney, Inc. of Cincinnati, Ohio, was retained by the City of Franklin to prepare the design and contract documents and to evaluate the project.

Ground-breaking ceremonies were held September 2, 1970, and the plant went into operation in mid-June 1971. The plant is being operated by the Black-Clawson Company, under a management contract with the City. Output of the plant consists of ferrous scrap that is sold to a nearby steel plant and paper fiber that is sold to a nearby roofing shingle plant.

The solid waste plant is the nucleus around which an environmental control complex is being built. Adjoining the solid waste plant is a new regional waste water treatment plant, designed, built, and operated by the Miami Conservancy District in Dayton, Ohio. When fully operable:

- The purified effluent from this plant will be the process and cooling water supply for the solid waste plant.

- The sludge from the municipal clarifier will be mixed with the non-recyclable organics of the solid waste plant and will be burned.

- The waste water from the solid waste plant will be treated in the new regional plant.

- The ash from the solid waste plant will be used as a settling agent in the regional plant's industrial clarifier.

Plans are underway to add a tank farm and blending station, where non-aqueous fluids, such as cutting oils, from the area will be stored, mixed, and disposed of safely in the facilities of the solid waste plant.

The solid waste processing plant's value was stated as being $2.4 million (Ref. 16). In July 1971, the Office of Solid Waste Management Programs authorized an addition to the solid waste plant. This plant will demonstrate a process developed by the Sortex Company of North America and the Glass

50

Container Manufacturer's Institute, to separate aluminum and color-sorted glass from the inorganic residues generated in the solid waste processing plant (Ref. 17). This addition is scheduled for startup in April 1973.

The system uses the output of the Hydrocyclone as its feedstock, separating the glass and aluminum by a series of steps involving air drying, magnetic separation, air classifying, and electrostatic separation of the glass from the aluminum (Ref. 18). Bench tests show that an aluminum that is 96-percent pure results.

The process recovers about 70 percent of the input glass in a size range large enough to color-sort using the Sortex process. Figure 15 shows a schematic diagram of the process. The cost of the glass and aluminum recovery process for the City of Franklin is $330,000.

Major operating problems encountered at the Franklin plant include Hydrapulper maintenance, the Hydrapulper noise level, erratic performance of the dewatering screws, grease in the fiber, and a low fiber yield. The latter two problems do not apply to the fuel recovery system, but their performance has been improved.

Normal maintenance of the Hydrapulper calls for removal of the rotor for refacing after every 100 hours of operation. The stator must be replaced after every 1200 hours of operation. The Hydrapulper cannot process bulky waste or tires. Hence, a separate bulky shredder would be needed in a complete processing facility.

An additional problem at the Franklin facility is the lack of provision for sufficient floor space in the receiving area. As a result, the front-end loader does not have room to maneuver after several trucks have dumped their loads.

AIR AND WATER POLLUTION CONSIDERATIONS

Some factors indicate that this process violates some of the newer, more stringent air and water pollution control laws. For instance, A. M. Kinney, Inc. has indicated that an additional $350,000 may be needed for dust collection equipment, over and above the original listing of equipment in their previous 1000-ton-per-day-system proposal. Earlier proposal and promotional material from the Black-Clawson Company indicated that no water pollution problem exists because of the closed loop of the system. However, this assertion was evidently based on the combustion of dirty water in the fluidized bed reactor. Considering a fuel-generating version of this process, the fluid bed reactor is omitted, and the polluted water must be treated in some manner that will keep the system in operation. As stated above under "Process Description," the waste water from the solid waste plant at Franklin, Ohio will be treated in a new regional plant. Therefore, it should be assumed that some treatment facilities will be required for the water used in this system.

It is also anticipated that sound-deadening construction will be required on the buildings housing this equipment.

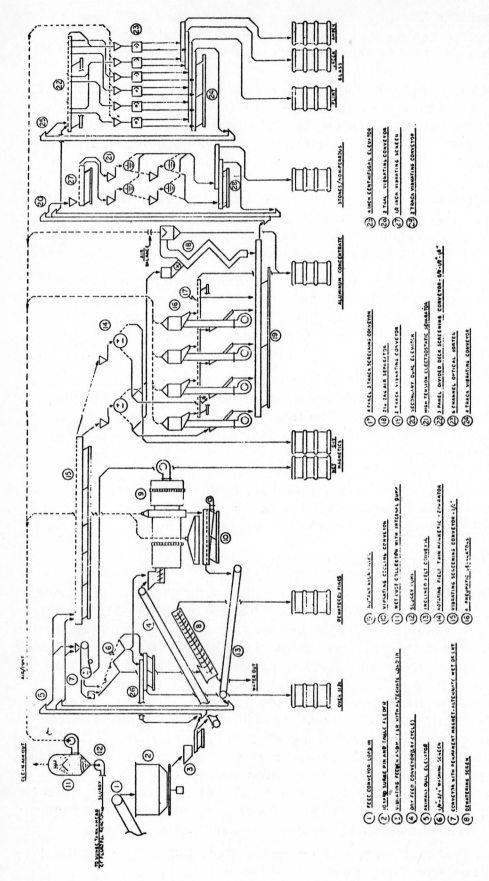

Figure 15. Schematic Diagram of Glass Container Manufacturer's Institute Glass Recovery Process (Source: Glass Container Manufacturer's Institute)

52

SCALING CONSIDERATIONS

Little difficulty is expected in scaling this operation up from the 150-ton-per-day pilot plant to the standard 1000-ton-per-day system under study. First, the pilot plant has been in operation and has attracted a great deal of attention for the past year and a half. Second, the apparatus to be used in the 1000-ton-per-day plant is simply multiples of the equipment used in the smaller pilot plant.

The problem appears to be in the logistics of the matter. The process involves a great deal of equipment, compared to some of the other processes, and the material handling and maintenance resulting from this complex process could prove to be the major consideration in scaling up to larger installations. The 1000-ton-per-day proposal submitted by A. M. Kinney, Inc. appears to be well thought out and realistic. A. M. Kinney has proposed a storage pit between the packer unloading area and the Hydrapulpers. This pit would conserve floor space but could cause construction, housekeeping, and maintenance problems because of the requirement for overhead cranes.

An even larger version of this process has been proposed to Hempstead, New York by the Black-Clawson Company. This proposed plant is designed to receive 2400 tons per day of refuse and will use four Hydrapulper lines, each having a capacity of 800 tons per day on a 24-hour-day basis. The plant will use the refuse fuel to generate electricity by means of a 36-megawatt steam powerplant, 7 megawatts of that output being used in the plant.

COST FACTORS

Capital and Operating Costs

Figures given by A. M. Kinney (Ref. 19) in January 1972 indicate estimated capital costs (Table 10) and operation costs (Table 11). Capital costs are for fuel processing only and do not include equipment for further separation of materials removed by the junker and liquid cyclones or for reclaiming of useful paper fibers. Costs are based on January 1971 prices.

A. M. Kinney, Inc. indicated an additional $350,000 may be needed for additional dust collection equipment to meet new, more stringent laws. Figures are based on burning in existing cyclone-fired boilers. For pulverized coal boilers (such as the Combustion Engineering tangentially fired units), an added cost of $136,000 is needed to cover additional stock bins, refuse burners, and boiler modifications. Additional drying equipment is also believed to be needed to reduce moisture content of the fuel from 50 to 25 percent. This equipment will add an additional $550,000 to the capital costs (Ref. 20).

This total differs considerably from a scale-down of the Black-Clawson Hempstead, New York proposal for a 2400-ton-per-day plant. When scaled

Table 10

SUMMARY OF CONSTRUCTION COSTS FOR A. M. KINNEY PROCESS
(1000-Ton-per-Day Plant)

Cost		Amount
Site work		$ 175,000
Architectural		500,000
Structural		730,000
Electrical		420,000
Mechanical		
Equipment	$1,830,000	
Piping and controls	430,000	
HVAC	70,000	
Plumbing	20,000	2,350,000
Subtotal		4,175,000
Contingencies, land acquisition, and engineering fees		950,000
Total estimated construction		$5,125,000

Table 11

SUMMARY OF ANNUAL OPERATING COSTS FOR A. M. KINNEY PROCESS
(1000-Ton-per-Day Plant)

Cost		Amount
Power cost		$ 105,000
Operating labor		350,000
Maintenance		140,000
Miscellaneous		
Building heat	$ 5,000	
Water	25,000	
Royalties	100,000	130,000
		$ 725,000
Reject disposal at $2/ton		150,000
		$ 875,000
Fixed charges		
Insurance	$ 25,000	
Amortization	550,000	
		575,000
Net annual cost		$1,450,000
Net cost/ton of refuse received*		$3.97

*These data are based on 1000 tons/day and 365 days/
year. Costs on the basis of a 250-day year will be higher.

54

down to 1000 tons per day, the Black-Clawson plant would cost $13.7 million.

Operating costs are based on wages and utility rates considered typical for similar operations in the Cincinnati, Ohio area. The total cost includes estimated royalties for the Black-Clawson Hydrapulper system. The actual royalties would have to be negotiated with the Black-Clawson Company. Amortization costs are based on 15-year municipal bonds at 6-1/2-percent interest.

Process Economics

Credit for Thermal Energy. The processed refuse, as delivered to the boiler feed system, has a net heating value of about 3350 Btu per pound. This value reflects the effect of the 50-percent water content of the fuel. For a 1000-ton-per-day plant, the fuel would return $670,000 annually, if the fuel could be sold for 25 cents per million Btu (which is very reasonable, compared to current fuel costs). The return would reduce the net cost to the city to $2.14 per ton of refuse received on a basis of 365 days per year. On a basis of 250 days per year, the net cost would be considerably higher.

Credit for Recovered Metal, Glass, and Paper Fiber. Ferrous metal recovery is now implemented at the Franklin, Ohio demonstration plant. The ferrous scrap produced is washed as part of the process and, as a result, is free of most paper labels and putrescible material. The scrap is sold for $14.25 per ton, delivered to the Armco Steel Corporation plant in Middletown, Ohio. Table 12 shows the effect of ferrous metal recovery on operating costs, assuming ferrous scrap is salable at $10 per ton, FOB, and fuel is salable at 25 cents per 10^6 Btu.

Installation of the glass and aluminum recovery process at the Franklin plant will produce color-sorted glass valued at $12 per ton, FOB Franklin, and aluminum valued at $200 per ton, FOB Franklin. Yield for the Franklin plant is not known; however, anticipated yields at the proposed Hempstead plant, based on a 2000-ton-per-day input are:

- Ferrous metal: 140 tons per day (7% of input)
- Color-sorted glass: 120 tons per day (6% of input)
- Aluminum: 15 to 20 tons per day (0.75 to 1% of input)

The aluminum figure will not yet apply to the State of Connecticut, because the use of aluminum cans is lower in Connecticut, where total aluminum content in solid waste varies from 0.35 to 0.55 percent.

CONCLUSIONS

The A. M. Kinney/Black-Clawson fuel recovery process uses components and techniques already demonstrated in the Franklin, Ohio plant. Costs ap-

Table 12

COST ESTIMATES ADJUSTED FOR FERROUS METAL RECOVERY
FOR A. M. KINNEY PROCESS

Cost	Amount
Construction Cost	
Construction cost for 1000-ton/day unit	$5,125,000
Cost for magnetic recovery equipment	50,000
	$5,175,000
Operating Costs and Fixed Charges	
Total annual cost for 1000-ton/day unit	$1,450,000
Revenue from sale of processed refuse	670,000
Net annual cost with thermal energy credit	$ 780,000
Increased operating cost with separation of ferrous metals	25,000
Increase in amortization and insurance	5,000
Reduced landfill cost (65 tons/day at $2/ton)	48,000
Revenue from sale of metal (65 tons/day at $10/ton)	240,000
Net annual cost	$ 522,000
Net disposal cost/ton of refuse with sale of thermal energy and ferrous metals	$1.43

Note: The net cost for processing refuse where thermal
energy and ferrous metals are recovered and sold
is about $1.43/ton of delivered refuse on the basis
of 1000 tons/day and 365 days/year. Net costs will
rise considerably for a shorter work year.

pear to be reasonable (Table 10) and, in fact, capital cost figures presented
by A. M. Kinney, Inc. for a 1000-ton-per-day plant work out to an astonishingly
low value of $5125-per-ton-per-day input capacity.

These figures may be open to question, however, when compared to the
Black-Clawson Hempstead proposal for a 2400-ton-per-day plant at a capital
cost of $34,000,000. Because this price includes a powerplant of 36,000-
kilowatt capacity, about $10,800,000 should be subtracted from the price if
only a fuel preparation plant is considered. (About $300/kw is the present

cost for powerplants in this size range.) This total leaves an approximate
cost of $23,200,000 for the fuel preparation plant (exclusive of land), which
works out to about $9700-per-ton-per-day input capacity. When scaled down
to a 1000-ton-per-day input size range, the cost will be in the $12,000,000 to
$15,000,000 range, or $12,000 to $15,000 per ton per day, much more than the
A. M. Kinney figure.

Another aspect of this process is the high moisture content of the resulting
fuel. A 50-percent moisture fuel will place too large a load on the induced
draft fans of a utility boiler if any appreciable supplemental fuel firing level
is used. Because 50-percent moisture is about the lowest moisture level
achievable with mechanical dewatering, it will be necessary to thermally dry
the fuel to a more acceptable level of 25-percent moisture.

The most logical source of thermal energy is the fuel itself. Assuming a
100-percent-efficient drier, calculations show that for the 1000-ton-per-day
plant, the drying process would consume 136 tons per day of fuel. The resulting
yield of 25-percent moist fuel would be 643 tons per day, with a heating value
of 5025 Btu per pound. The gross daily energy yield will then be 6.46×10^9 Btu
per day.

The power requirement for the process plant is estimated by A. M. Kinney
to be 2000 hp, which works out to 35.8 kilowatt-hours per ton, which is much
lower than the 179 kilowatt-hours per ton reported at the Franklin plant and
the 84 kilowatt-hours per ton predicted for the Hempstead plant. However,
assuming the 35.8 kilowatt-hour-per-ton figure to be correct, and assuming
the electrical energy to be generated at a 34-percent total efficiency, the power
consumption of the 1000-ton-per-day plant would be the equivalent of a heat
input of 0.36×10^9 Btu per day. The net energy output of the fuel plant would
be at most, 6.10×10^9 Btu per day as a result. As already noted, the added
cost of the drier would be on the order of $500,000 to $600,000 (Ref. 20).

The nature of the fuel indicates advantages and disadvantages. Uniformity
should be very good and should offer advantages in transportation and combustion. The smaller particle size produced by pulping should also be an
advantage in the combustion process, but may result in a larger fraction of the
ash becoming fly ash. The pulping process itself produces a larger fraction
of glass particles in a size range that is color-sortable, compared to dry
shredding.

On the other hand, the pulper is not capable of bulky shredding, and a
separate dry shredder will be necessary to handle bulky material. In a dry
shredding process, a bulky shredder could also serve as a standby primary
shredder. This advantage is not found in the wet pulping process.

Although the process uses relatively well proven equipment, the pulping
process is not so well defined that advances in the state-of-the-art come
easily.

The actual use of pulped fuel in an electric utility boiler has yet to be tried on a scale comparable to the St. Louis experiment. It is expected that the fuel will be quite comparable to the air-separated, dry-shredded fuel, and no major problems are foreseen for pulped fuel that do not also apply to the dry-shredded fuel.

In summary, the advantages and disadvantages are:

- Advantages

    Capital cost and operating cost are relatively low. (There is some doubt about these figures.)

    Consumption of fossil fuels is significantly reduced.

    Basic elements of the process have been proven at Franklin, Ohio.

    Material recovery is enhanced; the glass recovery potential is better.

    Fuel is uniform and of small particle size.

- Disadvantages

    The fuel has not yet been used as a boiler fuel with either coal or oil.

    Long-term boiler effects are not established.

    Air pollution effects are not established. The powerplant may be reclassified as well.

    The fuel is limited to boilers with bottom ash capacity. Ash disposal is required in significant amounts.

    A separate bulky shredder is necessary.

    A drier must be added to bring down moisture in the fuel.

    The energy yield is less than that for the St. Louis system.

    Economic projections seem optimistic.

    Waste water treatment is needed.

COMBUSTION POWER CORPORATION CPU-400 PROCESS (MENLO PARK)

The Combustion Power Corporation CPU-400 process is a package plant designed to consume 400 tons of waste per day. Steel, glass, and aluminum are separated out, and most of the balance is burned, with heat recovery accomplished by using a gas turbine generator. The process is modular in nature and can be expanded to larger sites with relative ease. For this study, the front end of the system, which prepares fuel for the gas turbine combustor, is described.

PROCESS DESCRIPTION

Process Flow

Figures 17 through 21 show the basic CPU-400 process and various modifications of it.

Figure 17 illustrates the front end of the system in functional diagram form, and Figure 19 shows a block diagram of the front end. This front end consists of a receiving area in which the municipal waste is unloaded from packer trucks. The waste is then pushed onto a slat conveyor by a front-end loader. The material is conveyed into the top of a vertical shaft shredder, which reduces it to a nominal 4- to 6-inch size. The shredded waste is then processed through a zig-zag air classifier.

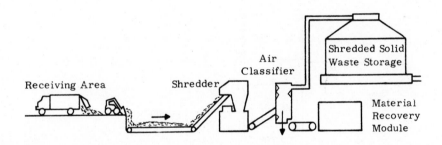

Figure 17. CPU-400 Process Front End (Source: Combustion Power Corporation)

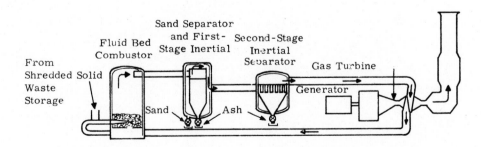

Figure 18. CPU-400 Process Heat Recovery System (Source: Combustion Power Corporation)

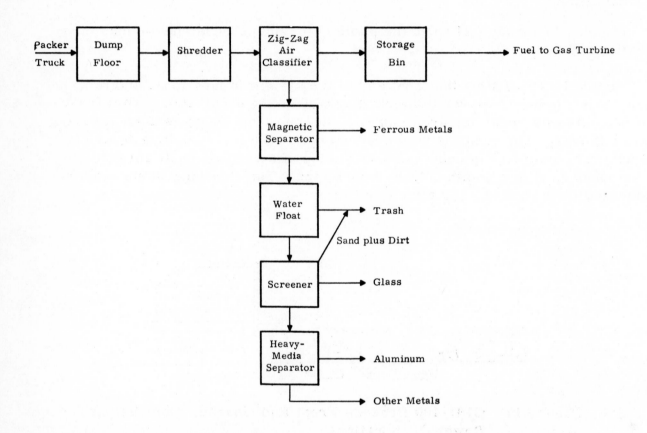

Figure 19. Block Diagram of CPU-400 Process Front End System

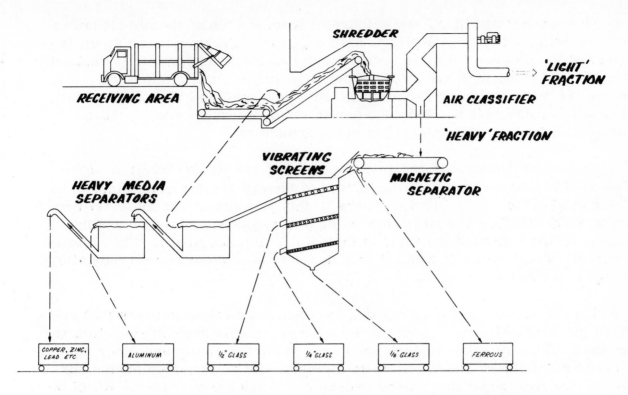

Figure 20. CPU-400 Process Material Recovery System
(Source: Combustion Power Corporation)

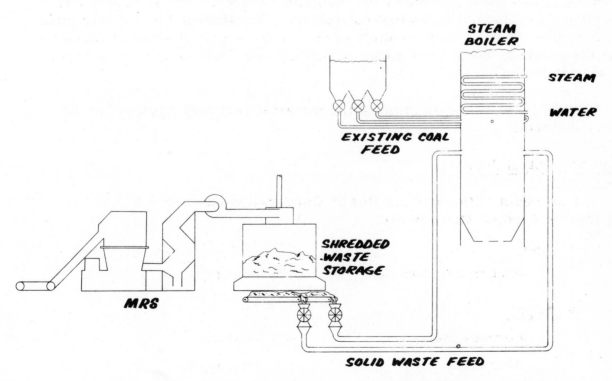

Figure 21. CPU-400 Process Suspension Fire to Existing Boiler
(Source: Combustion Power Corporation)

The low-density items are withdrawn from the top of the air classifier and are conveyed pneumatically to a storage bin. Cyclone dust collectors are used to meet air pollution requirements as the transport air is separated from the solid waste at the storage bin. The high-density items are transported from the bottom of the air classifier to a material recovery module. (The gas turbine heat recovery system is not of interest here, because only refuse fuel recovery processing is being described.)

Figure 20 illustrates the process flow of the dense fraction of the shredded waste that falls off the bottom of the zig-zag air classifier. The Combustion Power Corporation refers to this process as the material recovery system (MRS). The first step in the process is magnetic separation, which removes most of the ferrous material. An exception to this removed material is that which is trapped with nonferrous material or ferrous alloys that are not highly magnetic.

This first separation is followed by a set of vibrating screens that remove glass and dirt materials. A set of heavy-media separators follow the screens. These separators are used to recover aluminum and other nonferrous materials. The low-density material that is skimmed off the surface of the first separator can be recycled through the shredder. All of these items are currently available.

Figure 21 illustrates the use of an existing steam boiler to recover heat from the combustible portion of the municipal wastes. The shredded waste is extracted from conveyors that outfeed from the storage bin and introduce the shredded waste to burn in suspension in the boiler. It should be noted that the particle size from a single shredding operation is not optimum for most suspension burning boilers.

Figure 22 is a flow diagram of the material recovery system for the heavy material.

Input and Output Streams

On the basis of Combustion Power Corporation information (Ref. 21), the input and output streams are:

- Input

    Packer truck waste   1000 tons/day

- Output

    Ferrous metal   50 tons/day

    Aluminum   5 tons/day

    Other nonferrous metals   2 tons/day

62

Glass	30 tons/day
Solid fuel (dry basis)	519 tons/day
Total product	606 tons/day
Sand, dirt, and junk	75 tons/day
Ash content of fuel	75 tons/day
Total for disposal	150 tons/day
HHV of fuel (wet basis)	4506 Btu/lb
Gross total energy/day in fuel	6.74×10^9 Btu/day
Estimated power requirements based on 2000 hp at 34% generation efficiency	0.36×10^9 Btu/day
Net energy per day	6.38×10^9 Btu/day

Physical Requirements

Power. The annual cost of utilities (fuel, electricity, and water) is estimated by the Combustion Power Corporation to be $96,000. Total connected horsepower is estimated at 4030.

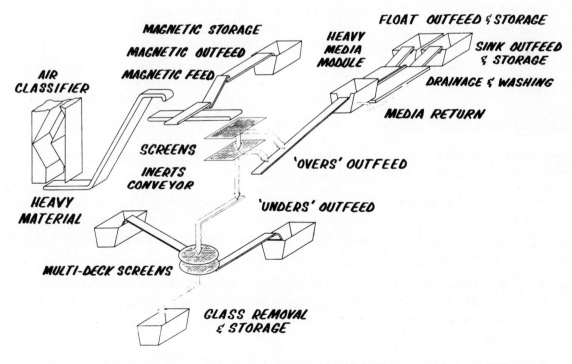

Figure 22. Flow Diagram of Heavy Material Recovery System
(Source: Combustion Power Corporation)

Space. Building requirements for the 1000-ton-per-day system are estimated to be:

Tipping area (for 2000 tons)	200 x 200 ft =	40,000 ft^2
Primary treatment	75 x 75 ft =	5,625 ft^2
Material recovery	75 x 125 ft =	9,375 ft^2
Storage	125 x 200 ft =	25,000 ft^2
Total	420 x 200 ft =	80,000 ft^2

Land area for a 1000-ton-per-day plant has not been supplied by the Combustion Power Corporation. However, a 477-ton-per-day unit requires 2.7 acres. Linearly scaling up to 1000 tons per day shows that 5.7 acres are required.

OPERATING HISTORY AND EXPERIENCE

The Combustion Power Corporation began research and development on its CPU-400 system in 1957 and has been working on it continuously since then. All the work has been sponsored by the Environmental Protection Agency, and approximately $6 million has been committed to the program to date. An 80-ton-per-day pilot plant at Menlo Park, California is scheduled for completion in early 1973, and extensive testing of the prototype is planned during 1973. Construction of a full-scale CPU-400 process prototype is scheduled to begin in mid-1973, with first operation to be late in 1974. The feasibility of disposing of nondewatered sewage sludge will also be investigated during the testing of the prototype. The Combustion Power Corporation claims that about 1000 tons of refuse have been shredded and separated in their pilot plant, and 50 to 100 tons have been run through the material recovery system (Ref. 22).

SCALING CONSIDERATIONS

Several considerations make the task of scaling this system up to a 1000-ton-per-day standard system difficult:

- The largest system that has even been attempted to date is an 80-ton-per-day pilot plant. Because it has had limited operation as a system, its abilities to meet Government regulations regarding safety and pollution cannot be certain.

- The proposals for a 400-ton-per-day system and an 800-ton-per-day system, which were released by the Combustion Power Corporation, have significant inconsistencies in the figures for equipment and facilities.

- A check with the component vendors, such as the shredder manufacturer, reveal that the Combustion Power Corporation has provided a very meager margin, above the basic equipment costs, to cover costs of system engineering, installation, contingencies, and such related expenses as shipment, taxes, and escalation.

● The job of scaling up is complicated by the fact that the 1000-ton-per-day standard system falls between the Combustion Power Corporation's combination of two CPU-400 systems to give an 800-ton-per-day overall installation and a combination of three of their standard modules to give a 1200-ton-per-day installation.

Most of the financial figures include the gas turbine heat recovery system, which does not enter into this technological assessment.

Figure 23 indicates an attempt to scale up the system to handle 1000 tons per day. Enough redundancy has been provided by using three complete lines

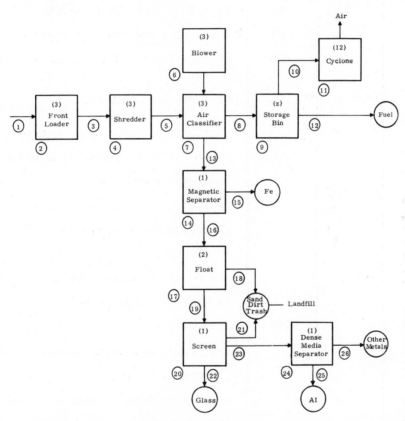

Figure 23. Combustion Power Corporation 1000-Ton-per-Day System

of equipment on the primary stream. Two 200-ton storage bins have also been used. Table 13 lists the equipment for a 1000-ton-per-day plant and indicates the estimated equipment cost.

AIR AND WATER POLLUTION CONSIDERATIONS

Opportunities for air pollution occur at the shredder and at several other devices downstream. The shredder is equipped with a hood, to minimize the

Table 13

1000 TON-PER-DAY SYSTEM EQUIPMENT

Item	Quantity	Cost ($K)		Description
		Each	Total	
1				Dump area
2	3	12	36	Front loader
3	3	78	234	Conveyor, shredder feed model 1000 (120 ft)
4	3	185	555	Shredder, Eidal SW 1000
5	3	8	24	Conveyor, classifier feed, model 1000
6	3	16	48	Blower drive, model 1000
7	3	24	72	Air classifier, model 1000
8	6	8	48	Conveyor, pneumatic, model 1000
9	2	270	540	Storage bin for shredded fuel, 200 tons
10 11 }	12	9	108	Cyclone dust collector
12	12	9	108	Conveyor, storage bin outfeed
13	3			Conveyor, air classifier outfeed (heavies)
14	1			Magnetic separator
15	1			Conveyor, magnetic material
16	2			Conveyor, float tank feed
17	2			Float tank, water
18	2			Conveyor, float rejects
19	2			Conveyor, screen feed
20	1		620	Screen
21	1			Conveyor, screening, sand, and dirt
22	1			Conveyor, glass
23	1			Conveyor, dense media separator feed
24	1			Dense media separator
25	1			Conveyor, aluminum
26	1			Conveyor, other metals
			2388	Total, based on scaled-up Combustion Power Corporation estimates

escape of dust, and much of the equipment downstream from the shredder is protected from air pollution by the fact that the system is enclosed to allow the blower (which sucks the air through the air classifier) to operate efficiently and also to convey the material pneumatically through the top of the storage waste bin. The cyclone dust collectors at the top of the storage bin may pose a problem, and the vendor should be consulted to make certain that these items are in compliance with federal and state regulations. Noise will be a consideration, and sound-deadening characteristics in the enclosing structures will be needed.

Water pollution can result from the material recovery system's float tanks and from floor drains or leachate from landfill operations if a small portion of the low-value material is disposed of in that manner. The flow tank problem could best be solved by a closed loop, with any required treatment included in the loop. The possibilities of closing the loop on the air transport system for the shredded waste should also be investigated.

COST FACTORS

Considerable discrepancy was found between the cost figures submitted by the Combustion Power Corporation in various documents (Ref. 23). Some of these variations may result from attempting to scale up from a very small pilot operation to a rather large installation. These variations are further complicated by the fact that the 1000-ton-per-day standard plant falls between the multiples resulting from the Combustion Power Corporation's 400-ton-per-day standard module.

The following cost figures were released to the General Electric Company by the Combustion Power Corporation as applicable to a 1000-ton-per-day system, with an anticipated economic life of 20 years (Ref. 21):

- Capital Investment

Solid waste primary treatment and storage and material recovery	$ 2,285,000
Auxiliary and support facilities	555,000
Structures	689,000
Total	$ 3,529,000

- Annual Operating Costs

Direct labor	$ 422,000
Purchased materials	52,000
Contract maintenance	
Solid waste subsystem	365,000
Controls	20,000

	Variable overhead	106,000
	Utilities	96,000
	Total	$ 1,061,000

- Anticipated Annual Income from Products

| | Quantity | | Price | |
Product	Percent	Price	(/ton)	Income
Ferrous	5	$ 18,300,000	$ 14	$ 256,000
Glass	3	10,950,000	12	131,000
Aluminum	0.5	1,825,000	200	365,000
Other metals	0.1	365,000	500	183,000
Sand and ash	15	54,750,000	3	164,000
Solid fuel (dry)	52	189,435,000	3	568,000
Total				$ 1,667,000

It is apparent, however, that capital investment costs would require adjustment upward to be consistent with good estimating practice. Contingency provision and installation increases are recommended (see Section 6, "Comparison of Front-End Systems").

CONCLUSIONS

The advantages and disadvantages of the CPU-400 front-end system are:

- Advantages

    Investment cost appears to be low.

    Many components are off-the-shelf items.

    The front-end system is simple.

    Consumption of fossil fuels is reduced significantly.

- Disadvantages

    Operating experience is lacking, especially for full-scale systems. Therefore, operating costs, purity of output streams, installation costs, and contingency estimates are highly speculative.

Single-stage shredding is unsatisfactory. The high maintenance of the shredding equipment on a single-stage device increases operating costs significantly. This is especially true when the output from the shredder must result in a particle size small enough for efficient suspension burning. A second stage, following the air classifier, would be desirable.

No experience using the fuel for boilers exists. Long-term corrosion effects and air pollution effects are not established. The powerplant may also be reclassified into a new emission category.

Boilers with bottom ash-handling capacity are required.

Significant inconsistencies exist in the various documents that have been released. These inconsistencies cause some doubt concerning the various statements that have been made concerning the abilities and capacities of the system.

A credibility gap exists as a result of the fact that a great deal of time and effort have been spent on this project, with very little apparent success. About $6 million in federal funds have been spent since 1967, and there has been relatively little operation to date.

COMPOSITION AND HEAT VALUE OF AVERAGE MUNICIPAL REFUSE

In this subsection, the various surveys on the material and energy contents of average municipal solid waste are reviewed, and an updated average is presented. This average is used as the standard in this study. Based on this standard waste composition and on a defined effectiveness of the refuse shredding and separating facility, details of proximate and ultimate analysis of both refuse and refuse fuel are derived.

REFUSE CHARACTERISTICS

The refuse considered here is an average municipal refuse. It is recognized, of course, that actual refuse composition varies widely between locations or communities and by seasons, days, and loads received at a collection-disposal site. Even for a fixed location, the proportion of the refuse constituents also varies on a long-range time scale. To have a common standard for discussion, an average composition has been proposed as shown in Table 14.

The main purpose in selecting a representative refuse analysis is to estimate the energy potential from the combustion of refuse as a fuel. Because the heat value of the refuse can be and is expressed in various manners, considerable confusion exists relative to this subject.

Various definitions of heat value in common use are presented below, with their interrelationships. To estimate energy potential, only the variation in

Table 14

SUGGESTED REFUSE COMPOSITION AS RECEIVED

Constituent	Weight (%)
Paper	37.0
Glass	9.0
Ferrous metals	7.6
Nonferrous metals	0.8
Plastics	1.4
Leather, rubber, textiles, and wood	6.0
Garbage and yard waste	10.0
Moisture	25.0
Miscellaneous	3.2
As received HHV (Btu/lb)	4700
Ash (total)	22.5

Basis: Data of the Midwest Research Institute (Ref. 2), the Envirogenics Company (Ref. 24), and W.R. Niessen (Ref. 25).

the proportion of combustible elements of the refuse is important; it is shown later that such variation is much less than the normal variation in moisture and noncombustible content.

HEAT VALUE OF REFUSE -- DEFINITIONS

Various methods of expressing heat value of refuse are:

- HHV (higher heating value, gross heat value)

- LHV (lower heating value, net heat value)
  LHV = HHV - 1040 × (moisture/lb of refuse)

- As received (or as discarded)

- Dry = $\dfrac{\text{As-received heat value}}{1 - \text{(per-unit moisture content)}}$

- MAF (moisture and ash-free) = $\dfrac{\text{As-received heat value}}{1 - \text{(per-unit moisture and ash content)}}$

Literature values are based on an as-received HHV basis or on an as-fired basis (this requires preprocessing data). The actual heat value would be defined by any combination of one of the methods; thus there are a total of six different methods in use.

70

The HHV basis is most commonly used in the United States, because all fuel purchases are made on this basis. The HHV of a fuel can be obtained easily and directly by testing; most refuse heat value data are usually derived on this basis, even though it is not always mentioned explicitly.

The LHV method allows for the loss of available heat energy in evaporating the moisture content of the refuse when the refuse is burned. This heat value practice is commonly used in Europe. For calculating available energy from burning of the refuse, the LHV basis is directly useful.

On the other hand, heat value is variously expressed on the basis of including or excluding some of the refuse constituents. Thus, as-received heat value of the refuse represents the available heat energy (on a HHV or LHV basis) when a pound of as-received refuse is burned. Other methods of expressing heat value, as defined above, assume that a particular refuse constituent is totally separated. This assumption is impractical in that both moisture and ash are usually intrinsic parts of most refuse constituents (paper, yard waste, etc.). Further, it is obvious from the above definitions that heat value expressed on the basis of excluding moisture content does not allow for moisture evaporation heat loss; such a distinction should be made by further specifying the HHV or LHV basis.

REFUSE COMPOSITION

Table 15 indicates weight percentages of refuse constituents, as reported by several prominent sources. The data by Kaiser (Ref. 26) published in 1964, are most complete and have been used widely. The Midwest Research Institute data are taken from their recent study conducted for the President's Coun-

Table 15

MUNICIPAL REFUSE COMPOSITION

Constituent	Weight (%)		
	Kaiser (Ref. 26)	MRI (Ref. 2)	Envirogenics (Ref. 24)
Paper	37. 8	33. 0	36. 0
Glass	5. 9	8. 0	12. 0
Ferrous metals	7. 8	7. 6	10. 0
Nonferrous metals	7. 8	0. 6	10. 0
Plastics, leather, rubber, textiles, and wood	5. 0	6. 4	4. 0
Garbage, yard waste	13. 8	15. 6	9. 6
Moisture	20. 7	27. 0	25. 3
Miscellaneous	9. 0	1. 8	3. 1
	100. 0	100. 0	100. 0

MRI = Midwest Research Institute

cil on Environmental Quality (Ref. 2). The data in the last column are abstracted from a comprehensive study, conducted by the Envirogenics Company for the Environmental Protection Agency and published in November 1971 (Ref. 24). These data were based on an averaging of Kaiser's data and other data compiled by the Environmental Protection Agency.

It should be noted that the percentage weight values for the refuse constituents are given on a moisture-free basis, and that total moisture content is given separately. For instance, the moisture in the paper is included under "Moisture." Each constituent also contains certain noncombustibles along with volatile matter.

A further ultimate analysis of the refuse can be carried out to determine its composition in terms of basic chemical elements. Table 16 presents such

Table 16

ULTIMATE ANALYSIS AND HEAT VALUE

Composition	Weight (%)			
	Kaiser (Ref. 26)	St. Louis (Ref. 10)	Chicago NW (Ref. 27)	Envirogenics (Ref. 24)
Carbon	28.00	--	26.50	25.5
Hydrogen	3.50	--	3.50	3.4
Oxygen	22.35	--	22.00	21.7
Nitrogen	0.33	--	--	0.5
Sulfur	0.16	0.14	0.25	0.1
Noncombustibles	24.93	19.9	33.25	21.7
Moisture	20.73	25.7	14.50	27.1
Heat value (HHV) (Btu/lb)				
As received	4917	4280	4360	(4800)
Dry	6203	(5780)*	(5100)	(6600)
Moisture and ash-free	9048	(7720)	(8350)	9410
No. samples	--	68	5	10+
Sampling period	1950-1962	July 1972	May 1971	1968-1970

*Values given in parentheses are computed by the General Electric Company.

composition data from various sources. Comprehensive ultimate analysis data are given in References 28 and 29. The data for St. Louis are the most recent (presented in the Union Electric seminar in October 1972). The refuse composition data presented in the seminar were expressed on the basis of samples received at the research laboratory of the Ralston-Purina Company (Ref. 10); thus, the refuse in the samples had already passed through the shredding and magnetic separation stages before it was analyzed at the laboratory.

The data presented in Table 16 are given on an as-received basis and have been corrected for magnetic separation. The Chicago data are used for the determination of operating efficiency of the Chicago Northwest Waterwall incinerator recently installed.

The ultimate analysis data can be used to estimate the as-received HHV of refuse by using Dulong's formula:

$$\text{HHV (as received)} = 14544C + 62028 \left(H - \frac{O}{8}\right) + 4050 \, S$$

where C, H, O, and S are expressed on a per-unit weight basis. As explained below under "Ultimate Analysis of Refuse and Refuse Fuel," however, the caloric value of refuse calculated on this basis does not prove to be satisfactory, in most cases, compared to test data.

SUMMARY OF REPORTED HEAT VALUE DATA

The data available for various major solid waste heat recovery systems are compiled in Table 17. In each case, the reported heat value and associated

Table 17

EVALUATION OF REPORTED HEAT VALUE DATA

Source	Reported Heat Value (Btu/lb)	Sample Condition	Reference	Calculated Heat Value, As-Received, HHV Basis
St. Louis (Union Electric Co.)	4631	As-received basis after shredding and magnetic separation	10	4280
Chicago Northwest	4360	As fired, no separation	27	4360
A.M. Kinney, Inc.	3350	As fired, 50% moisture, unpulpable material (approx 20%) rejected	19	3685
Monsanto Co.	4600	Shredding, no separation	30	4600
Ecology, Inc.	6620-8500	Compost process output, 12% moisture, ferrous metal removed, 640 lb fuel-out/ton-in	31	2120-2720

major preprocessing stages are mentioned. The last column of Table 17 lists the heat values calculated on a uniform as-received HHV basis, by using the relations previously described. The A.M. Kinney process output is slightly lower in heating value, apparently because some combustibles are lost with the metal and glass that are separated. The compost product heating value is lower yet, because the composting process itself uses up the heating value of the refuse. Section 5, "Composting Processes," describes the comparable loss in heating value for all of the high-rate compost processes studied.

SUGGESTED REFUSE COMPOSITION

Because of inherent variations in refuse composition and in heat value, by location and time of testing, it is believed that a comprehensive gathering of data to characterize refuse in fine details will be of marginal value. Instead, the best available data (given in Tables 14 through 17 and in Ref. 25) will be examined, and a representative refuse composition will be developed, as given in Table 14. In the absence of any better source of data, the data in Table 14 will be used for all future energy recovery calculations.

73

As reported by Niessen and Alsobrook (Ref. 25), paper content of municipal refuse is lowest in summer; spring, fall, and winter refuse have increasing paper content. Thus, viewing the data in Table 16 from this perspective, the refuse composition data for St. Louis and Chicago will not be a representative average for the whole year. The suggested composition of Table 14 is selected to be a representative yearly average of an average municipal refuse.

Niessen and Alsobrook also analyze the refuse composition changes (particularly yard waste) for various regions; their data indicate that yard waste for northern states follows the all-region average values for all seasons.

SEASONAL VARIATION IN HEATING VALUE

The data on seasonal heat value variations are scarce. The Envirogenics Company estimate of the variation (Ref. 24) is shown in Figure 24. Note that the moisture and ash-free (MAF) heat-value basis does not indicate as significant a variation as is evident in the as-received value.

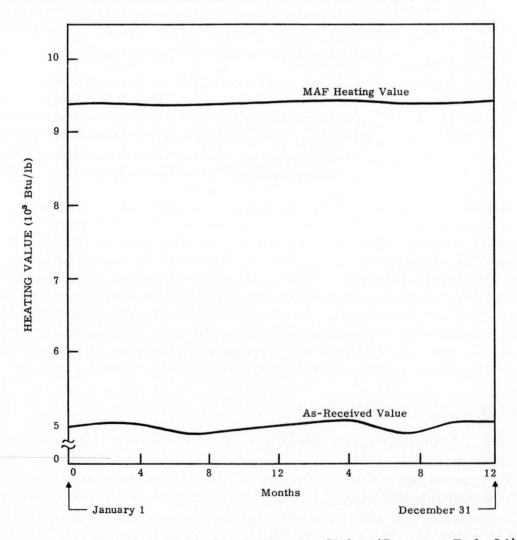

Figure 24. Seasonal Variation in Heating Value (Source: Ref. 24)

PROJECTED COMPOSITION CHANGES

Various methods have been used to project the changes in municipal refuse composition until the year 2000. The results are presented in Figure 25.

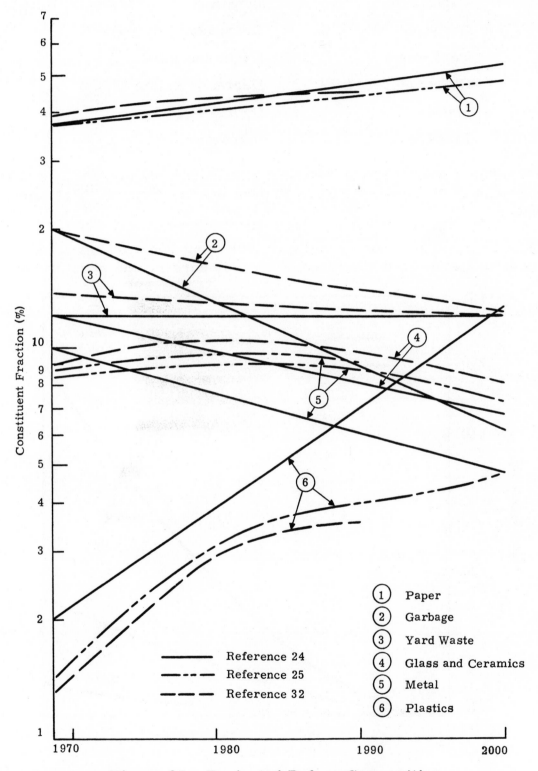

Figure 25. Projected Refuse Composition

Despite the fact that various estimates differ in their details, the following trends are evident:

Constituent	Trend (% weight)
Paper	Increase
Garbage	Decrease
Yard waste	Slight decrease
Glass and ceramics	Decrease in late 1990's
Metals	Decrease
Plastics	Increase

Of major interest here is the effect of such composition changes in the heating value of the refuse. Figure 26 presents the projected heat value variations.

The projected increase by the year 2000 in as-received heat value of refuse is estimated to be in the range of 18 to 50 percent; the moisture and ash-

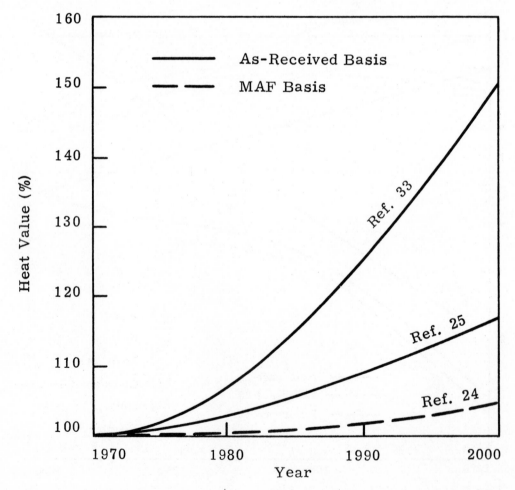

Figure 26. Projected Heating Value of Mixed Municipal Refuse

free (MAF) heat value projected by the same source indicates a 4-percent change during the next 30 years.

Despite the differences in heat value projections, the net effect of increasing heat value would be that a given facility designed to handle a certain energy release rate from refuse would be able to burn a lower quantity of higher Btu value in the future.

ULTIMATE ANALYSIS OF REFUSE AND REFUSE FUEL

The suggested composition and heat value of average municipal refuse presented above under "Suggested Refuse Composition" can be used to estimate energy potential from the combustion of refuse. For certain detailed combustion calculations, it is further necessary to derive an ultimate analysis of the prepared refuse. This subsection presents data on ultimate and proximate analyses, both for the average municipal refuse and for the as-fired refuse fuel, after it passes through certain fuel preparation stages.

The recent Union Electric Company seminar (Ref. 10) presented test results of the composition of refuse and refuse ash associated with its Merramec powerplant, which is burning coal and refuse. The paper presents a statistical distribution of the quantities and proximate analysis of refuse (moisture, sulfur, chloride, and ash); however, it does not include similar statistical data for the elements of ultimate analysis.

Two papers (Refs. 28 and 29) presented at the National Incinerator Conference of the American Society of Mechanical Engineers present comprehensive details of ultimate analysis of each constituent of municipal refuse. These data are in general agreement in both cases; the data of the 1970 paper (Ref. 29) are used here. Using the refuse composition described above under "Suggested Refuse Composition," and using data presented here, an ultimate analysis for the average municipal refuse is derived, as shown in Tables 18 and 19.

While passing through the fuel preparation stage, a certain fraction of each of the constituents of the incoming refuse is separated from the process output. The exact separation effect is directly related to the particular process; the assumed separation efficiency is indicated in Table 20 (Ref. 34). In a manner similar to that used for raw refuse analysis, the ultimate analysis for prepared refuse fuel is computed as shown in Tables 18 and 21.

The following observations are made, based on its composition, when the heat value of any refuse is estimated:

- Using the dry heat value of each constituent and its weight percentage, the heat value contributed by each of the refuse constituents, and hence the total heat value, can be computed. This value, however, does not check with reported total heat values for both References 28 and 29.

- Using ultimate analysis data and the Dulong Formula, the heat value can be computed. These computed values also are in strong disagreement with reported heat values of References 28 and 29. E. R. Kaiser (Ref. 28) and M. B. Owen (Ref. 35), however, indicate that such correlation is not expected to be satisfactory for estimating the refuse calorific value.

- E. R. Kaiser (Ref. 36) indicated that the moisture and ash-free (MAF) heat value of the organic content of municipal refuse is in the range of 8800 Btu per pound to 9100 Btu per pound; he had used 8800 Btu per pound in his paper. The heat value in this report is estimated on the basis of an 8800-Btu-per-pound heat value of organic content.

Table 18

ANALYSIS OF REFUSE

Category	Raw Refuse	Refuse Fuel
Ultimate Analysis of Refuse (% Weight)		
Carbon	26.18	33.13
Hydrogen	3.51	4.46
Oxygen	22.08	28.65
Nitrogen	0.58	0.56
Sulfur	0.10	0.11
Chlorine	0.08	0.11
Noncombustibles	22.47	9.23
Moisture	25.00	23.75
Proximate Analysis of Refuse		
Moisture	25.00	23.75
Volatile matter	45.89	58.99
Fixed carbon	6.64	8.03
Noncombustibles	22.47	9.23
Heating Value		
Organic	4622	5858
Partial oxidation of metal	78	8
Total Btu/lb	4700	5866

Table 19

DETAILED ULTIMATE ANALYSIS OF DRY RAW REFUSE
(% Weight)

Category	Carbon	Hydrogen	Oxygen	Nitrogen	Sulfur	Chlorine	Fixed Carbon
Metal (8.4%)	0.39	0.05	0.36	0.004	0.0008	--	0.042
Paper (37%)	16.80	2.26	15.58	0.111	0.0444	--	4.180
Plastics (1.4%)	0.84	0.12	0.27	0.014	0.0042	0.084	0.070
Leather and rubber (0.9%)	0.63	0.08	0.034	0.031	0.0135	--	0.058
Textiles (0.8%)	0.37	0.05	0.33	0.018	0.0016	--	0.030
Wood (4.3%)	2.07	0.26	1.82	0.013	0.0047	--	0.606
Food waste (4.0%)	1.67	0.23	1.10	0.112	0.010	--	0.212
Yard waste (6%)	2.95	0.39	2.17	0.174	0.02	--	1.158
Glass (9%)	0.047	0.006	0.03	0.003	--	--	0.036
Misc (3.2%)	0.42	0.06	0.38	0.096	--	--	0.240
Total	26.18	3.51	22.08	0.58	0.10	0.084	6.636

Table 20

EFFICIENCY OF FUEL PREPARATION STAGE

Refuse Constituent	Initial Average (%)	Fraction Retained in Fuel (%)	Distribution of Fuel Output (%)
Paper, cardboard, bags, etc.	37	95	55.64
Glass	9	10	1.43
Ferrous	7.6	5	0.60
Nonferrous	0.8	15	0.19
Plastics	1.4	80	1.77
Leather, rubber, textiles, and wood	6.0	60	5.70
Garbage and yard trimmings	10	50	7.92
Water	25	60	23.75
Misc.	3.2	60	3.00
Total	100		100

Note: Approximately 63% of the raw refuse material content is retained in the output fuel fraction.

Table 21

DETAILED ULTIMATE ANALYSIS OF PREPARED REFUSE FUEL
(% Weight, Dry Basis)

Category (percent)	Carbon	Hydrogen	Oxygen	Nitrogen	Sulfur	Chlorine	Fixed Carbons
Metal (0.79)	0.037	0.005	0.035	0.0004	0.0001	--	0.004
Paper (55.64)	25.26	3.39	23.42	0.167	0.067	--	6.287
Plastics (1.77)	1.06	0.15	0.336	0.018	0.005	0.11	0.090
Leather and rubber (0.8)	0.56	0.07	0.030	0.028	0.012	--	0.051
Textiles (0.8)	0.37	0.05	0.333	0.018	0.0016	--	0.031
Wood (4.1)	0.98	0.25	1.74	0.012	0.0045	--	0.578
Food waste (5.5)	2.29	0.32	1.52	0.154	0.014	--	0.292
Yard waste (2.4)	1.98	0.16	0.87	0.07	0.008	--	0.463
Glass (1.43)	0.007	0.001	0.005	0.0004	--	.	0.006
Miscellaneous (3.0)	0.39	0.06	0.36	0.09	--	--	0.225
Total	33.13	4.46	28.65	0.558	0.112	0.11	8.027

Very little information is available pertaining to the physical characteristics of solid waste, other than densities. As part of a program to develop a more rational approach to solid waste size reduction (the present design approach for shredders can be characterized as "brute force") the University of California (Ref. 37) has been conducting a study of mechanical properties of refuse components under Environmental Protection Agency Grant No. R-801218. Only preliminary results are now avaiable. Table 22 shows stress versus strain data for slowly applied loads to various materials common to solid waste. It is interesting to note that cardboard and steel have comparable rutpure energies, and aluminum has a rupture energy about three times as great.

Table 22

STRESS VERSUS STRAIN DATA[*]

Material	Container Type	Container Shape and Specimen Locations	Specimen Thickness (in.)	Ultimate Strength (psi)	Ultimate Strain (in.-in.)	Rupture Energy (ft-lb/in.^{3})
Steel	12-oz can, beverage	Cylinder -- spec cut from side, axially and circumferentially	0.007	82,000	0.005	9.4
Aluminum	12-oz can, beverage	Cylinder -- spec cut from side, axially and circumferentially	0.006	31,000	0.012	26.5
Carboard	Box, laundry detergent	Rectangular box -- spec cut from front and back panels	0.025	6400	0.025	8.3
Paper	Bag, brown paper	Grocery-type bag -- spec cut at various locations	0.009	4000	0.025	5.1
Plastic, poly-vinyl chloride	Bottle, liquid soap	Sculpted molding -- spec cut from front and back panels	0.19 to 0.026	4000 to 5000	0.360 - $\dot{e}$ = 0.1 0.130 - $\dot{e}$ = 1.0 0.060 - $\dot{e}$ = 10	111 - $\dot{e}$ = 0.1 44 - $\dot{e}$ = 1.0 19 - $\dot{e}$ = 10
Plastic, poly-ethylene	Bottle, shampoo	Cylinder -- spec cut as in cans, above	0.028 to 0.036	1000	0.80 for $\dot{e}$ = 0.1 0.84 for $\dot{e}$ = 1.0 0.90 for $\dot{e}$ = 10	56 - $\dot{e}$ = 0.1 60 - $\dot{e}$ = 1.0 66 - $\dot{e}$ = 10

[*]Where $\dot{e}$ = elongation rate, all materials tested at $\dot{e}$ = 0.1, 1.0, and 10 in/min, with the plastics showing the effects of elongation rate as given above. Materials other than the plastics showed no elongation rate effects.

Impact tests were also performed in which various containers were impacted with a known weight falling through a known distance. The energy required to effect a given volume reduction could then be deduced. The containers were placed in various orientations for testing.

Impact test results are shown in Figure 27 as dashed lines. Steady-state tests, where forces are slowly applied to crush the containers, are shown as solid lines.

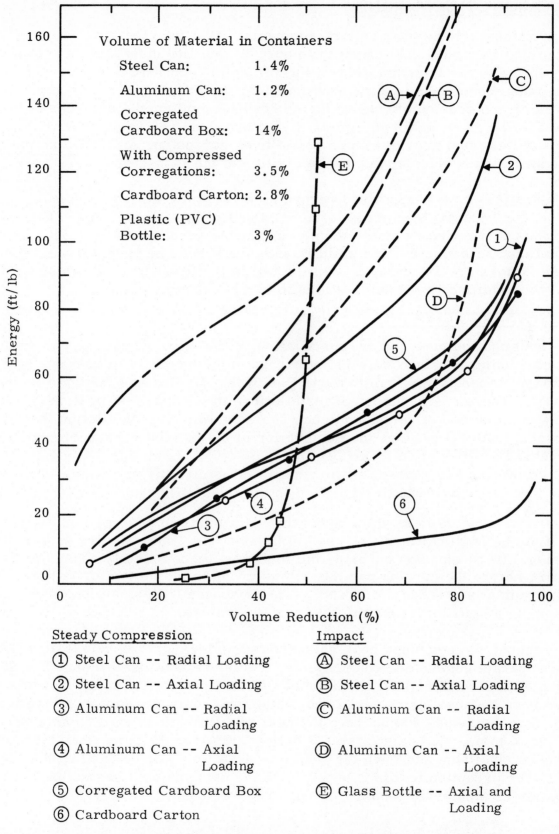

Volume of Material in Containers

Steel Can: 1.4%

Aluminum Can: 1.2%

Corregated
Cardboard Box: 14%

With Compressed
Corregations: 3.5%

Cardboard Carton: 2.8%

Plastic (PVC)
Bottle: 3%

Energy (ft/lb)

Volume Reduction (%)

Steady Compression

① Steel Can -- Radial Loading

② Steel Can -- Axial Loading

③ Aluminum Can -- Radial
                   Loading

④ Aluminum Can -- Axial
                   Loading

⑤ Corregated Cardboard Box

⑥ Cardboard Carton

Impact

Ⓐ Steel Can -- Radial Loading

Ⓑ Steel Can -- Axial Loading

Ⓒ Aluminum Can -- Radial
                   Loading

Ⓓ Aluminum Can -- Axial
                   Loading

Ⓔ Glass Bottle -- Axial and
                   Loading

Figure 27. Energy Volume Reduction for Various Containers
Under Quasi-Steady-State and Impact Conditions

APPLICATION CONSIDERATIONS

If refuse is to be used as a supplementary fuel for electric utility boilers, there must be a sufficient capacity to burn it in utility installations conveniently located near the points where the refuse is generated and processed. Also, when considering refuse as a fuel, careful attention must be given to corrosion and erosion effects on the combustion equipment, and consideration must also be given to air pollution. Using prepared refuse as a 10- to 30-percent auxiliary fuel provides some relief from corrosion, erosion, and particulate emission problems, due to the dilution of the refuse products.

Questions have been raised regarding synergistic effects of burning oil and refuse together, compared to the coal and refuse burning at the City of St. Louis fuel recovery plant. Some experience does exist in Europe, where oil and unshredded municipal solid wastes are burned in the same installation. These factors are discussed in more detail in the following text, along with desirable additions and modifications to fuel processing plants.

EUROPEAN EXPERIENCE

Investigation revealed one installation, in Stuttgart, Germany, where fuel oil and municipal refuse were being burned simultaneously to heat one stream of steam (Appendix I, "Trip Reports Describing Munich and Stuttgart Incinerators"). There are a number of installations in Europe that heat one stream with a combination of refuse and coal; there is even one installation that burns refuse and natural gas (see the description of the Munich North incinerator in Appendix I). None of these installations, however, including the one at Stuttgart, can provide a record of experinece that is directly applicable to the problems that might arise in firing refuse in oil-fired utility boilers.

First, none of these units burns the solid waste in suspension, whereas all of the boilers that might be used in Connecticut would require suspension burning. (In suspension burning, the fuel must be completely consumed by the fire before the burning particle either drops to the bottom or is blown out of the radiant section of the boiler.) The European boilers all have moving grates of various forms on which to burn the refuse.

Second, the combined coal/refuse-fired boilers and oil/refuse-fired boilers have separate combustion chambers for the refuse and fossil fuel. The utilization of separate combustion chambers for the refuse and fossil fuel was decided upon because the required heat output was much greater than that available from the refuse and because the heat requirement existed on an around-the-clock basis. This arrangement better optimized the time when there was no refuse available and when the total heat load would be carried by the fossil fuel. The exception to this rule was a boiler using natural gas as the fossil fuel, where a common combustion chamber was used.

The unique aspect relative to these installations in Europe is the common ownership of the power station and central heating station and refuse incinerating plant. One city agency is charged with being the utility arm of the city. This common ownership and management allows unity of funding, planning, installation, and management that is not usually possible in the United States.

U.S. EXPERIENCE WITH OIL AND REFUSE FIRING

One installation in the United States is known to burn refuse and oil in the same combustion chamber. This unit is an industrial installation, located in Rochester, New York, that consists of a tangentially fired Combustion Engineering, Inc. boiler similar to the Merramec powerplant of the Union Electric Company. This unit burns oil, industrial sludge, and solid waste; the boiler has a design rating of 185 tons per day of solid waste and 134 tons per day of sludge. At rated loadings, 60,000 to 70,000 pounds per hour of steam can be attributed to the sludge plus solid waste and 40,000 to 50,000 pounds per hour can be attributed to the oil. Steam conditions are 400-psi saturated steam (446°F) resulting in metal temperatures much lower than those in a utility boiler. Stack temperatures are about 600°F, compared to 200°F to 300°F in a utility boiler, so effects of condensation in the stack and precipitator are not directly comparable either. However, it is felt that both the Stuttgart and Rochester units will give some insight into problems of operating with oil and refuse firing.

It is believed necessary, however, to run a full-scale experimental test in an oil-fired utility boiler to definitely establish the feasibility of applying the St. Louis fuel recovery process in such boilers.

AIR POLLUTION

In any waste disposal system involving the combustion of the waste material, four products would ideally be produced: heat, water vapor, carbon dioxide, and a compact sterile ash. Any materials other than these four are very likely to cause either air or water pollution; in some cases, the heat may even be considered pollution.

In the case of a utility burning refuse as a fraction of its fuel stream, the primary pollution elements causing concern at this time are particulates (small solid ash particles entrained in the flue gas), sulfur dioxide, or trioxide, chlorine gas or hydrochloric acid, and various oxides of nitrogen.

Federal and State standards have been set for particulates and sulfur oxides. The possible reaction of refuse burning on these two emissions is therefore considered in some detail in this subsection.

Particulate Emission

The particulates in the flue gas generated by burning refuse come from two sources: the ash content of the burned materials and, equally important, the small particles of ashes, sand, floor sweepings, and broken glass remaining in the refuse fuel as a result of incomplete separation.

All calculations dealing with mixtures such as refuse must be based on an assumed analysis; the assumed analysis used here will be that given in Table 14.

The problem of particulate emission needs careful examination for a mixed burning situation because:

- Existing utility plants are required by State air pollution control regulations to limit particulate emissions to 0.2 pound per million Btu of heat release, whereas plants modified for mixed-burning would be required to limit particulate emission to 0.1 pound per million Btu.

- The particulate collection efficiency of electrostatic precipitators for such situations is uncertain because:

  Existing precipitations are designed to remove ash particles from high-sulfur coal. The sulfur content of both the refuse and the oil is significantly lower.

  The ash content of the fuel oil is very low; however, the combination of sticky oil ash and fluffy refuse ash could seriously affect the collection efficiency, if it causes even partial blockage of the flow paths in the precipitator.

On the basis of the above potential problems, and in the absence of first-hand experience in cleaning gases from a mixed burning operation, utilities have expressed serious concern that the particulate collection efficiency for a mixed burning situation could drop drastically from its design efficiency of about 99 percent for coal to levels as low as 60 percent for mixed oil and refuse. (The current state-of-the-art for new electrostatic precipitations is 99.5-percent efficiency for coal-burning boilers and 95-percent efficiency for oil-burning units.)

Technical discussions (Ref. 38), however, with German engineers from Apparatebau Rothemühle, a West German manufacturer of electrostatic precipitators and the only company in the world having built a precipitator for an oil and refuse system (at Stuttgart), indicate that the problem may prove to be easier to handle. The opinion is advanced on the basis that:

- If unburned paper does not enter the precipitator, the refuse-and-oil ash is easier to handle than coal ash.

- Because refuse contains moisture, it provides enough conductivity for the ash particles, and the high sulfur content is not really required. About 20- to 25-percent moisture in flue gas would ensure good collector performance; about 6- to 10-percent moisture in the flue gas would be a lower limit.

- Gas flow rates higher than the designed rates would reduce collection efficiency. As a rough rule of thumb, an 18-percent increase in gas flow rate would double the grain loading of the gas leaving the precipitator.

Hence, in the opinion of Rothemühle engineers, precipitators designed for coal with a 99-percent efficiency would give about 95-percent collection efficiency for a mixed refuse-and-oil burning situation, with 20-percent heat input from the refuse. This opinion is based on the assumption that the existing precipitator is in a proper operating condition (with 99-percent efficiency for coal) and that mixed burning would have approximately a 40-percent higher gas flow rate than the design flow rate for the precipitator.

It is a common experience that the efficiency of an electrostatic precipitator designed for coal drops drastically when it is used with oil burning units. Efficiencies in the 50- to 60-percent range are reported. This efficiency drop is primarily due to the fact that coal ash precipitators are designed on the basis of the sulfur content of coal, whereas oil ash precipitators are designed for high carbon carryover (and relatively low sulfur content) in the ash. However, in the opinion of Rothemühle engineers, the addition of refuse ash and associated moisture in the oil ash would bring the precipitator efficiency to the 95-percent efficiency range.

Other factors affecting the particulate emission level are described in the following paragraphs.

Noncombustibles in Processed Refuse. For material recovery, the solid waste will be shredded and then passed through several material separation stages. Such preprocessing of waste will also prepare the refuse in a form suitable for utility furnace firing.

Ideally, the material separation system should separate the total noncombustible fraction (excluding the bound ash, which is distributed in the organic fraction) of the incoming refuse. However, any practical separation system would retain some noncombustibles in the outcoming fuel fraction and would also carry over certain combustible content in the reject fraction. Table 20 illustrates operating characteristics of a hypothetical separation system; in this case, the initial noncombustible content (excluding 3.7% bound ash) of 18.7 percent is reduced to 2.7 percent, equivalent to an overall noncombustible removal efficiency of approximately 86 percent.

As indicated in Table 20, only 63 percent of the raw refuse material content is retained in the output fuel fraction. On this basis, the total noncombustible content (including bound ash) in the output fuel is 9.3 percent. Material separation also increases the per-pound heat value of output material, in this case, from 4700 Btu per pound for raw refuse to 5870 Btu per pound for the refuse fuel fraction.

As a sample case of an ideal material separation system, it is assumed that all noncombustible content of the incoming refuse is separated from the output fuel fraction, without any loss of the combustible fraction. It is assumed that 60 percent of the input moisture will appear in the fuel fraction, as was assumed in Table 20. For this case, the output fuel fraction will have only the bound ash, which amounts to 5.4 percent of the output fuel. The heat value for the fuel fraction in this case is 6425 Btu per pound. Thus:

Noncombustible Removal Efficiency (Excluding Bound Ash) (%)	Total Noncombustible Output (%)	Heat Value of Output Material (Btu/lb)
0 (no separation)[*]	22.5[**]	4700
86	9.3	5870
100	5.4	6425

Fly Ash Versus Bottom Ash. It is estimated that for suspension burning of pulverized coal in a dry-bottom furnace, 80 percent of the coal ash ends up as as fly ash (Ref. 40). The percentage of total noncombustibles of refuse resulting as fly ash will depend on the particle size of the shredded refuse, the point of the refuse feeding in the furnace, and the gas velocity in the furnace section. In the absence of firm numbers, it is assumed that 75 percent of the noncombustibles in prepared refuse will become fly ash. This estimate is believed to be very conservative.

Percent of Heat from Refuse. In the absence of extensive operating experience, it is estimated that up to a maximum of 20 percent of the total heat release in the furnace can be supplied from refuse without any noticeable effect on boiler performance. From the viewpoint of maximizing net fuel income (or, at worst, for minimizing net disposal costs), it would be preferable to burn larger refuse quantities, if the actual operating experience with mixed burning proves encouraging. It is expected that the refuse heat would be limited to 20 percent during the first two years of mixed-burning furnace operation.

Particulate Loading in Stack. Because uncertainties exist in determining the magnitude of several factors outlined above, a single number representing particulate loading for the mixed burning case cannot be used. The net stack particulate loading is calculated based on two assumptions. Figure 28 is based on the assumption that the fly ash from the oil is negligible compared to that from the refuse fuel. Figure 29 assumes that a fly ash loading of 0.2 pound per million Btu heat input at the precipitator inlet section is associated with 0.5-percent sulfur oil operation. For any specific application, if the assump-

---

[*]See Ref. 39 for the assumed refuse composition and heat value.
[**]Aside from glass and ferrous and nonferrous metals, other constituents of refuse (paper, yard waste, etc.) contain a certain amount of ash, which is termed "bound ash."

86

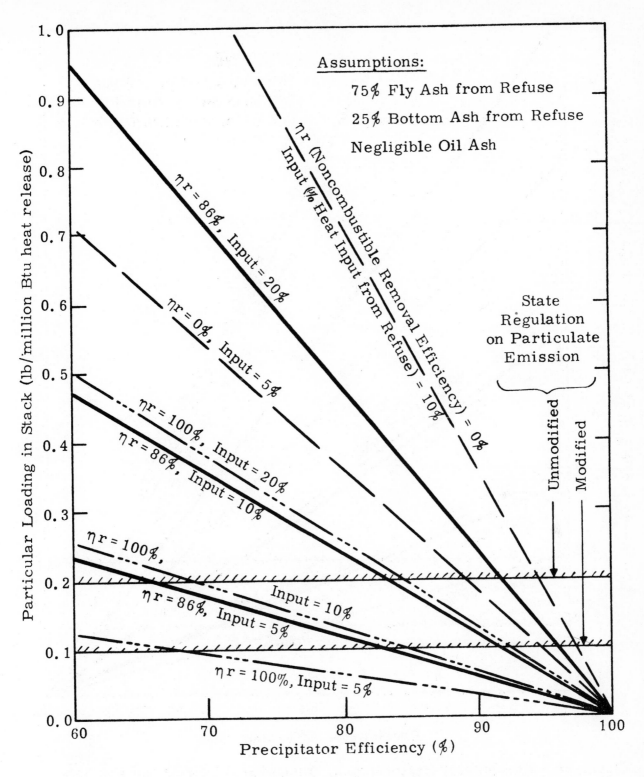

Figure 28. Precipitator Performance, Assuming Negligible Fly Ash

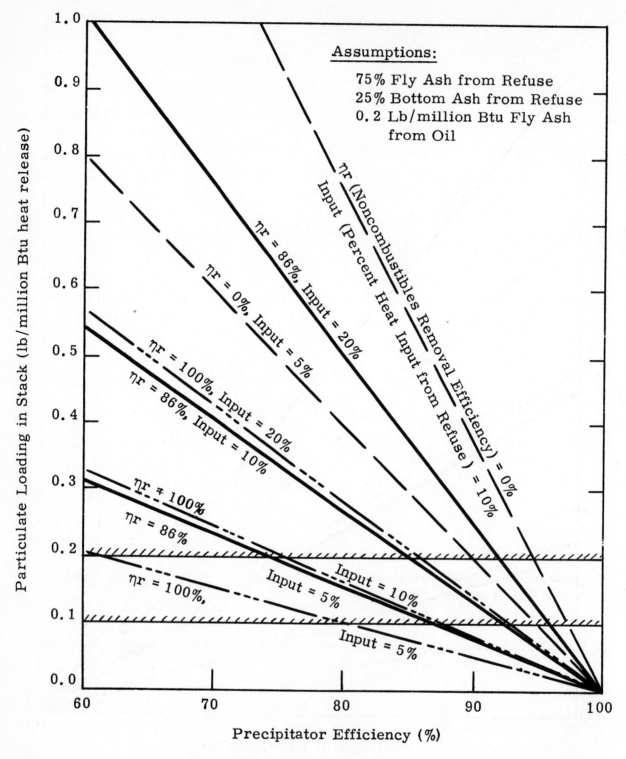

Figure 29. Precipitator Performance, Assuming Oil-Produced Fly Ash of 0.2 Lb/Million Btu of Oil Input

88

tions of Figure 29 are not correct, the data of Figure 28 can be added to the specific oil ash content (with the precipitator collection effect) to obtain total stack perticulate loading. As mentioned earlier, the calculations of Figures 28 and 29 are based on the assumption that 75 percent of the total noncombustibles from the refuse ash ends up as fly ash.

Based on Figure 29, the following observations can be made regarding the constraints imposed by State air particulate emission control regulations:

- Even with a 100-percent noncombustible removal efficiency, a 20-percent refuse heat release rate would require a minimum particulate collection efficiency of 93 percent to meet the most stringent code. A collection efficiency of 81 percent will allow only a 5-percent refuse heat release rate.

- If the noncombustible removal efficiency is the more realistic 86-percent value, a particulate collection efficiency of at least 96 percent will be required for 20-percent refuse heat input, whereas even a 5-percent heat input rate will require a collection efficiency of a minimum of 87 percent.

- If the particulate collection efficiency is only 80 percent, evan a 5-percent heat input from refuse (with 100-percent removal) will exceed applicable particulate emission limits.

The above observations change somewhat if a lower fly ash/bottom ash split occurs. For instance, with a 50/50 split, and a noncombustible removal efficiency of 86 percent, a particulate collection efficiency of 94.3 percent will be needed for a 20-percent refuse heat release rate.

Rothemühle engineers predict about 95-percent efficiency when original coal-burning precipitators are used for mixed burning duty. From Figure 29, it can be seen that 95-percent efficiency would produce particulate loading in a stack well below the 0.2-pound limit, but slightly above the 0.1-pound limit for a suggested fuel preparation system and 20-percent heat input from refuse.

Comparison with Particulate Emission from Grate-Burning Incinerators. In order that the particulate emission from the mixed burning process be evaluated in comparison with an alternate method of burning the refuse in an incinerator, an estimate is made for particulate emission levels typical for such systems. Estimates based on various assumptions are summarized as follows:

- For grate-burning incinerators (without heat recovery):

    9.6 pounds per million Btu, if 20 percent of all noncombustibles become fly ash (upper limit).

    2.2 pounds per million Btu, if all glass and metals are excluded from the fly ash (lower limit).

    5.3 pounds per million Btu, based on 50 pounds of fly ash per ton of refuse.

89

- For modern incinerators (grate-burning, with heat recovery):

  Test data for the Chicago Northwest incinerator indicate

    30 pounds of fly ash per ton of refuse
    3.44 pounds of fly ash per million Btu
    Fly ash = 4.5 percent of total noncombustibles

  Test data for the Dusseldorf incinerator indicate that 13 percent of the total noncombustibles are fly ash.

- For modified utility boilers: 0.1 pound per million Btu of heat input.

- For existing incinerators, average 0.4 pound per 1000 pounds of gas (corrected to 50-percent excess air). Assuming 3.8-pounds of stoichiometric air per pound of refuse, the regulation translates to 0.57 pound per million Btu of heat input.

- For new incinerators, 0.08 grain per scf (12% CO_2), maximum, two-hour average: the regulation translates to 0.22 pound per million Btu of heat input.

Note that the particulate loads are estimated on the basis of excluding any emission control system.

For grate-burning refractory furnace incinerators, it is estimated that 20 percent of the total ash results as fly ash (Ref. 40). Because grate-burning systems do not require any size reduction, the refuse from packer trucks is directly fed to the furnace, without any material separation. As an upper limit, all noncombustibles are assumed to be in the form of ash. For estimating the lower limit of fly ash quantity, it is assumed that all the glass and ferrous and nonferrous metals are completely trapped in the bottom ash and that only 20 percent of the remaining noncombustibles become fly ash. Reference 41 indicates that the furnace dust emission for a high-performance incinerator is 35 pounds per ton of refuse and that it could be up to 60 pounds per ton for some incinerators. Most of the old incinerators are likely to have the emission rate of about 50 pounds per ton of refuse.

To further estimate the emission problem for modern heat recovery incinerators, the test data obtained for the Chicago Northwest incinerator (Ref. 27) were analyzed. The specific design of the water-walled incinerator indicates very good combustion characteristics. The boiler tube banks of the heat recovery section will impinge somewhat on the dust separation effect for the combustion gas passing through them. The fly ash quantities for such a modern facility are variously expressed above. The data for the Dusseldorf incinerator (Ref. 24) are also presented.

Trade-Offs in Particulate Emission. The utility boilers modified for mixed refuse burning would be subject to a tougher particulate control regulation than is now applicable to them. Because the same refuse can also be burned in an incinerator, the applicable regulations are translated on a uniform basis, as illustrated above under "Comparison with Particulate Emission from Grate-Burning Incinerators." For identical purposes of burning refuse, the mixed burning approach requires more stringent controls than the incineration.

In the State of Connecticut, particulate emission limits for powerplants are 0.2 pound per million Btu for existing plants and 0.1 pound per million Btu for new or modified powerplants. State of Connecticut limits on existing incinerators are equivalent to 0.57 pound per million Btu, and for new incinerators, the limit is 0.22 pound per million Btu input. An argument could be made that a utility plant burning supplementary refuse fuel is serving the function of an incinerator as well as a powerplant. If the plant is treated on the basis of an existing incinerator, for its emissions due to refuse burning, and as an existing powerplant, for its emissions due to oil burning, the resulting particulate emission limits would result:

Refuse Firing Rate on Heat Release Basis (%)	Allowable Particulates Due to Oil (lb/million total Btu)	Allowable Particulates Due to Refuse (lb/million total Btu)	Total Allowable Particulates (lb/million total Btu)
0	0.2	0	0.200
5	0.19	0.0285	0.218
10	0.18	0.0570	0.237
20	0.16	0.114	0.274
30	0.14	0.171	0.311
40	0.12	0.228	0.348

In the following paragraphs, the total allowable particulates noted above will be referred to as the "semiincinerator limit," bearing in mind that the term has no official sanction.

Considering the case where the refuse input firing rate is 20 percent and the noncombustible removal efficiency, ηr, is the realistic value of 86 percent, Figure 30 shows this case on an expanded scale. Here it is seen again that the modified boiler requirement is met by a 96-percent efficient precipitator, while the unmodified boiler limit requires 92-percent efficiency. The semiincinerator limit only requires a precipitator efficiency of slightly over 89 percent.

Emission from Existing Incinerators. If the refuse handled by the mixed burning approach is instead burned in existing incinerators, the corresponding particulate emission rates should be compared. Because no test data for existing incinerators are available at this stage, only preliminary estimates can be made. Using data from Table 14 as a guide, a fly ash rate of at least 4.5

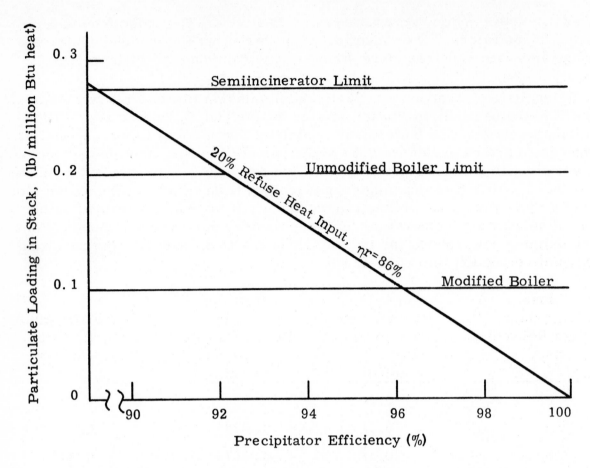

Figure 30. Precipitator Performance for 20-% Refuse Burning

pounds per million Btu can be used. Because none of the incinerators uses high-efficiency particulate control equipments (electrostatic precipitators, bag filters, etc.), an average 60-percent collection efficiency is assumed. Thus, approximately a 1.8-pounds-per-million-Btu emission rate can be considered typical performance for existing incinerators.[*]

It thus appears that the particulate control standards applicable to the mixed burning approach should be reviewed and revised.

Sulfur Emissions

Because sulfur in refuse is like all other constituents that are randomly variable, a single hard percentage value cannot be allocated to it. Figure 31

---

*Some data available for the air compliance mathematical model are summarized in Ref. 42, which indicates about 9 lb of particulates/ton of refuse, or about 1 lb/million Btu. However, these data are not based on actual stack emissions and were derived in 1970 by using certain EPA multipliers. Also, the multiplier appears to be uniformly applied irrespective of the existence and nature of particulate control equipment.

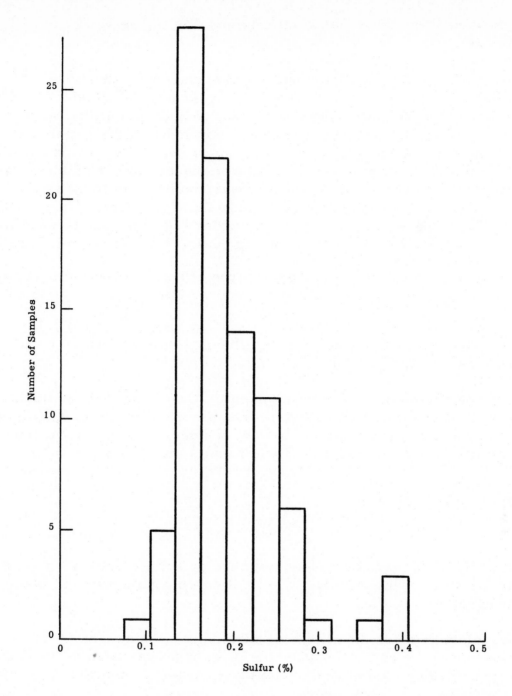

Figure 31.   Sulfur Content in Refuse Fuel (St. Louis Test Results)

shows the statistical distribution of sulfur content measured in St. Louis refuse fuel (Ref. 10) over a limited period of time.

A number that is typical of many analyses (Refs. 24 and 39) is 0.15 percent by weight.   Using an average value of as-received refuse heating, of
4700 Btu per pound, derived above under "Refuse Composition," the result
would be 0.31 pound of sulfur per million Btu.   This figure compares to 0.27
pound of sulfur per million Btu in 0.5-percent fuel oil, indicating that the

average as-received refuse has a sulfur content in the same range as low-sulfur fuel oil.

Separation of the refuse into combustibles and noncombustibles will upgrade the heating value considerably (to the order of 5000 to 6000 Btu/lb). Because there is no definitive data as to what components of the refuse contain the sulfur, it is not possible to predict how much of the sulfur will remain with the combustible stream and how much will be separated out with the noncombustibles. If it is assumed that the percentage of sulfur in the noncombustible stream is the same as that in the combustible (fuel) stream, the prepared refuse fuel would have a sulfur content of 0.255 pound per million Btu. For the refuse fuel described above under "Ultimate Analysis of Refuse and Refuse Fuel," the sulfur content would be 0.187 pound per million Btu.

To assure removal of one obvious source of sulfur, it would make sense to tune the separation process to remove rubber items such as heels and pieces of tire.

The problem of sulfur content of the flue gas may be further reduced by the retention of the sulfur in the fly ash and bottom ash.

Kaiser has stated that in refuse burning, 25 percent of the sulfur as sulfur dioxide in the refuse is released in flue gas, 25 percent remains in the fly ash, and 50 percent remains in the bottom residue. These observations were made on grate burning incinerators, and results may differ for suspension burning.

CORROSION AND EROSION

Corrosion

Any study of the various methods of burning solid wastes must involve a very careful study of the possible effects of corrosion on the total plant equipment involved.

Most information on the effects of burning solid wastes has been learned by observation of the corrosion after the investment has been made in the plant. This is true whether the particular plant has been an incinerator or a more sophisticated heat recovery system.

Because solid waste appears to be a logical and inexhaustible, if not unlimited, source of energy and generally requires the expenditure of effort to dispose of it, it is very important to fully understand the problems that have arisen whenever its energy has been utilized as a source of heat. Some of these problems, of course, exist if the waste is burned as a method of volume reduction, even if no attempt to recover the heat is involved. Corrosion is a term generally applied to the removal of a useful material by chemical or electrochemical action, as contrasted to erosion, which is the removal by mechanical, abrasive action.

Corrosion experience, laboratory testing, and analysis point to a number of involved factors:

- Alternating oxidation/reduction.

- Reaction with chlorine and hydrochloric acid.

- Reaction with alkali metals, such as lead, and with sulfur compounds.

Major published literature on corrosion has been surveyed to define factors affecting corrosion attack, various corrosion theories based on field and laboratory investigations, and the impact on a mixed burning approach. An important summary statement in a recent report for the Environmental Protection Agency (Refs 24 and 43), made on the basis of an extensive survey, field visits, and tests, sums up the present position: "The...comments should counteract the unfortunate impression, still widespread in the U.S., that high-performance, refuse fired steam generators in Europe continue to experience intolerable corrosion problems." A similar statement was also made in a recent paper (Ref. 44) by a German investigator.

The conditions known to be associated with corrosion attack are:

- Fireside corrosion, which is reported to occur when:

    Radiant superheater tube metal temperatures exceed 950°F.

    Waterwall tube metal temperature exceeds 500°F to 600°F.

    The surface is under flame front impingement.

- Convection surface corrosion which usually occurs:

    On the upstream side of superheater tubes in the first few rows.

    At a threshold metal temperature of 850°F to 950°F for oxidizing atmosphere, and about 600°F under reducing conditions.

- Cold-end corrosion, which is caused by condensation of corrosive gases (at gas temperatures below 450°F). Almost all forms of corrosion take place in the presence of ash deposits on tubes.

The corrosion potential for the mixed burning approach is:

- Experience so far is for grate burning of unprocessed refuse. Units experiencing severe corrosion had reducing conditions.

- Corrosion in refuse boilers is mainly a nuisance, at present, rather than a critical problem; this corrosion poses no serious threat, in Europe, to existing installations (Refs. 24 and 43).

- For a mixed burning approach, corrosion could be even less severe if:

    Preprocessing stages will significantly reduce noncombustibles in refuse fuel and in flue gas.

    Suspension burning of refuse will not create a reducing atmosphere.

Refuse particle size will keep most noncombustibles in the bottom ash.

A small ratio of refuse to oil will minimize ash deposits on the tubes.

Background. The published literature on corrosion, surveyed here, is based on the European incineration experience and on some field tests and laboratory studies conducted by both European and U.S. investigators. In spite of their extensive operating experience, "no one in Europe really knows the exact mechanisms of refuse boiler corrosion, and the proposed explanations are based upon surmise" (Refs. 24 and 43). This situation exists because early corrosion studies were primarily based on chemical models, and the operating conditions responsible for deposit formation or tube corrosion were not adequately reported. Two recent papers presented at the National Incinerator Conference of the American Society of Mechanical Engineers, in 1972 also significantly differ in their understanding of controlling corrosion mechanisms.

The information given in this section therefore serves to present knowledge despite the inability to certify the identification of important corrosion mechanisms, the probable roles played by certain constituents of refuse ash, and by certain operating conditions affecting corrosion rates. This information will be useful in specifying preprocessing stages for refuse fuel preparation and in development of an optimum mixed burning system.

It is generally established that in most cases corrosion takes place in the presence of ash deposits on the tube metal. Thus, conditions promoting fouling of tubes also tend to enhance corrosion activity.

The corrosion mechanisms are first discussed below in relation to the role played by operating conditions and constituents of refuse ash; the corrosion experience is also presented in terms of the zones of combustion and the flue gas path. The European approach in corrosion control is then described, along with certain recommendations regarding steps to minimize corrosion. Finally, the potential for corrosion in the mixed burning approach and certain desired features of new plant construction for refuse burning are presented.

Relation between Corrosion and Fouling. In most cases, buildup of certain constituents in ash deposits enhances tube corrosion. Corrosion at high temperatures appears to be strongly dependent on the existence of ash in contact with tube surfaces, both for reducing and oxidizing conditions. Acceleration of the normally slow gas attack appears to be due to the presence of lead, sodium, and potassium in the ash deposit (Refs. 24 and 43). In one European experience, furnace tube corrosion was minimized by studding the tube surface and covering it with a 0.4-inch-thick refractory layer to prevent slag buildup; similarly, corrosion of convection surfaces was minimized by covering the tubes with shields made of Sicromal.

According to Reference 45, chemical analysis indicates that fouling is due to the formation of sulfates of the alkali metals, zinc, and lead, in some combination.

Based on past experiences with various coals, the Foster Wheeler Corporation (Ref. 45) suggests that the combined alkali content (sodium and potassium, expressed as Na_2O) in the ash should not exceed 0.4 to 0.7 percent on a dry fuel basis, if fouling is to be minimum. Above 0.7 percent, the fouling becomes severe and uncontrollable. It is also indicated in Reference 45 that fouling problems occur, particularly for high-alkali-content ash, if hot gases entering the convection pass are not cooled to at least 1850°F to 1900°F.

Corrosion Theories.   The following three corrosion theories exist:

- Corrosion by Reducing Atmosphere.   The reducing atmosphere is casued by a lack of proper excess air in certain combustion zones, by inadequate turbulence, or by insufficient residence time necessary for perfect combustion. The reactions are very slow below a metal temperature of 760°F (Ref. 43). The speed of reactions is enhanced by the alternating of oxidizing and reducing gas conditions. During the period of oxidizing conditions, a protective layer of iron oxide is formed on the metal but is removed during the next reducing period. This continual removal of the protective coating accounts for the speed of the corrosive action.

- Corrosion by Chloride Compounds.   The sources of formation of hydrogen chloride are:

    Polyvinyl chloride at temperatures from 446°F.

    Hydrolysis of alkali chlorides at 754°F.

    Reaction between acid sulfates in the deposits and alkali chlorides at temperatures from 392°F.

Some chlorine can occur by the reaction:

$$4\ HCl + O_2 \underset{1830°F}{\overset{\substack{Catalyst \\ 480°F \text{ to } 930°F}}{\rightleftharpoons}} 2\ Cl_2 + 2\ H_2O$$

Hydrogen chloride or chlorine, in the presence of deposits, diffuses to the tube surface to form iron chlorides; these iron chlorides diffuse to regions of higher temperature within the deposit and are decomposed to iron oxides and hydrogen chloride and chlorine in gaseous form. The freed gaseous chlorine can then return to the tube metal surface and remove more iron in a closed cycle.

- Corrosion by Sulfur Compounds.   This theory is based on observations made in coal-firing units. Two types of sulfur-associated corrosion mechanisms occur. A Pyritic attack is usually associated with reducing conditions resulting from flame impingement on waterwall tubes in

97

the furnace, with the temperature ranging between 600°F and 800°F. A complex alkali iron sulfate corrosion is usually associated with metal surface temperatures of 950°F to 1100°F. This corrosion process is of a cyclic or regenerative nature. It frequently occurs on the leading edge of the first few rows of superheater tubes and is strongly dependent on the existence of a liquid phase. Sulfate formation depends directly on the SO_2 concentration in the gas stream.

Corrosion in Boiler Furnace Zones. Conditions found to influence corrosion attack in various parts of the furnace and boiler tubes are listed above. While the threshold metal temperatures for corrosion proposed by several investigators are in broad agreement, the controlling mechanisms are not. Two recent papers (Refs. 46 and 47) present results of extensive field and laboratory studies on fireside corrosion and suggest a corrosion model, which indicates

- Chlorides and chlorine are reactive with the tube metal in a regenerative manner.

- Sulfur forms low melting pyrosulfates or biosulfates, and a regenerative loop involving Na_2SO_4 is possible.

- Zinc and lead salts serve to lower the melting points of the mixtures on the metal surface.

Another paper, based on a European experience (Ref. 44), presents the view advanced by the VGB Laboratory, based on its own work: "The often assumed influence upon the corrosion by alkalisulfates, alkalipyrosulfates, or alkalibisulfates could not be proved in any case." According to the VGB Laboratory, corrosion is caused by a reaction between the tube material or the iron oxide protective scale and hydrochloric acid.

The rate of corrosion is also in some dispute. Reference 46 indicates that high initial corrosion rates taper off with time and reach an asymptotic limit (in their tests, the carbon steel corrosion waste rate reached this limit after about 600 hours). F. Nowak (Ref. 48) has found that the material loss in furnace tube corrosion continues approximately linearly (his data indicate this to be true at least up to 5000 hours of operation). The Dusseldorf experience (Ref. 44) for superheater tube corrosion supports Reference 46.

For corrosion at high temperatures, a liquid phase may be essential to promoting corrosion under oxidizing conditions, although this restriction does not appear to hold under reducing conditions. The reaction between metal and combustion products in gaseous forms is slow in an oxidizing atmosphere. Hydrogen chloride will be less damaging in the gaseous forms at a high temperature than when it is mixed with moisture to act as hydrochloric acid during the shutdown period. The presence of a liquid phase is critical in this respect.

Metal temperature limits suggested by three firms for 100-percent refuse burning include:

- The Babcock and Wilcox Company suggests an 825°F limit for acceptable life, 600°F as a conservative limit.

- Combustion Engineering Inc. suggests 850°F steam temperature (approximately 950°F to 1000°F metal temperature).

- The Foster Wheeler Corporation indicates an 850°F metal temperature limit.

Corrosion in Mixed Burning Systems -- Potential and Methods of Control. The existing field experience is for grate burning of municipal refuse without any separation or size reduction. As indicated above under "Corrosion," the severe corrosion experience during the 1960's in Europe is no longer considered serious, even though the grate burning practice is still in existence. In most cases, this improvement was achieved by suitable boiler modifications (Ref. 44) to eliminate reducing environment zones, to achieve perfect combustion, and to promote better turbulence and mixing.

In a mixed burning approach, the fuel preprocessing stages will have the following effects:

- If higher efficiency in the separation of combustible and noncombustible fractions is achieved, only the bound ash and a small fraction of metals and glass will enter the furnace.

- Because of the probable role played by alkali metals in corrosion, refuse constituents with significant alkali content should be removed from the combustible stream, even if it is not an economical recycling step. (Glass is a prime example in this category.)

- The sulfur problem in a mixed burning process is likely to be less severe. The oil used now has 0.5-percent sulfur, and 95 to 100 percent of the sulfur is released as oxides in the flue gas. The refuse has a slightly lower content of sulfur than of oil, on a Btu basis; however, it is found that only 25 percent of the refuse sulfur is released as oxides in the flue gas, whereas 25 percent is discharged in fly ash and 50 percent is discharged with residues. The refuse sulfur distribution is presumably for a grate burning situation and could be different for suspension burning.

  If the sulfur content in the flue gas is less in the mixed burning situation, it is expected that fouling, but not necessarily corrosion, due to the alkalis will occur (Ref. 45). This expectation is based on the assumption that corrosion takes place as a result of the formation of complex alkali sulfates and pyrosulfates.

- The ratio of fly ash to bottom ash is generally higher for suspension burning than for grate burning (Ref. 49); the ratio is influenced by particle size and furnace flow characteristics. While a smaller particle is preferred for combustion purposes, the aerodynamics would favor larger particle sizes to minimize fly ash problems. Refuse preprocess-

99

ing stages should therefore aim at reducing combustible particle size, leaving noncombustible spillover as large in size as possible. (This ideal may be difficult to achieve.)

Tests made at the current experiment of the Merramec powerplant of the Union Electric Company in St. Louis, Missouri, where refuse and coal are burned together, indicate that the sodium content in fly ash had not changed after the unit was modified for mixed burning; this situation is believed to be the result of the fact that most of the sodium is contained in the glass that falls out as bottom ash.

- In the mixed burning of oil with refuse, the ash loading will be diluted; however, if the sticky oil-ash combines with the refuse ash and generates uneven or heavy ash deposition on critical tube surfaces, the corrosion potential could be significant. Lacking a definitive set of balanced chemical equations, the quantitative corrosion effects cannot be clearly stated, yet they are not likely to be severe.

- In the mixed burning of oil with refuse, the possibility also exists for burning particles of refuse to adhere to the sticky oil-ash. This situation will produce local hot spots and will tend to produce a localized reducing atmosphere of limited duration.

- It is suggested (Refs. 46 and 50) that the minor constituents of refuse (sodium, potassium, zinc, tin, and lead) are significant in the potential corrosion attack on boiler tubes. From the data on ultimate analysis of refuse constituents (Ref. 29), it is seen that effective removal of metallic content from the refuse fuel would tend to remove zinc and lead, because these elements are usually found as alloys in magnetic metals. In addition to metals, other refuse constituents containing zinc are leather and rubber.

In view of the above uncertainties, the specific extent and nature of potential corrosion effects cannot be identified with precision. If the tests that will be conducted on the first utility boiler to fire a mixture of oil and refuse indicate high corrosion rates, additional steps may be taken to reduce it:

- Provide better fuel preparation to remove offending materials such as chlorine-rich polyvinyl chloride or sulfur-rich rubber and sodium-rich glass.

- Possibly shield critical tube elements with strips of steel or a layer of silicon carbide, a practice successfully employed in Europe.

It is not believed that the use of additives in the refuse fuel will be a successful answer to corrosion. This approach has been tried in European installations and was abandoned. Use of additives helped to ease removal of slag deposits, but was expensive and did not reduce corrosion.

Finally, where it is possible to consider designing a new steam plant that will burn refuse as a supplementary fuel with oil, corrosion can essentially be eliminated by:

- Firing the refuse and oil in separate furnaces.

- Limiting the metal temperature of its boiler in the refuse firing unit to the values listed above under "Corrosion in Boiler Furnace Zones."

- Careful fuel preparation.

Erosion

Erosion in boilers is primarily the result of abrasive particles of fly ash and residual solids that become entrained in the high-velocity portions of the flue gas and, by impinging on the exposed portions of the water and superheater tubes, wear away the metal.

This effect must be controlled, when boilers are first designed, by controlling the velocity of ash-laden streams. In existing installations, care should particularly be taken to limit the volume of fly ash and entrained particles for refuse fuel by properly screening or air classifying to remove the grit from the combustibles before firing. In some severe cases, steel strips have been used in front of the tubes to be eroded instead of the tubes themselves.

As noted at the beginning of this section, under "St. Louis Fuel Recovery Process," erosion is also a severe problem in the refuse fuel feeding lines at the Merramec powerplant. This problem is believed to be the result of the presence of highly abrasive glass and nonferrous metal particles in the refuse fuel. The addition of air classification to the fuel preparation process will remove these heavy abrasive particles and will alleviate the feed-line erosion problem as well as the boiler erosion due to the particles in the fly ash. It is believed that both the A. M. Kinney/Black-Clawson fuel recovery process and the Combustion Power Corporation CPU-400 process should prove effective in removing the abrasive particles, because they both provide separation of heavier materials.

HYBRID FUEL PROCESSING SYSTEM

It is apparent that any refuse fuel preparation system should remove as many of the noncombustibles as possible. Segregation of the noncombustibles from the input stream not only produces a better fuel, but also provides a ready-made feedstock for a material separation system to recover materials such as aluminum, glass, and ferrous metals. The economics of material recovery from the noncombustible stream must of course be checked before reaching a firm decision.

From a fuel processing standpoint, both the dry shredding and wet pulping processes described above would require modification. Both of the dry shredding processes use only one stage of shredding -- a high-power-consumption high-maintenance approach, if the end-product fuel size is to be 1 inch or smaller.

The wet pulping process produces a fuel that is too moist and will require additional thermal drying before it is acceptable as a utility boiler supplemental fuel. There is also some question that the fuel from a wet pulping process may be ground so fine that a larger fraction of the resulting ash becomes fly ash. Because no wet-pulped fuel has yet been burned in quantity in utility boilers, the latter argument remains conjecture. The wet pulping process is more gentle to glass than the dry shredding processes, providing about three times the glass yield in color-sortable sizes. At this point, however, the economic advantages of yielding color-sorted glass have not yet been established and may prove marginal for the State of Connecticut application.

As a result of these various considerations, a hybrid system seems to offer many advantages. One such system is described in Section 6, "Comparison of Front-End Systems." The fuel preparation train described there would utilize:

- Primary Shredding. Reduction to a dimensionless 6 to 8 inches will require less power and shredder maintenance, while still producing an air-classifiable product.

- Air Classification. The shredded material will be separated into a light fraction and a heavy fraction. The light fraction will consist of predominantly combustible materials, such as paper and plastic films. The heavy fraction will predominantly consist of glass flour, dirt, metals, ceramics, and other combustible materials. Some of the heavier combustibles, such as solid chunks of leather, rubber, or moisture-laden vegetable scraps, will appear in the heavy stream while, of course, some scraps of aluminum foil and other noncombustible materials will appear in the combustible stream. The processing of the noncombustible stream is described in Section 6.

- Secondary Shredding. This shredding step is performed on a feedstock that has already lost most of the metals and glass and other inorganic materials that shorten hammer life and cause high power consumption in the shredder. Size reduction is to a maximum dimension of less than 1 inch. The 1-inch maximum dimension assumes that the piece is really a platelet rather than a solid body. The important consideration is really the time required to burn the piece completely, and experience in the powerplant may require adjustment of this dimension downward or upward. Unfortunately, with the variety of materials existing in municipal solid waste, simple dimensional specifications are not really adequate. It is possible, however, that the air classifier will drop out the chunks of heavy, slow-burning materials, thereby protecting the furnace from unburned pieces that would drop to the bottom.

- Secondary Screening. The secondary shredder output may still contain fine abrasive particles, such as glass flour, that escaped the primary air classifier. Screening is one effective way to remove these particles. A last-resort alternative is to use a secondary air classifier to remove noncombustible carryover, but it is not believed that this step will be necessary.

Experience will also determine whether sufficient ferrous metal pieces, such as tin can shreds, will escape the primary air classifier. If this carryover is significant, a magnetic separator will be needed between the primary air classifier and the secondary shredder, in order to protect the secondary shredder. Magnetic separation before the primary air classification is not recommended, because the large quantities of ferrous scrap at that point in the process stream will entrain significant amounts of paper into the ferrous scrap stream.

- Fuel Shipment and Storage. Fuel shipment to the utility user can be accomplished by truck, by barge, or by pneumatic transport (if the utility plant is nearby). Barge hauling offers the potential for economical transport if the geography is appropriate, and the barges can also serve as short-term buffer storage.

Storage and handling of dry-shredded or pulped fuel requires more experimentation. The physical characteristics of shredded or pulped fuel are not yet established. Data pertinent to how these fuels tend to mat with time are needed, in order to develop a more rational approach to design of storage, mechanical handling, and pneumatic conveying facilities.

PYROLYSIS PROCESSES

To include all processes, "pyrolysis" is defined here as a process in which solid waste is heated in the absence of air or with deficient air, to produce gas, liquids (oils), char, and ash in varying quantities. Some of the pyrolysis products have economic value as fuel.

The processes considered in this section, along with a brief description of their products are:

Process	Description
Garrett Research and Development Company flash pyrolysis system	Produces fuel oil and char and recovers glass and ferrous scrap. The process can be altered to produce fuel gas instead of oil.
Monsanto Landgard system	Produces char residue, and the heat recovery unit produces steam. This process is primarily a high-grade incineration process.
Union Carbide Corporation oxygen refuse converter	Produces a fuel gas and frit residue.
Urban Research and Development Corporation system	Produces a fuel gas and frit residue.
Torrax Services, Inc. system	Produces frit residue and steam. This process is also a high-grade incineration process.
Bureau of Mines pyrolysis research	Determined basic parameters of solid waste pyrolysis (bench research study).
Hercules Powder Company system	Produces internally used process gas (pyrolysis used in a subprocess of a compost plant).
Battelle Memorial Institute system (Pacific Northwest Laboratories)	Bench scale study.
Institute of Gas Technology system	Produces fuel gas (a proprietary process).
Regional resource recovery system (Charleston, West Virginia)	Produces fuel gas (now in the bench scale stage).

Process		Description
American Thermogen process		Produces heat (basically a slagging incinerator with heat recovery capability). This process is not truly a pyrolysis process, but is included here because of the similarity of products.

Pyrolysis processes can be either exothermic or endothermic and involve process temperatures ranging from 900°F to 3000°F.

Various programs on pyrolysis processes have been progressing for long periods. The processes, with their available operating experience, are listed in Table 23.

Table 23

PROCESS STATUS

Process	Pilot Plant	Estimated Operating Time	Proposed Plant
Torrax	75 tons/day	3-4 yr	--
American Thermogen	120 tons/day	2-3 yr	--
Regional Resource Recovery	2-ft dia	2-3 yr	--
Garrett	4 tons/day	2 yr	200 tons/day (San Diego)
Monsanto	35 tons/day	2 yr	1000 tons/day (Baltimore)
Union Carbide	5 tons/day	1-2 yr	200 tons/day (Charleston)
Battelle	3 tons/day	6 mo	--
Urban Research and Development	120 tons/day	6 mo	--
Hercules	--	--	1500 tons/day (New Castle)
Bureau of Mines[*]	--	--	--
Institute of Gas Technology[**]	--	--	--

[*]Research only.
[**]Proprietary pilot plant work only.

Each of the approaches representing a given pyrolysis process is slightly different from the others; however, all of these systems can be grouped into three broad categories:

- Refuse conditioning required

    Garrett Research and Development Company flash pyrolysis system

    Hercules Powder Company system

    Monsanto Landgard system

    Regional Resource Recovery system

- Pyrolysis on as-received refuse (no conditioning)

    American Thermogen process

    Torrax Services, Inc. system

    Union Carbide Corporation oxygen refuse converter (bulky refuse requires size reduction)

    Urban Research and Development Corporation system

- Undefined

    Battelle Memorial Institute system

    Bureau of Mines pyrolysis research

    Institute of Gas Technology system

The processes shown in the undefined category include those where the information available to date is incomplete. These three processes are in the early research phase. The projects may not have advanced far enough at this stage to determine whether processing of input refuse is required.

GARRETT PYROLYSIS SYSTEM

The Garrett Research and Development Company has a resource recovery system designed to recover salable synthetic heating fuels, glass, and magnetic metals from mixed municipal refuse. The system is an outgrowth of nearly four years of intensive research into methods of production of synthetic fuels. The heart of the Garrett system is the flash pyrolysis process. The pyrolysis process has been studied in considerable detail in a laboratory reactor system for over a year.

COMPONENTS AND PROCESS

A simplified flowchart of the Garrett process is shown in Figure 32, and a schematic diagram is shown in Figure 33. The Garrett system incorporates the following basic functions:

107

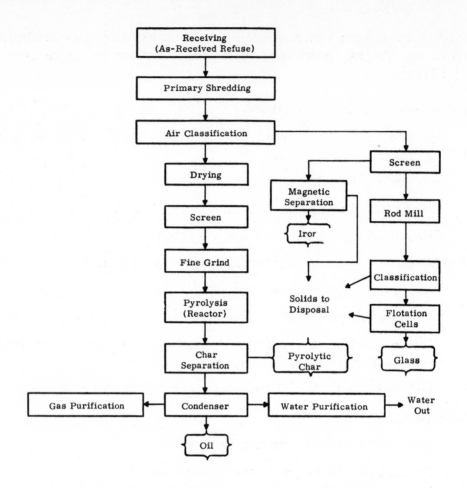

Figure 32. Flowchart of Garrett System

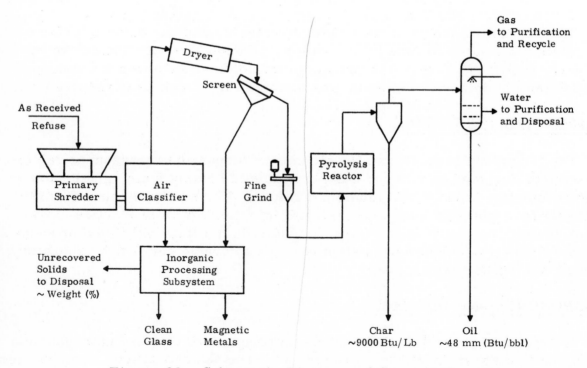

Figure 33. Schematic Diagram of Garrett Process

- Primary shredding of the raw refuse to approximately 2 inches.

- Air classification to remove most of the inorganics from the feed material.

- Drying of the shredded refuse to about 3-percent moisture.

- Screening of the dry material to reduce the inorganic content to below 4 percent by weight.

- Recovery of magnetic metals and clean glass products.

- Secondary shredding of the organics to a -20 mesh.

- Flash pyrolysis of the organics.

- Collection of the pyrolytic products.

These functions can be grouped into three major operations associated with the Garrett system (Refs. 51 through 53):

- Refuse preparation

- Pyrolysis and fuel recovery

- Material recovery

Refuse Preparation

The primary purpose of this operation is to deliver dry, finely divided, essentially inorganic refuse to the pyrolysis reactor. An important byproduct of this step is the preparation for magnetic metal and glass recovery.

The Garrett system is aimed at processing normal municipal refuse, generally the household variety. Because no operating facility of the order of magnitude of 1000 tons per day exists, the extrapolation for such a refuse processing facility is based on the Garrett estimates, based on their 4-tons-per-day pilot plant.

The preferred plant size of the Garrett Company is actually 2000 tons per day; the following discussion pertains to that size range. Figures will subsequently be converted back to a 1000-ton-per-day range for comparison with other processes.

In the 2000-tons-per-day Garrett system, a tipping area is planned that will provide storage for 48 hours of input. From the tipping area, three parallel streams of conveyors, loaded presumably by front-end loaders, carry the refuse into primary shredders. Each parallel line is designed to carry refuse at a rate of 1000 tons per day. (One line serves as a standby line.) Two vertical-shaft Eidal shredders provide coarse size reduction (Ref. 54). Each primary shredder accepts as-received household garbage and reduces it to a 2-inch nominal size. The estimated power requirement for a primary shredder would be 1000 hp

One significant function of the refuse preparation operation is to minimize the inorganic content in the pyrolysis feed material. The Garrett Company claims that their pyrolysis process itself is virtually unaffected by the inorganics. Yet, the Garrett Company is concerned about the presence of agglomerating-type material (e.g., glass in the pyrolysis stream), because it contaminates the fluidized bed (Refs. 51, 55, and 56). It is true, however, that the inorganics degrade the quality of the residual char and increase maintenance and operating costs for secondary shredding.

A zig-zag, vertical-column air classifier, similar to the one offered by the Combustion Power Corporation, is used for first-stage separation. Shredded waste (2-in. nominal) discharged from the primary shredder is carried on mechanical conveyors into the inlet throat of the air classifier. The conveyance from the shredder into the classifier should utilize adequate sealing at the interfaces between the classifier and the conveyor. It is estimated that air velocities through the classifier reach approximately 2000 feet per minute.

Preliminary work done by the Garrett Company on air classification showed that air classification alone could not yield the desired degree of separation between inorganic and organic components. Therefore, a two-stage screening operation, following air classification, is incorporated (Refs. 55 and 57).

In the early phases of the pilot plant program, the Garrett Company felt that the purity of the pyrolysis feed material could be enhanced substantially by drying coarsely shredded refuse before air classification. One estimate indicates that the inorganic content of the feed stream to the reactor can be reduced to less than 4 percent by using this approach (Ref. 51). However, the proposed configuration for the commercial plant incorporates drying after air classification but before screening.

Material separated by the air classifier forms two streams:

- Mostly organics for the pyrolysis reactor.
- Mostly inorganics for material recovery.

The organics stream is fed into a rotary drum dryer after being discharged from the air classifier. Mechanical conveyors can be used for this purpose. All the air through the air classifier is processed through appropriate filtering devices before discharge into the atmosphere. For the pilot plant, a sophisticated pulse-jet heat source generating sonic shock waves within the fluidized bed was used to reduce the moisture content from 25 to 3 percent. However, a more conventional rotary drum is proposed for the commercial processing plant.

The dryer is sized to reduce the moisture from 25 to 3 percent. When the moisture content of the incoming refuse exceeds 25 percent, the type of provision made in the dryer to accommodate this variation is not clear. The options available are to:

- Reduce the through-put to allow greater heat transfer.

- Increase the hot gas temperature in the dryer.

The drying operation greatly facilitates further removal of inerts by screening, improves handling characteristics, and apparently reduces horse-power requirements during secondary shredding. In addition to these bene-fits, it is essential to reduce the moisture content during pyrolysis, in order to lower the heat load in the reactor and also to minimize subsequent oil-water separation operations.

Dryer output, consisting mainly of the organics from the air classifier and with moisture content reduced to around 3 percent, is fed to a two-stage screen-ing device. The first stage is a 1/4-inch screen and the second successive stage is a 14-mesh screen. It is estimated that the inorganics in the organic stream can be reduced to about 2 percent by this successive screening process. About 14 percent of the organics pass through the screens and are ultimately returned to the pyrolysis stream by subsequent material recovery operations.

The most critical element in the refuse preparation operation is the sec-ondary shredder. The Garrett Company feels it is highly desirable to feed the pyrolysis reactor a finely size-reduced organic stream. This fine stream increases oil yields at atmospheric pressure. During pilot-plant operation, it was found that a hammermill would require 40 hp for a 1000-pound-per-hour organic stream, to shred it from the coarsely shredded size to a -20 mesh (approximately 0.015 in.) (Ref. 51). Extrapolating this to a 2000-ton-per-day facility, a secondary shredder would require approximately 1000 to 1500 hp to process 600 tons per day, operating 24 hours per day. There are three sec-ondary shredders proposed for this facility. One major concern is to locate a commercial piece of equipment available for this application. None of the hammermill shredder manufacturers responds optimistically to such a severe size reduction.

The screened organic stream output is then conveyed to the secondary shredder. (Mechanical conveyors should be adequate for this purpose.) Again, the shredder must reduce the coarse stream to a 0.015-inch nominal size. Finely ground output is then conveyed pneumatically to the pyrolysis reactor feed mechanism. Approximately 60 percent of the input stream (by weight) is fed to the pyrolysis system.

Pyrolysis and Fuel Recovery

The Garrett pyrolyiss process involves rapid heating of finely shredded organic materials in the absence of air. A proprietary heat exchange system is utilized, and a single reactor is used, the reactor being fed from a surge storage bin that is served by the three refuse preparation lines. The Garrett philosophy is to provide only two hours of surge storage upstream of the reactor (Ref. 54). It also has been found that a single reactor is preferable, even in the 2000-ton-per-day plant, in order to achieve economy of scale.

The Garrett process is flexible in that it can be tuned to produce either a fuel oil or a fuel gas, depending on the operating temperature of the pyrolysis unit. For fuel gas production, temperatures on the order of 1400°F are maintained and the process is usually endothermic by about 800 to 1200 Btu per pound (Refs. 51 and 54). The resulting fuel gas product has a heating value of 770 Btu per cubic foot (Ref. 54). It contains 23-percent carbon dioxide, which can be scrubbed out to produce pipeline-quality gas of 900-Btu-per-cubic-foot heating value (Ref. 58).

The energy yield and the economies, however, are more attractive for the oil process than for the gas process; therefore, the remainder of this discussion of the Garrett process is devoted to the oil process. The oil process is an atmospheric-pressure, 900°F process that is slightly exothermic. A 550-Btu-per-cubic-foot byproduct gas is also produced and entirely used, in the process, for heat. (Some Garrett literature quotes 500 Btu/ft^3, and other Garrett reports quote 600 Btu/ft^3 for this gas. These numbers were rounded numbers; 550 Btu/ft^3 is the correct value.)

The pyrolysis reactor is a vertical unit. Solid waste is carried into the reactor by recycled gas. Rapid mixing occurs within the reactor as the solid waste passes upward under turbulent flow conditions, and high heat transfer rates are obtained during the short residence time. This flash pyrolysis reaction, conducted at about 900°F, minimizes secondary thermal decomposition (degradation) of the volatilized compounds, and high liquid yields are thus assured. After removal of char in cyclones, the hot gases pass to a standard oil recovery collection train. In the oil recovery collection train, the hot, char-free products leaving the reactor cyclone are rapidly cooled to 200°F by a venturi quench system using recirculated product oil (Ref. 59). Preferred moisture content of the oil is 14 percent (Ref. 60). The liquid fuel oil is separated through a settling process. Water in the settling tank is removed for further purification, prior to its discharge. (The BOD of this waste water is about 100,000 before treatment; however, it is biodegradable.) The gas fraction is used in the process as auxiliary fuel.

The Garrett Company claims that pyrolysis of chlorinated plastics, such as polyvinyl chloride, in municipal refuse results in the production of methyl chloride instead of hydrogen cloride.

In its present configuration of 1000-ton-per-day commercial plant, the entire process is to have one stack venting to the atmosphere. All combustible gases are burned in a process heater, and all of the air streams (e.g., classifier and dryer) are passed through the process heater combustion chamber.

The stack gases are then cooled to 350°F and are vented through a bag filter. It should be noted that when methyl chloride burns, it produces hydrogen chloride, which may prove to be a problem in the stack and bag filter.

About 80,000 cubic feet of gas per ton of waste processed is released from the system. About two-thirds of this gas is the air used only for pneumatic conveyance and drying, while the remainder represents combustion products from the process heater. It is estimated that another 600,000 Btu per ton of waste is available from these gases if additional process heat is required.

Material Recovery

The Garrett Company believes that the most important aspect of their recovery system is the quality of the products. To help achieve this objective, the Company has plans to incorporate several components in the 1000-ton-per-day commercial plant to enhance the quality of recovered materials. Another obvious plus in the Garrett material recovery system is that materials with sales potential are recovered before becoming residue. The Garrett Company does not believe in pyrolyzing these materials when they can be recovered more readily in the shredded refuse form.

The Garret glass reclamation process is proprietary (Ref. 55), but they claim recovery of 70 percent of all glass contained in municipal refuse.

A block diagram of the glass flotation system is shown in Figure 34, and a schematic diagram is shown in Figure 35. Inorganics separated from the

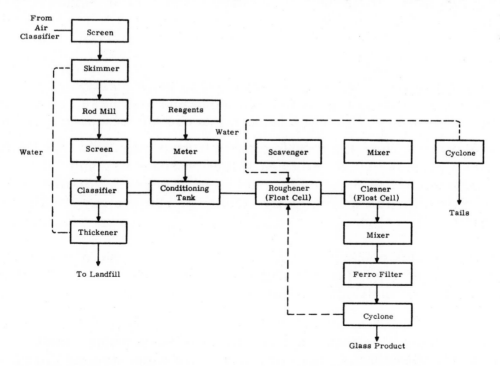

Figure 34. Block Diagram of Garrett Glass Recovery Plant

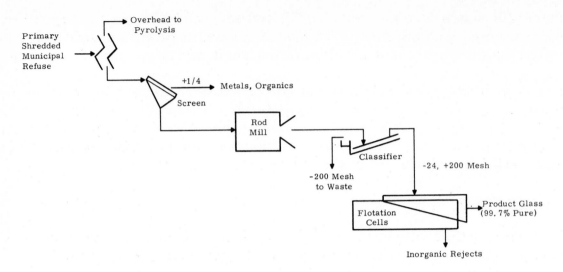

Figure 35. Schematic Diagram of Garrett Glass Recovery Process

air classifier under the flow and screens are processed through a series of skimmers, conditioning tanks, mills, flotation cells, and cyclones.

The significant steps of the Garrett glass recovery process are:

- Screening. A +1/4-inch screen is used, which is expected to contain 95 percent of the glass.

- Coarse milling. The glass-rich fractions are pulped in water, and floating paper and wood are removed. The sink material is ground in a rod mill. After one pass through the mill, the +8-mesh fraction is removed. This fraction consists mostly of metals, with some plastics and rubber.

- Fine milling. The -8-mesh material of ground glass, stones, ceramic, and ash are milled to -32 mesh and are classified to remove the -200 mesh. The material is thickened and disposed of as waste.

- Flotation. The -32, +200-mesh fraction is repulped, conditioned with a proprietary set of glass selective reagents, and fed to a flotation cell. The flotation product is recleaned by flotation once or twice and then is magnetically cleaned to remove any tramp iron particles.

The float and sink products are then dewatered, and the glass is separated as a clean recovered material. Analysis of the tailings is stated as showing no phosphate values. Tailings are discarded to the landfill.

The +8-mesh fraction from coarse milling is 90-percent metallic. This fraction is combined with the +1/4-inch screen fraction, and the mixture is processed to recover magnetic metal.

114

Input and Output Streams

The Garrett Research and Development Company flash pyrolysis system is designed to process normal municipal solid refuse that is typical of packer truck pickup. The limiting components in this system are the primary and secondary shredders and the pyrolysis reactor.

Items objectionable to a ring-grinder-type or hammermill-type primary shredder include:

- Automobiles
- Scrap steel in bulk quantities
- Reinforced concrete
- Explosives
- Wire

The ultimate size reduction requirement on the output of the secondary shredder (0.015 in.) severely restricts the input to this component. Among the types of material suitable for input to a secondary shredder are organics, preferably those which are soft and easy to shred. These materials should not include resilient materials (e.g., rubber and leather), which would be difficult to size-reduce to this extreme. From the standpoint of adverse effects on the pyrolysis reactor, the feed material should be free of agglomerating substances, especially glass, which would contaminate the fluidized bed.

Moisture content of the incoming refuse is another input stream variable. Material entering the reactor is dried to a 3-percent moisture level. Sizing of the dryer is based on an input stream moisture content nominally at 25 percent. As this percentage is increased, the throughput of the dryer would be reduced, and system capacity would be correspondingly downgraded.

Figure 36 shows the product yield and power input for a 2000-ton-per-day Garrett flash pyrolysis system (Refs. 54 and 57), and Figure 37 shows a more detailed material and power flow for a 150-ton-per-day system (Ref. 54) and also shows process water and air flows. In terms of the percentage output yields, both systems are similar. In terms of electric power consumption, the 150-ton-per-day unit consumes 176 kilowatt-hours per ton of input, while the 2000-ton-per-day unit requires 127 kilowatt-hours per ton. The fuel oil produced by the Garrett Company has a higher heating value of 10,500 Btu per pound, according to Garrett tests (Ref. 51), although later tests at Combustion Engineering, Inc. (Ref. 60) quote a heating value of 8490 Btu per pound.

This discrepancy is partially explained by the fact that the higher Garrett figure is for oil with no moisture content, while the Combustion Engineering

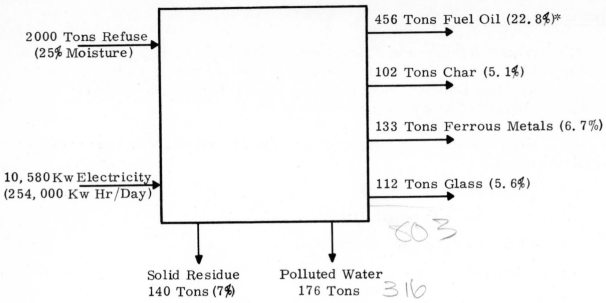

2000 Tons Refuse
(25% Moisture)

10,580 Kw Electricity
(254,000 Kw Hr/Day)

456 Tons Fuel Oil (22.8%)*

102 Tons Char (5.1%)

133 Tons Ferrous Metals (6.7%)

112 Tons Glass (5.6%)

Solid Residue
140 Tons (7%)

Polluted Water
176 Tons

*Percentages are based on incoming refuse weight.

Figure 36.   Product Yield and Power Input for 2000-Ton/Day
Garrett System (Sources: Refs. 54 and 57)

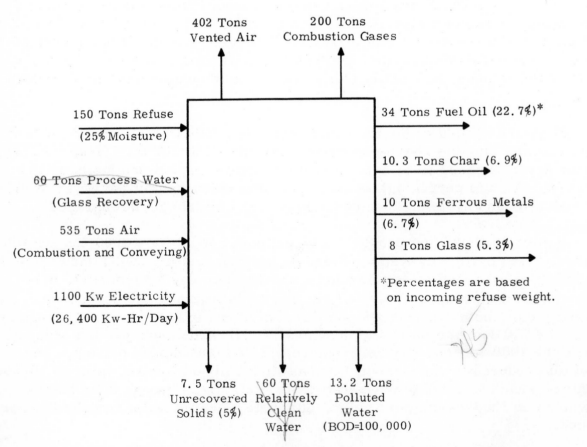

402 Tons
Vented Air

200 Tons
Combustion Gases

150 Tons Refuse
(25% Moisture)

60 Tons Process Water
(Glass Recovery)

535 Tons Air
(Combustion and Conveying)

1100 Kw Electricity
(26,400 Kw-Hr/Day)

34 Tons Fuel Oil (22.7%)*

10.3 Tons Char (6.9%)

10 Tons Ferrous Metals
(6.7%)

8 Tons Glass (5.3%)

*Percentages are based
on incoming refuse weight.

7.5 Tons
Unrecovered
Solids (5%)

60 Tons
Relatively
Clean
Water

13.2 Tons
Polluted
Water
(BOD=100,000)

Figure 37.   Material and Power Flow for 150-Ton/Day Garrett
System (Source: Ref. 54)

results are for oil with the preferred moisture content of 14 percent. When the Combustion Engineering results are corrected for moisture content, the Garrett and Combustion Engineering results agree within 5 percent.

Scaling the 2000-ton-per-day figures linearly down to the 1000-ton-per-day size, which is being used as a standard of comparison, and using the higher value of 10,500 Btu per pound for the heating value of the oil, the input/output stream will be:

- Input

Packer truck waste	1000 tons/day
Process water	800 tons/day

- Output

Fuel oil (0% moisture)	228 tons/day
Char	51 tons/day
Ferrous metals	66. 5 tons/day
Glass	56 tons/day
Total product	401. 5 tons/day
Solid residue	70 tons
Ash in oil	4. 8 tons
Total for disposal*	74. 5 tons
HHV of fuel oil (0% moisture)	10, 500 Btu/lb
Gross total energy/day in fuel oil	4.8×10^9 Btu/day
Power requirement based on 127 kwh/ton generated at 34% efficiency	1.3×10^9 Btu/day
Net energy per day	3.5×10^9 Btu/day

Physical Requirements

The Garrett Company has conducted preliminary studies on sizing land manpower, and operating requirements for a 2000-ton-per-day plant (Refs. 56 and 61). This work reflects an extrapolation from the 200-ton-per-day demonstration plant proposed for San Diego County and approved as a grant by the Environmental Protection Agency.

From the information generated for a 2000-ton-per-day pyrolysis plant, the corresponding estimates for a 1000-ton-per-day facility have been derived.

The total land requirement for this facility is six acres, to accommodate the equipment and to adequately provide for storage of raw refuse. Proper enclosures are required to protect the material from the weather, especially rain and snow. Total land requirements are gross estimates. For improved planning the following areas must be properly identified:

---

*If char is considered a waste product, the tons for disposal rises to 125. 8 tons/day.

117

- Planned storage capacity for raw and shredded refuse.

- Redundancy in the pyrolysis system.

- Storage provision for pyrolysis products.

- Storage provision for glass and magnetic metals.

The major components of the Garrett system that demand high power input are:

- Primary shredder

- Secondary shredder

- Dryer

- Pump

- Glass froth flotation system

- Pneumatic conveyance

- Mill

- Classifier

The power requirement, based on 127 kilowatt-hours per ton, is 127,000 kilowatt-hours per day. A water flow of 15 gpm is required.

Manpower requirements have been furnished by the Garrett Company for 500- and 2000-ton-per-day plants and are listed below, with a General Electric estimate of staffing for a 1000-ton-per-day plant:

Function	500 Tons/Day	2000 Tons/Day	1000 Tons/Day
Feed preparation			
Weighmasters	1/day	1/day	1/day
Operators	2/shift	6/shift	4/shift
Glass recovery			
Operators	2/day	2/shift	2/shift
Pyrolysis system			
Operators	2/shift	4/shift	3/shift
Product Loaders	1/day	2/day	1/day
Supervisors	5 total	5 total	5 total
Off-site administration, maintenance, etc.	10/day	13/day	10/day
Total	35	69	54

Operation is based on 24 hours per day, 7 days per week, and 350 days per year.

OPERATING HISTORY AND EXPERIENCE

G. M. Mallan and C. S. Finney, in a recent paper (Ref. 51), stated that the Garrett Company initiated a new coal conversion program about four years ago (in 1968). This program employed a flash pyrolysis process, operating essentially at atmospheric pressure and requiring no catalyst and a modest capital investment. In 1969, the Garrett Company found that a similar process could readily convert the organic portion of municipal solid waste into good quality synthetic fuel oil. The early experiments showed that about one barrel of pyrolytic oil could be obtained from one ton of refuse. Since this discovery, an integrated solid waste recovery process has been developed. The Garrett Company claims that this process converts or reuses over 90 percent of the raw materials in the refuse.

A 5-pound-per-hour reactor was built and tested during 1970 and 1971. This reactor was a continuously operating unit. The results from this laboratory model were presumably encouraging to the Garrett Company, in terms of tentative oil yields and offered recovered products. A 4-ton-per-day pilot plant was constructed in 1970 and has been in use for two years.

Experience with the 4-ton-per-day pilot plant did not include primary shredding. The Garrett Company received shredded waste from the Combustion Power Corporation. The waste was then run through the Garrett pulse-jet drier, which reduced the moisture content from 25 to 3 percent. The dried material then went to a Garrett-designed zigzag air classifier, sending the heavy fraction to the glass and magnetic metal recovery process. The light fraction went to a set of Sweco screens, to remove glass and metal particles, and was then sent to the Reitz RD-12 secondary shredder. This shredder is a 40-hp, 1000-pph-capacity machine, which produces a 20-mesh output. The output is then air-transported to the holding bin in the pyrolysis unit from which the feedstock is fed into the pyrolysis unit by a positive displacement feed.

In addition to shredded waste supplied by the Combustion Power Corporation, the Garrett Company also processed pulped municipal refuse supplied by the A. M. Kinney/Black-Clawson plant in Franklin, Ohio. Currently, the plant is processing a variety of agricultural feedstocks, including rice hulls and tree bark, under a Government contract.

The operation of this pilot plant, using prepared municipal refuse, yielded the following pertinent data:

- Pyrolytic fuel oil can be generated continuously at essentially atmospheric pressure, requiring neither a catalyst nor hydrogen.

- Over 70 percent of the glass, with purity higher than 99.7 percent, can be recovered.

- More than 90 percent of the raw materials contained in the municipal waste can be recovered and recycled.

Using the 4-ton-per-day pilot plant, the Garrett Company evolved a system for a 200-ton-per-day plant. This system plan was submitted to the Environmental Protection Agency as a demonstration project for San Diego County and was approved in August 1972. Funding will be:

Environmental Protection Agency grant to San Diego County	$2,962,710
San Diego County funds	600,000
Garrettt Company funds	300,000
San Diego Gas and Electric Company funds	150,000
Total	$4,012,710

Plant design is now nearly complete, with construction slated to begin in early 1976. The schedule calls for completion of construction in 20 months, 4 months for startup, and 1 year of operation by the Environmental Protection Agency. The facility will convert all of the wastes of two communities (Escondido and San Marcos) into fuel oil that will be used in a San Diego Gas and Electric Company steam powerplant.

The San Diego system is expected to handle about 200 tons of solid waste per day. Anticipated revenues are between $200,000 and $300,000 from the sale of facility products.

SCALING CONSIDERATIONS

The Garrett Research and Development Company proposed flash pyrolysis system has been evaluated for a 4-ton-per-day pilot plant for more than two years. While the experiment has been successful in demonstrating recovery of products, it has been a laboratory-type demonstration under controlled conditions.

Three major areas with many unknowns emerge as the critical areas when scaling a pilot operation to a 1000-ton-per-day commercial facility:

- Feed material preparation for flash pyrolysis
- Glass recovery process
- Pyrolytic reactor

This concern relative to scaling could have been one of the considerations made when the Garrett Company proposed a 200-ton-per-day demonstration plant for an Environmental Protection Agency grant.

Extreme size reduction requirements of input material for the pyrolysis reactor (0.015 in.) would create a considerable challenge for shredder manufacturers. To date, no equipment of this type seems to be commercially available to handle the order of magnitude of feed material expected in a 1000-ton-per-day facility.

The reactor feed material should be dried to a nominal 3-percent moisture content. A dryer to perform this task must be flexible, to handle the input with incoming moisture content varying from 5 to 60 percent, as encountered in municipal refuse. At the same time, the dryer should maintain a constant throughput and a fairly constant moisture content of output.

Because details of the ractor are not known, an accurate assessment of its scalability is not possible. It is believed, however, that the Garrett scale-up from 4 tons per day to a single-unit, 200-ton-per-day reactor for the San Diego project should uncover most scaling problems. Because economy of scale can be achieved only by use of a single reactor, plants larger than 200 tons per day will require an additional scale-up, which will, of course, involve some risk.

The sophisticated proprietary froth-flotation glass recovery system is quite complex, at best. Adding scale-up considerations causes the system to become quite unwieldy. To date, the Garrett Company has not even built a glass recovery system compatible with their 4-ton-per-day pilot plant. The recovery plant tested by the Garrett Company was rated at 50 pounds per hour.

The Garrett approach in assimilating various components into their flash pyrolysis system appears to follow their parent company's experience. The configuration begins shaping up as a refinery outgrowth.

In general, the Garrett Company feels it will have determined the critical design parameters for a 2000-ton-per-day plant 20 months after the start of the San Diego project.

AIR AND WATER POLLUTION CONSIDERATIONS

The Garrett flash pyrolysis system, as configured for a 1000-ton-per-day facility, will have only a single stack venting to the atmosphere. All air streams, such as those from the classifier and dryer, will pass through the process heater combustion chamber to oxidize odors and to burn any fine particles not removed by process cyclones. Stack gases are colled to 350°F and are vented through a bag filter. The anticipated efficiency of this filter is 95 percent.

Total particulate emission is expected to be no more than 0.06 gram per cubic foot (Ref. 52). Environmental Protection Agency standards call for 0.08 gram per standard cubic foot of dry gas, corrected of 12 percent carbon dioxide.

The flash pyrolysis process requires approximately three pounds of air per pound of refuse processed. While it might be feasible to attain required air quality standards, it appears that the water pollution control task is a sizable one. One of the products of pyrolysis, out of the reactor, is water. Water represents 13 percent of the pyrolysis products, by weight. The chemicals present in this water are:

- Acetaldehyde
- Acetone
- Formic acid
- Furfural
- Methanol
- Methylfurfural
- Phenol

The waste water has a BOD of 100,000 and is biodegradable. It is estimated that for a 1000-ton-per-day facility, the amount of water from pyrolysis would be on the order of 1000 gallons per hour (88 tons/day). A substantial quantity of water (400 tons/day) will also be discharged continuously from the froth flotation glass recovery process. Planning is required for adequate purification of this water before it is discharged into any sewers.

Another pollutant in the discarded waste would be the sink from the thickener and the tails, both nonusable byproducts of the glass recovery process.

In looking at the overall pollution control solution, some positive action on the disposition of these byproducts, along with waste water purification, must be incorporated. The Garrett Company has touched on these points; their solution will be worked out later.

COST FACTORS

Following are Garrett data for the 200-ton-per-day, 2000-ton-per-day, and (where available) 500-ton-per-day plants:

Factor	200 Tons/Day	500 Tons/Day	2000 Tons/Day
Capital investment (no land)*			
Structures	--	--	$ 3,200,000
Equipment	--	--	10,720,000
Total	$3,860,000 (Ref. 54)	--	$13,920,000 (Ref. 57)
Capital cost per ton/day capacity	$ 19,313	--	$ 6,960

*A General Electric estimate for a 1000-ton-per-day facility is $25.4 million. No provision is made for effluent water treatment.

122

Factor	200 Tons/Day	500 Tons/Day	2000 Tons/Day
Land area required (acres)	--	--	4
Economic life (yr)	15	15	15
Operating cost factors (Ref. 54)			
Operating employees	11	25	56
Administrative employees	--	10	13
Hr/day	24	24	24
Days/yr	350	350	350
Primary shredders	1	--	3
Dryers	1	--	3
Secondary shredders	2	--	3+
Flash pyrolysis units	1	1	1
Liquid collection train	1	1	1

Projected revenues are shown in Table 24 as provided by the Garrett Company. Prices are delivered prices and may be optimistic. For instance,

Table 24

PROJECTED REVENUES FROM SOLID WASTES

Recovered Commodities	Weight (approximate % of as-received refuse)	Available for Disposal (%)	Commodity Value ($/ton)	Estimated Revenue ($/ton refuse)
Magnetic metals	7	95	20.00	1.33
Glass, mixed color	8	70	14.00	0.78
Oil at 4.8 mm (Btu/bbl)	24	95	15.35	3.50
Char at 9000 Btu/Lb	15	34	4.50	0.23
Gas at 600 Btu/Ft^3	6	0	4.00	0.00
Water	33	34	-0.25	-0.03
Inorganic wastes	7	85	-1.50	-0.09
Net revenue/ton				5.72

Source: Ref. 51

the $15.35-per-ton price for the oil becomes $.90 per million Btu at a heating value of 8490 Btu per pound. A value of $.50 to $.60 per 10^6 Btu appears more reasonable at 1973 prices.

GARRETT SYSTEM PRODUCTS

The Garrett Company system is very complex. It does, however, offer a product flexibility that is not available in other pyrolysis processes. It also offers a storable fuel, when the oil process is used. Due to the desirability of the types of products produced by the Garrett system, a closer look at the actual products and their applicability is in order.

Pyrolytic Oil (Garboil)

The composition of the oil obtained by pyrolysis of solid waste at the Garrett pilot plant depends on both the quench temperature and on further treatment. Pyrolytic oil is made of hundreds of compounds, varying in molecular weight from 32 to perhaps 10,000, with boiling points ranging from 55°C to greater than 360°C. About half the material is not capable of being distilled.

Some state of oxygenation is always present, so the carbon-to-oxygen ratio usually ranges between one and two for most fractions. Specific gravities from 1.1 to 1.4 are typical of organic fluids that are oxygenated.

The oil is darkly colored, having the red-brown hue typical of compounds derived from the destructive distillation of carbohydrates of polysaccharides. The oil has a pH of 1.7 to 2.5, measured in either water or methanol, and owes its acidity to carboxylic acids such as glycolic, maleic, acetic, or acrylic acid. Aqueous solubilities are highest in caustic soda or potash solutions.

A comparison of the oil with a typical No. 6 fuel oil is given in Table 25 (Ref. 51). The pyrolytic oil properties shown were obtained from the bench test output, which differs somewhat from the pilot plant output.

Table 25

TYPICAL PROPERTIES OF NO. 6 FUEL OIL AND PYROLYTIC OIL

Property	No. 6 Oil	Pyrolytic Oil
Carbon weight (%)	85.7	57.5
Hydrogen	10.5	7.6
Sulfur	0.7-3.5	0.1-0.3
Chlorine	--	0.3
Ash	<0.5	0.2-0.4
Nitrogen	2.0	0.9
Oxygen	2.0	33.4
Btu/lb	18,200	10,500
Specific gravity	0.98	1.30
Lb/gal	8.18	10.85
Btu/gal	148,840	113,910
Pour point (°F)	65-85	90
Flash point (°F)	150	133
Viscosity SSU at 190°F	340	3150
Pumping temperature (°F)	115	160
Atomization temperature (°F)	220	240

Results of a chemical analysis and of physical tests on Garrett Garboil samples produced by the 4-ton-per-day pilot plant are given in Table 26. Complete details of these tests, conducted by Combustion Engineering, Inc. appear in Reference 60.

Table 26

GARRETT GARBOIL ANALYTICAL DATA
(Combustion Engineering Inc. Tests)

Report Basis	As Received	Ash Composition	
Total moisture	14.1	SiO_3	14.2
Volatile matter	74.3	--	--
Fixed carbon	9.5	Al_2O_3	13.1
Ash	2.1	Fe_2O_3	37.4
Total	100.0	CaO	9.4
HHV			
Btu/lb	8,490	MgO	5.2
Btu/gal	94,000	--	--
Moisture	14.1	Na_2O	4.1
Hydrogen	4.9	--	--
Carbon	48.7	K_2O	2.7
Sulfur	0.3	TiO_2	1.4
Nitrogen	1.3	--	--
Oxygen	28.1	P_2O_5	2.3
Chlorine	0.5	--	--
Ash	2.1	SO_3	6.1
Total	100.0	Present not assayed	
Ash fusibility (°F)	--	--	4.1
Reducing atmosphere	--	--	
IT	1,930	Zn, Ba, Mn, etc. --	4.1
ST	2,230	total = 100.0	
FT	2,660		
Specific gravity 25/25	1.33	Base/acid ratio	2.0
Distillation			
Ibp	105°C	Fe_2O_3/CaO	--
280°C	17.4%	Ratio	4.0
Residue	82.6	--	--
Total	100.0	--	--
Sample frothed at 290°C, stopping distillation			--
Flash point	133°F	--	--
Fire point	Not obtained	--	--
Pour point	77°F	--	--

Saybolt viscosity (14% moist Garboil)

170°F	3100 SSU	190°F	1100 SSU	210°F	450 SSU
180°F	1700 SSU	200°F	670 SSU	215°F	350 SSU

IT - Initial deformation temperature   FT = Fluid temperature

ST - Softening temperature   IBP = Initial boiling point

These tests consumed five barrels of Garboil at firing rates up to 18 gallons per hour. Comments that follow are a combination of extracts from the Combustion Engineering report and comments made by C. S. Finney and G. M. Mallan (Ref. 54):

- The oil is corrosive. Because about half of the pilot plant consisted of mild steel and the other half was stainless steel, a 2.1-percent ash, very high in iron, was observed in the oil. Tests performed on an all-glass bench unit produced a 0.2-percent ash content.

- The recommended pumping temperature is 170°F, and the atomization temperature is 230°F for Garboil, at the recommended moisture content of 14 percent. Higher moisture content reduces combustion efficiency, and lower moisture content unduly raises viscosity. The flash point of 133°F is considered safe. Temperatures below 120°F would be dangerous.

- The oil degrades at temperatures above 200°F. In particular, localized hot spots at the oil heater surfaces must be located. These hot spots can exist even though the bulk oil temperature may be well under 200°F. Also, any bypass-type pressure regulator loops downstream of the atomizing oil heaters must be avoided.

- The Garboil swells considerably with temperature increases, going from a specific gravity of 1.36 at 110°F to a specific gravity of 1.05 at 190°F. Storage tanks should be sized accordingly.

- Steam atomization works well with Garboil; air atomization does not.

- Garboil is compatible with the San Diego No. 6 fuel oil (an Alaskan oil) and was found compatible with two South American mixes of No. 6 oil. Here, compatibility is defined as not causing a chemical change in either fuel when mixed. Garboil was found to be incompatible with the low-sulfur, blended No. 6 fuel oil normally used at the Combustion Engineering Windsor Locks powerplant. The Combustion Engineering oil may be typical of oil used in Connecticut utility plants.

- Blends of Garboil and compatible No. 6 oils are actually mixtures that tend to settle out over a period of days, due to specific gravity differences between the two components. Garrett recommends blending on-line, just prior to firing. This procedure requires 100-percent stainless steel for the Garboil fuel portion of the system.

- Ignition stability of properly atomized Garboil, or blends of Garboil with San Diego No. 6 fuel oil, was equal to that obtained on No. 6 fuel oil alone. Tests were run at Garboil/No. 6 oil ratios of 100/0 percent, 75/25 percent, 50/50 percent, 25/75 percent, and 0/100 percent.

- Plugging from Garboil or Garboil blends is a problem. The purging of the fuel handling system with No. 6 oil immediately before shutting down the system is necessary to avoid plugging of lines and/or spray tips.

- Stack emissions showed negligible amounts of unburned carbon when burning Garboil and Garboil blends at excess oxygen levels above 2 percent.

- Sulfur dioxide in the flue gas during 100-percent Garboil firing was nearly that expected for a 0.3-percent sulfur fuel (250-300 ppm). NO_x was on the order of 400 ppm, somewhat less than that calculated, based on the 1.3-percent nitrogen in the oil. Sulfur dioxide and NO_x values are corrected to 3-percent excess oxygen.

- The Combustion Engineering conclusion is that Garboil or blends of Garboil and compatible No. 6 fuel oils can be successfully burned in a utility boiler with a properly designed fuel handling and atomization system.

- The Garrett Company has made an in-house determination that Garboil is not a satisfactory gas turbine fuel, due to its high ash content.

Pyrolytic Gas

Little information has been released on the characteristics of the gasification process product other than the 770-Btu per cubic foot heating value. However, the byproduct gas produced by the liquefaction process may provide some clues. When run at optimum liquefaction temperatures of 900°F to 1000°F, the gas yield is about 27 percent, based on the dry pyrolysis feed to the reactor. All of this gas is reused in the process. The heating value of this gas is about 550 Btu per cubic foot, and its typical chemical composition is shown in Table 27. The chloride content of this gas is of considerable significance.

Table 27

TYPICAL GAS ANALYSIS FROM LIQUEFACTION
CONDITIONS AT 950°F*
(Garrett Process)

Component	Volume (%)
Water	0.1
Carbon monoxide	42.0
Carbon dioxide	27.0
Hydrogen	10.5
Methyl choride	<0.1
Methane	5.9
Ethane	4.5
C_3 to C_7 hydrocarbons	8.9

*Feed material supplied by the Black-Clawson Company. Gas yield, 27% (by weight); heating value 550 Btu/ft^3.

127

Repeated analyses by mass spectroscopy have failed to detect any hydrogen chloride in any pryolytic gas sample. Only methyl chloride has been detected in these tests, and this potential pollutant is quite easy to remove with simple water scrubbing. Removal results in essentially no chloride air pollution when this gas is burned for process heat.

Pyrolytic Char

The Garrett pyrolysis pilot plant effort conducted at La Verne, California to date has relied upon municipal solid waste feedstock that has previously been coarsely shredded and air-classified. Under these conditions, the pyrolysis feed has contained at least 10-percent inerts, and the yield of pyrolytic char has been about 20 percent by weight. The properties of a typical char are summarized in Table 28.

Table 28

CHAR CHARACTERISTICS

Characteristic	Nominal	Experimental Range
Heating value	9000 Btu/lb	6000-9700
Ash content	31.8% (by weight)	21.8-50.0
Ultimate analysis		
Carbon	48.8% (by weight)	37.1-61.3
Hydrogen	3.9	2.0-4.0
Nitrogen	1.1	1.0-1.5
Oxygen	12.7	12.2-17.2
Sulfur	0.28	0.28-0.65
Chlorine	0.22	0.22-0.45
Moisture content	0.6% (by weight)	0.0-1.44

The calorific value of the chars produced is equivalent to that of a low-grade coal, largely because of the quite high oxygen content and the considerable amount of ash present. The oxygen content, of course, reflects the cellulosic nature of many of the constituents that make up municipal refuse. Ash content is high, and further process development is needed to reduce it to below 30 percent. The Garrett Company had hoped to upgrade the char into a low-grade, activated carbon, but has found this development to be impractical using the char from municipal solid waste. It is practical, however, with char from several agricultural waste feedstock. The char from municipal waste is, at best, a marginal product.

Glass

The Garrett Company claims that their proprietary process enables recovery of over 70 percent of all glass in the incoming refuse (Ref. 55). The recovered glass is a sand-sized, mixed color product with a purity of better than 99.7 percent. The high purity is due to a froth flotation process. Approximately two/thirds of the impurities are carbonacious. These impurities presumably present little problem upon remelting.

Laboratory melt studies conducted by Owens-Illinois, Inc., on sample cruets show none of the normal imperfections often found in reclaimed glass. The Garrett Company is continuing efforts to color-sort recovered glass, using a high-intensity magnetic separator. Thus far it has yielded excellent results in the laboratory, but is very expensive from a power standpoint. Another check on glass quality is made by comparing its chemical composition with that of a typical production green. Owens-Illinois, Inc. made this comparison and found the two to be very similar.

Metals

Most of the ferrous metal recovered from municipal trash will be tin-coated steel cans that have been tightly crumpled during a preliminary shredding in the Eidal mill. Miscellaneous large pieces of scrap will be periodically kicked out of the Eidal mill and will be mixed with the cans. The metal fraction will contain some 4 percent of organic and nonmetallic refuse.

Nonferrous metals are presently not being recovered in the residual fraction that must be landfilled. If most of these nonferrous metals were recovered, not only would the small residual waste fraction be reduced, but the gross income of the process would be increased by about $1 to $2 per ton of as-received refuse. Development work on several promising recovery processes will soon begin at the Garrett Research and Development Company.

MONSANTO LANDGARD SYSTEM

The Monsanto Company has been developing the Landgard system for several years. The Landgard system is designed to be a totally enclosed, self-contained operation. The system encompasses all operations for receiving, handling, shredding, and pyrolyzing waste; for quenching and separating the residue; for generating steam from waste heat; and for purifying the off-gases. Potential products are steam, carbon char, glassy aggregate, and ferrous metal.

COMPONENTS AND PROCESS

The Landgard system represents a coupling of three major subsystems:
- Refuse preparation
- Pyrolysis and energy recovery
- Residue materials recovery

Each of the subsystems comprises functional grouping of various components and options. Figure 38 shows the flow diagram for the Landgard system.

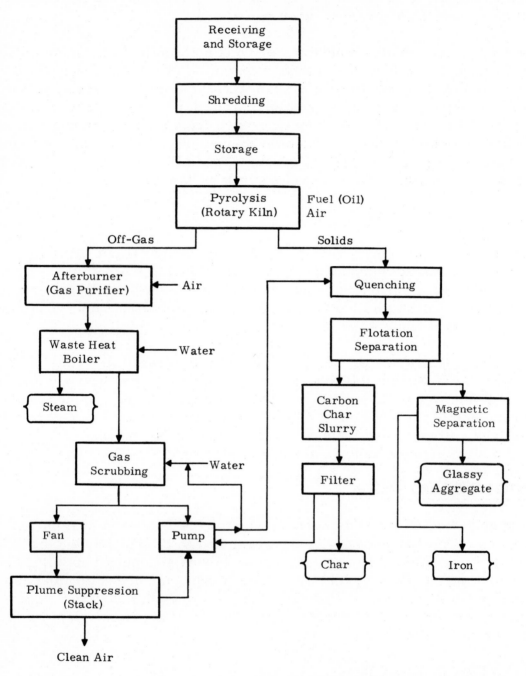

Figure 38. Flow Diagram of Monsanto Landgard System (1000 Tons/Day)

Refuse Preparation

Refuse conditioning and preparation prior to pyrolysis includes:

- Receipt and storage of incoming raw waste.

130

- Size reduction of the refuse.

- Storage and outfeed of shredded waste for pyrolysis.

The receiving area for a 1000-ton-per-day facility is sized to accept raw refuse over a six-hour period to correspond to an eight-hour collection cycle. Planning of the truck dumping area will allow for each truck to enter the facility, unload, and move out in no more than five minutes, on a sustained basis, during the six-hour refuse dumping period.

Provision for raw refuse storage should include adequate space to accommodate at least one day's normal collection. This quantity would require a holding capacity of 1000 tons. The receiving and storage facility should be enclosed and protected from the weather.

Shredding action is accomplished by Eidal-type shredders. It is presumed that two parallel size reduction streams are utilized; however, shredded refuse streams are combined into one storage bin. For a 1000-ton-per-day processing plant operating 16 hours per day, two Model 1000 Eidal mills are required. Size reduction of the waste is to -6 inches, to provide an optimum time-temperature profile through the pyrolysis rotary kiln (Refs. 30 and 62).

Metered raw refuse out of live-bottom hoppers is conveyed mechanically to the shredder. The shredded discharge stream again utilizes mechanical conveyors for storage of shrddded refuse into a bin. From a housekeeping standpoint, and for protection from the weather, both the charging conveyors and the shredded refuse conveyors must include a protective cover.

Atlas-type enclosed conical bins are utilized for storage of shredded refuse. The bins should have the capacity to hold at least 1000 tons of shredded refuse. A single-discharge metered feed of shredded refuse is conveyed mechanically into the pyrolysis kiln.

Pyrolysis and Energy Recovery

In the basic pyrolysis process, shredded waste is heated in an oxygen deficient atmosphere to a temperature in the vicinity of 1800°F (Refs. 30, 62, and 63). This environment pyrolyzes organic matter into gaseous products (mostly CH_4). Pyrolysis also forms a residue consisting mostly of ash, carbon, glass, and metal.

Shredded waste is continuously fed into a rotary kiln. The kiln is refractory-lined to withstand high operating temperatures. Shredded waste feed and direct-fired fuel (oil) enter opposite ends of the kiln. Countercurrent flow of gases and solids exposes the feed to progresively higher temperatures as it passes through the kiln, so that first drying and then pyrolysis occur. The kiln is designed to uniformaly expose solid particles to high temperatures (Ref. 64). This exposure is intended to maximize the pyrolysis reaction.

Sealing at the inlet and discharge ends of the rotary kiln does not appear to be a problem, perhaps because of the negative pressures maintained in the pyrolysis chain.

Pyrolysis gases from the kiln are drawn into a refractory lined after-burner (gas purifier). These off-gases are mixed, in the gas purifier, with incoming air and are burned. The design of the gas purifier is aimed at preventing discharge of combustible gases (e.g., CH_4) to the atmosphere. High temperatures in the purifier thermally destruct off-gases and control odors.

High-temperature gases discharged from the gas purifier are passed through a water tube boiler. This heat exchanger is designed to produce up to 200,000 pounds per hour of saturated steam at 250 psig (Refs. 62 and 63). Exit gases from the boiler are further cooled and particulate matter is removed while gases pass through a water-spray scrubbing tower.

An induced draft fan is utilized to move the exhaust gases from the rotary kiln through the entire system and out the exhaust stack. Gases, after scrubbing and temperature reduction, are introduced in the fan. The scrubbing action saturates these gases with water. To minimize and suppress plume formation, gases leaving the fan are passed through a dehumidifier. By using ambient air, the gas temperatures are dropped further, and condensed water is removed and recycled. The cooled gas stream is then heated by ambient air and is discharged to the atmosphere. Suitable pumps are utilized to recirculate scrubber-discarded water, which is processed through appropriate filters, after using it for quenching of the rotary kiln residue.

Residue Materials Recovery

The hot residue from the rotary kiln is discharged at 1800°F into a water-filled quench tank. In the basic system, the residue is conveyed out of the quench tank, and no further processing of this mixture of ash, carbon, glass, and metal is done.

The Monsanto Company also offers a resource recovery version of the Landgard system. In the resource recovery system, however, additional equipment is utilized to remove iron and to separate the char. In this version, the residue is elevated, on a conveyor, out of the quench tank and into a flotation separator. Here low-density materials floats off as a carbon char slurry, is thickened, and is then filtered to remove the water. This slurry is then conveyed to a storage pile and is transported away from the site.

The heavy material from the bottom of the flotation separator is conveyed to a magnetic separator for recovery of iron. The remaining heavy material, defined by the Monsanto Company as glassy aggregate, is passed through screening equipment and is stored for removal from the site. Solids from the scrubber are removed by diverting the sediment-laden part of the recirculated water to a thickener. Underflow from the thickener is transferred to the quench tank, and the clarified overflow stream is recycled to the scrubber.

Input and Output Streams

The Monsanto Landgard system is designed to process normal municipal refuse. The system is not capable of handling bulky waste. The limitations on the type of refuse seem to be due to the shredding and size reduction equipment capability. Among the items unacceptable to the shredder are:

- Automobiles
- Bulk quantities of scrap steel
- Reinforced concrete
- Chemicals
- Oil
- Wire

Also, such items as explosives, concrete, and certain chemicals are objectionable in the pyrolyzing rotary kiln.

High moisture content, greater than 50 percent, could result in a reduction of the throughput processed by the shredder. Also, a high moisture content could adversely influence size reduction by the shredder, because of a sticky paste formed by squeezing the refuse instead of shredding it.

There are two versions of the Landgard system:

- Basic plant generating gas and steam.
- Complete system with ferrous metal recovery potential.

Looking at a plant with resource recovery, the Monsanto Company estimates that for a 1000-ton-per-day normal municipal refuse input, the input-output stream will be:

- Input

    Packer truck waste   1,000 tons/day

- Output

    Ferrous metal   70 tons/day

    Glassy aggregate   170 tons/day

    Carbon char   80 tons/day

    Steam   200,000 lb/hr

133

- <u>Output</u> (Cont'd)

Gross energy value of steam	5.51×10^9 Btu/day
Energy of fuel oil used	1.00×10^9 Btu/day
Estimated power requirement based on 50 kwh per ton generated at 34% efficiency	<u>0.18 $\times 10^9$ Btu/day</u>
Net energy produced	4.33×10^9 Btu/day

From the standpoint of any recovery: the char is wet, it will not burn, it is highly alkaline, and its heating value is 2400 Btu per pound. Therefore, no financial benefits can be derived from the char in this form, as discharged from the plant (Ref. 64).

The glassy aggregate might offer some potential as a roadbed material, but it might require further drying before any transporting could take place.

Physical Requirements

The Monsanto Company has comlpeted preliminary sizing of the land requirement for a 1000-ton-per-day plant. This preliminary sizing is perhaps extracted from their demonstration **p**lant study for the City of Baltimore.

The requirements are defined in terms of: 1) a basic plant without resourcⵉ recovery and 2) one with resource recovery. According to the Monsanto Company, a basic plant would require three to four acres; a resource recovery plant using pit storage would require up to five acres (Ref. 30, 62, and 63). If a level floor receiving area is used, six acres will be required.

The above estimates appear to be gross. To arrive at refined estimates, the following data must be known:

- Planned storage capacity for raw and shredded refuse.
- Extent of redundance desired.
- Storage capacity for char and other residue.
- Capacity for on-site processing of recirculating water

Major components of a 1000-ton-per-day system, requiring a high power input, would be:

- Shredder
- Rotary kiln
- Boiler
- Scrubber

- Pump

- Induced draft fan

- Conveyor

- Shredded refuse storage area

Of these components, the shredder is known to require a 1000-hp motor. There are at least two shredders required for this size facility. Based on the power requirements for a smaller capacity plant, the power requirement for a 1000-ton-per-day system can be estimated on an order of magnitude basis. Gross numbers of 3000 to 5000 hp seem adequate to supply the power needs for the above facility.

OPERATING HISTORY AND EXPERIENCE

E.D. Stewart (Refs. 30 and 62) stated that the Landgard system was developed at the Monsanto Company in Dayton, Missouri. A one-ton-per-day pilot plant was constructed before 1969 to demonstrate feasibility of resource recovery through solid waste pyrolysis (Ref. 30). A 35-ton-per-day pilot plant achieved operational status in St. Louis County in the fall of 1969. This prototype was operated for two years, and the Monsanto Company claims that through this plant, the Company:

- Demonstrated continuous direct fire pyrolysis of typical unclassified municipal waste.

- Gained operational experience in solid waste handling.

- Obtained performance data for optimizing the process and scaled up for large facilities.

- Compiled guidelines on a performance guarantee.

- Performed development work on preliminary resource recovery from residue.

The Environmental Protection Agency has selected the Landgard system to be a 1000-ton-per-day demonstration plant. The total project will cost $13 million, the Environmental Protection Agency's share being $6 million. The system, which will be constructed in Baltimore, Maryland, is expected to be operational in 1975.

In the 35-ton-per-day prototype, the Monsanto Company claims to have run up to 25 percent, by weight, of various plastics and tires and up to 30 percent, by weight, of solids and sewage sludge. Apparently these quantities ran satisfactorily, with no significant variations in effluents. The key problem pertained to the handling of liquids, rather than to pyrolyzing.

No sale of a commercial Landgard system has yet been made (Ref. 65). However, the Monsanto Company is currently negotiating with several major cities, presumably the result of aroused interest emanating from the recent Environmental Protection Agency grant.

The Monsanto system primarily utilizes commercially available components; however, some of these components have not been previously used for this specific application, on a commercial scale (e. g., the Atlas storage bin and the Eidal shredder). The Monsanto Company indicates that over 100 rotary kilns, similar to those used in the Landgard system, are in operation within the chemical industry. Some of these kilns are more than 300 feet long.

SCALING CONSIDERATIONS

Most of the elements of the Monsanto Landgard system have been utilized in other applications that process throughputs representative of 1000 tons per day. Rotary kilns of capacities larger than 1000 tons per day have been operating commercially, but not in the pyrolyzing of municipal waste. Shredders of the type considered by the Monsanto Company for their full-scale system have been in operation, grinding municipal refuse.

The Monsanto pilot plant capacity was 35 tons per day. Company plans for a 1000-ton-per-day plant in Baltimore will probably lead to an operational system by 1975. The areas of concern, when scaling data from a 35-ton-per-day pilot plant to a 1000-ton-per-day full-scale plant include:

- Waste heat boiler: scaling effects and commercial versus demonstration operation.

- Gas scrubber: scaling factors and commercial plant life.

- Water recirculation components: scaling factors and life of pump, filter, and flotation separation equipment.

- Storage bins: effect of greater heights on packing of shredded refuse.

In determining resource recovery potential for a 1000-ton-per-day plant, the Monsanto Company seems to have used very optimistic scaling factors from their 35-ton-per-day unit. Energy recovery in the form of steam is given as more than 60 percent of the wet refuse heating value. Considering losses through the kiln, the afterburner and conversion efficiency of the waste heat boiler on overall efficiency greater than 60 percent is highly optimistic. (This percentage may be partially offset by the fact that fuel oil is used to heat the pyrolysis kiln.) The same situation is true for the ferrous metal recovery potential, described in excess of 95-percent recovery from the refuse, after pyrolysis. These areas require further assessment when scaling up from from the pilot plant.

AIR AND WATER POLLUTION CONSIDERATIONS

The Monsanto Company has completed stack sampling of their 35-ton-per-day prototype plant. The Monsanto data specify ultimate characteristics of the refuse (4600 Btu/lb) at the time of stack sampling (Refs. 30 and 62). It is therefore assumed that normal household (municipal) refuse of the packer truck type was pyrolyzed.

Based on the information furnished by the Monsanto Company, the stack gas analysis by volume consists of:

Gas	Volume
Nitrogen	78.7%
Carbon dioxide	13.8%
Water vapor	1.8%
Oxygen	5.7%
Hydrocarbons	10 ppm
Sulfur dioxide	150 ppm
Nitrogen oxides	65 ppm
Chlorides	25 ppm

The estimates furnished on particulate loading in the stack, using the Environmental Protection Agency method, are 0.02 grains per standard cubic foot of dry gas, corrected to 12-percent carbon dioxide. The allowbale particulate emission levels established by the Environmental Protection Agency are 0.08 grains per standard cubic foot, corrected to 12-percent carbon dioxide. The data from the prototype plant indicate that the plant should meet the Environmental Protection Agency standards. In extrapolating to a full-scale 1000-ton-per-day plant, however, gas scrubber sizing and efficiency would be a prime factor.

It is conceivable that low particulate emission levels can be achieved in the proposed system. Following the path of the gas stream from the rotary kiln, the off-gases from the kiln are burned in the afterburner and are conveyed to the waste heat boiler. From there, all exiting gases pass through a wet gas scrubber. To minimize objectionable plume, some plume suppression device similar to a condenser and heat exchanger is utilized. The gases then pass through a plume suppressor before discharging into the atmosphere.

While air pollution aspects appear satisfactory, water pollution potential cannot be ignored. The Monsanto Company mentions that all water is recirculated and clarified in a closed-loop system, and no discharge is planned. However, there is a substantial amount of water, in the wet char, that is lost from the system. The filtering loop of the char slurry makes the outflow water highly alkaline, and water removed from the gas scrubber is highly corrosive. A schematic diagram of the closed-loop water system, as visualized by the Monsanto Company, is shown in Figure 39.

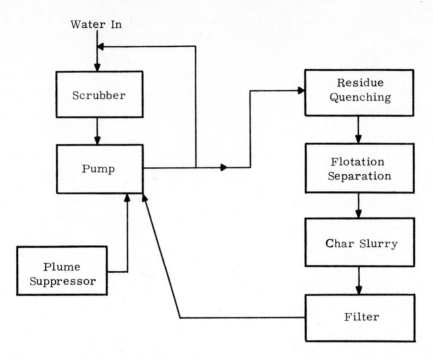

Figure 39. Schematic Diagram of Monsanto Landgard Water Circulation System

Two major questions, relative to water, that must be resolved satisfactorily are:

- Does the water in wet char make this material unsatisfactory for landfill?

- How is component life affected by recirculating highly corrosive water effluents from the scrubber, char slurry, and the plume suppressor?

COST FACTORS

The following costs pertain to a middle east coast city (probably Baltimore) and include all construction costs, but exclude land, taxes, and insurance (cost figures were furnished by the Monsanto Company):

Factor	250 Tons/Day*	500 Tons/Day*	1000 Tons/Day*
Capital costs			
Basic plant			
Total	$3,200,000	$5,400,000	$8,500,000
Capacity/ton/day of solid waste	$ 12,800	$ 10,800	$ 8,500

*The system could handle up to 30% sewage sludge by cutting back solid waste figures to 70% of the throughput values quoted.

138

Factor	250 Tons/Day*	500 Tons/Day*	1000 Tons/Day*
Capital costs (Cont'd)			
Resource recovery plant			
Total	$4,700,000	$7,600,000	$11,600,000
Capacity/ton/day of solid waste	$ 18,800	$ 15,200	$ 11,600
Expected life	25 yr	25 yr	25 yr
Operating costs			
Basic plant ($/ton)	$6.98	$5.54	$4.27
Resource recovery plant	$7.72	$6.11	$4.75
Operating cost factors			
Shredders	1	2	2
Operating manpower	13	15	19
Water (gpm)	125	225	450
Annual plant capacity (tons)	77,500	155,000	310,000

According to the information furnished by the Monsanto Company, "these costs do not take into consideration any credit for the sale of recovered materials. They include fuel oil, power, water, manpower, maintenance, and miscellaneous costs." The plant operates 24 hours per day, 7 days per week.

As a cross check on the above figures, the actual numbers for the Monsanto 1000-ton-per-day plant to be built in Baltimore under a $6 million Environmental Protection Agency demonstration grant are:

Cost	Amount
Capital cost	$13 million
Total operating cost/ton	$9.61
Revenues recovered	$4.67
Net cost/ton	$4.94

UNION CARBIDE SYSTEM

The Union Carbide Corporation oxygen refuse converter system for recovering the resources from solid waste is characterized by:

- Acceptance of mixed municipal refuse without size reduction or sorting.

---

*The system could handle up to 30% sewage sludge by cutting back solid waste figures to 70% of the throughput values quoted.

- Use of oxygen instead of air for pyrolysis.

- Recovery of clean-burning fuel gas from the pyrolysis furnace.

- Totally inert residue.

COMPONENTS AND PROCESS

A process flow diagram for a 1000-ton-per-day plant proposed by the Union Carbide Corporation is shown in Figure 40. This plant, with a daily capacity of 1000 tons of mixed municipal refuse, would include:

- Three shaft furnaces.

- Three gas cleaning trains.

- Oxygen generation plant.

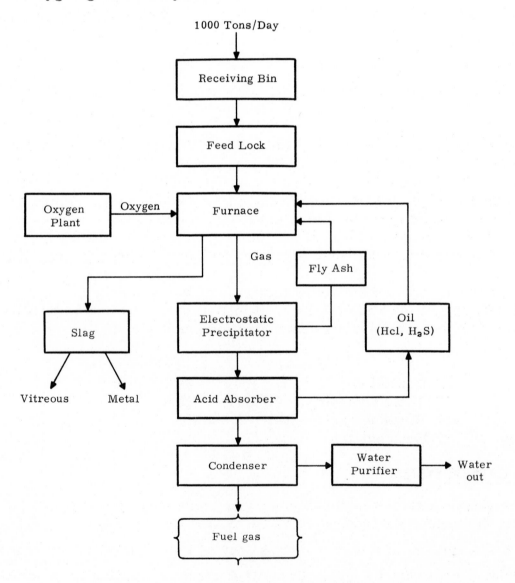

Figure 40. Flow Diagram of Union Carbide System

A normal grouping of the major components in this facility would consist of:

- Refuse preparation equipment

- Pyrolysis equipment

- Fuel gas cleaning equipment

Refuse Preparation Equipment

It is claimed that the Union Carbide system accepts mixed municipal refuse as-received, without any prior conditioning or preparation. The only constraints on the refuse pertain to its physical size and to the requirement that most of it should be solid waste. The size limitation is set by the flow-through opening in the feed lock mechanism. Acceptable refuse can be up to 4 feet in each dimension. Bulky combustible and noncombustible items in the refuse require a size reducer. This reducer can be either a shear or any of the numerous types of shredders (e.g., a vertical-shaft or horizontal-shaft hammermill). A 1000-hp shredder is required for this application.

The flow path of the refuse in an operational facility set up to process 1000 tons daily would be as follows. Mixed municipal refuse consisting of household garbage, sofas, white goods, and other bulky materials would be dumped into the receiving area by incoming packer trucks. Manual sorting of the refuse would enable separation of objects larger than four square feet. These separated items would be accumulated in a separate area for later shredding and coarse size reduction.

The feed lock acceptable refuse is then loaded onto each of the three conveyors. Each conveyor discharges into a receiving bin. (Overhead cranes and a grapple system could also be considered for this step.) The refuse from the bin enters a feed lock. Each feed lock is coupled with one of the three shaft furnaces. The lock feeds intermittently into the top of the furnace.

While in a pilot plant type of operation, the front-end equipment requirements can conceivably be simplified: the refuse can be discharged directly through the feed lock into the furnace or, alternately, it can be fed from an intermediate storage bin. This bin can be intermittently loaded to maintain the natural flow into the furnace; however, for a commercial facility operating continuously, it is essential to have this segment of refuse handling mechanized as described above.

Storage of raw refuse should include adequate space provision to hold at least one day's normal refuse collection. The area should be enclosed and protected from rain and snow. Accessibility and maneuverability are important for determining how high the refuse should be piled. In general, the height should be restricted to about 6 feet.

For a 1000-ton-per-day plant, this subsystem includes three shaft furnaces, an oxygen generating plant, and three molten slag containers. The furnaces are similar to the blast furnaces commonly used in the iron and steel industry or in the chemical industry. Each furnace is vertically oriented, refractory lined, and is designed to withstand high temperatures (during pyrolysis) and the effects of corrosive gases (e. g., chlorides and sulfides evolved from the refuse). Incoming refuse is fed intermittently into the top of the furnace, through respective feed locks.

Gaseous oxygen is fed continuously into each furnace through tuyeres near the bottom. The oxygen is generated on the site. For every ton of refuse processed, 0.2 tons of oxygen are required (Ref. 66). The oxygen travels upward through the furnace bed at the low velocity of 1 foot per second. For a 1000-ton-per-day facility, the oxygen generating plant alone would require a 3000-kilowatt power input. Total power required by the plant is 4850 kilowatts.

The solid products of pyrolysis accumulate in the bottom of the furnace as molten slag. The slag is tapped continuously, into a separate tank with water, and is quenched. This procedure is normal for the pilot plant and is recommended for the 1000-ton-per-day facility. However, if metal recovery is a consideration, then the molten slag offers potential. In the furnace, the pool of residue consists of an upper vitreous layer and a lower metallic layer. These layers could be tapped separately.

The normal operating mode is to hold the oxygen flow constant and to maintain the bed level between a minimum and maximum height. The bed-height tolerance is fairly wide and can vary from 8 to 20 feet without affecting the process in a 333-ton-per-day furnace.

The gas produced in the furnace, from the combustible portion of the refuse, exits at a low temperature (170°F to 200°F) near the top of the furnace. This gas contains significant moisture and some oils and fly ash. The Union Carbide Corporation predicts that the particulate loading in this gas would be about 0.08 grain per cubic foot, corrected to 12 percent carbon dioxide (Refs. 67 and 68). The volume of this gas is only 5 to 10 percent of the gas volume produced in a conventional incinerator, due to the absence of nitrogen and excess air. Production of oxides of nitrogen is also very much retarded for the same reason.

Fuel Gas Cleaning Equipment.

The furnace flue gas is ducted to an electrostatic precipitator, where the oil and fly ash are removed from the pyrolysis fuel gas. The oil is then condensed and returned to the furnace, where it is cracked to gaseous products.

The fly ash settling from the gas in the electrostatic precipitator is also re-cycled back to the furnace, where it gets another chance to settle to the bottom and get tapped out as molten residue. The gases then pass into an acid ab-sorber, where a neutralizing solution removed hydrochloric acid, hydrogen sulfide, and organic acids. The aqueous solution containing the above salts is continuously bled from the recycled absorber liquid and is fed to the fur-nace. In the furnace, the salts are eliminated with the slag.

The moisture remaining in the gas is removed in a condenser, aft of the acid absorber. The saturated gases entering the condenser leave it with a moisture content of about 6 percent, by volume. Water accumulated in the condenser requires purification before discharge into a sewer.

The fuel quality, as estimated by the Union Carbide Corporation, is:

- Moisture: 6%.

- Particulates: 0.008 grain per scf corrected to 12% carbon dioxide.

- Heating value: 300 Btu per scf.

- Flame temperature and heat transfer characteristics similar to those of natural gas.

Input and Output Streams

The Union Carbide oxygen refuse converter system can accept mixed municipal refuse as received. Packer truck refuse does not require any size reduction prior to pyrolysis. However, the Union Carbide Corporation esti-mates that 10 percent of the mixed refuse would be bulky waste, requiring shredding. In a 1000-ton-per-day plant, approximately 100 tons daily would have to be size-reduced. Therefore, a shredder capable of accepting bulky items (e.g., sofas and water heaters) would process this type of refuse. Most of the incoming refuse is processed through the pyrolysis furnaces, without any size reduction or sorting of the combustibles from noncombustibles.

The Union Carbide system is relatively tolerant of variations in input ma-terial composition. Tests on the pilot plant show that 2-to-1 changes in mois-ture content had no effect on oxygen consumption, on the temperature of the gas leaving, and on refractory performance. Tests with plastic contents ranging from 0 to 25 percent and inert contents ranging from 0 to 40 percent produced similar results.

The input and output streams, for a 1000-ton-per-day plant are:

- Inputs

    Packer truck and bulky refuse   1000 tons/day

    95 percent pure oxygen   200 tons/day

- Outputs

Fuel gas*	600 tons
Solid residue	<u>220 tons</u>
Total	820 tons
Heating value of fuel gas	300 Btu/scf
Gross energy per day in gas based on 4700 Btu/lb of refuse	7.05×10^9 Btu/day
Power requirement, based on 4850 kilowatts generated at 34% efficiency	<u>1.17×10^9 Btu/day</u>
Net energy per day	5.88×10^9 Btu/day

The energy available from a 1000-ton-per-day facility, either as fuel gas or as electric power produced in an on-site gas turbine generator, is estimated by the Union Carbide Corporation in Table 29, based on 5000 Btu

Table 29

UNION CARBIDE SYSTEM USE
OF AVAILABLE ENERGY
(1000 Tons/Day)

Energy	Btu/Hr	Percent
Available energy in refuse*	416,000,000	100
Energy losses in conversion process**	70,000,000	17
Energy available in fuel gas	346,000,000	83
Fuel gas uses		
Process steam	16,000,000	4
Building heating	10,000,000	2
Energy to maintain auxiliary combustion chamber at operating temperature	7,000,000	2
Net energy available in fuel gas	313,000,000	75
	Kilowatts	
Electric power generation	21,000†	
Electric power used in plant	7,000††	
Electric power available for export	14,000	

Source: Union Carbide Corporation

*Based on a refuse heating value of 5000 Btu/lb, this figure is calculated as (5000) (2000) (1000)/24.

**Includes latent heat of moisture in refuse, sensible heat of fuel gas, heat content of molten slag and metal, and heat leak.

†Based on combustion of the net fuel gas in a gas turbine generator, set with 24% thermal efficiency for the gas turbine and 96% electrical efficiency for the generator.

††Power consumption to produce fuel gas at 35-psig pressure is 4850 kw.

*Actually produces 700 tons, but 100 tons are reused in the process.

per pound of refuse. This optimistic estimate shows that approximately 83 percent of the energy in the incoming refuse is made available in the fuel gas. About one-tenth of this available energy is needed for operation of the facility. In the case of the gas turbine powerplant, 7000 kilowatts are required to run the oxygen plant, the electrostatic precipitator, the condenser, auxiliaries, the bulky waste shredder, and the gas compressors, to compress the fuel gas to 200 psig for firing in the gas turbine. If the plant is only to produce fuel gas, the product gas need only be compressed to 35 psig, and the required power drops to 4850 kilowatts.

Table 30 shows the gas composition and tolerance band expected from the Union Carbide system. Table 31 shows an estimated comparison of combustion

Table 30

COMPOSITION OF UNION CARBIDE FUEL GAS

Gas Composition	Average Value	Range of Values
CO (%)	50	40-50
H_2 (%)	30	25-30
CO_2 (%)	14	10-20
CH_4 (%)	4	3-6
C_2H_2, C_2H_4, C_2H_6 (%)	1	0.5-2
N_2 (%)	1	--
HHV (Btu/ft^3)	300	270-320

Table 31

COMPARISON OF REFUSE CONVERTER FUEL GAS
WITH NATURAL GAS

Characteristic	Refuse Converter Fuel Gas	Natural Gas*
Heating value (Btu/scf)**	300	1000
Air required for combustion		
Scf air/scf gas	2.4	9.5
Scf air/thousand Btu	8	9.5
Combustion products[†]		
Scf/thousand Btu	9.4	10.5

Source: Union Carbide Corporation

*Assumed to be methane.
**Scf = standard cubic feet, measured at 60°F and 1 atmosphere.
[†]Calculated for complete combustion with stoichiometric volume of air.

characteristics between the fuel gas from the oxygen refuse converter and natural gas. The Union Carbide Corporation has included the air requirements and the volume of combustion products in addition to heating values. On an equal heat release basis, less air is required to burn the fuel gas. This means that the air blowers in a given utility boiler for natural gas would have sufficient capacity to meet the requirements of the fuel gas, a very important feature. The volume of combustion products formed for a given heat release is also less. Therefore, there should be no equipment handling problems in units designed for handling natural gas as the fuel. The only change needed, to use fuel gas in an existing utility boiler, would be an increase in the size of the burner ports.

Physical Requirements

The Union Carbide oxygen refuse converter system requires a minimum of front-end equipment ahead of the pyrolysis blast furnaces. The refuse storage requirements for incoming raw refuse would have to be sized for at least one-day capacity. The Union Carbide Corporation recommendation is for two-day storage capacity for a 1000-ton-per-day facility.

A projection, by the Union Carbide Corporation, of the major equipment for a plant processing 1000 tons of mixed municipal refuse is given in Table 32. To install the equipment, provide accessibility, and provide an enclosed

Table 32

EQUIPMENT PROJECTIONS FOR 1000-TON/DAY PLANT

Equipment	Quantity
Storage area for two-day refuse (2000 tons)	--
Overhead cranes and grapple system	3 (optional)
Front-end loaders	3
Raw feed conveyors	3
Bulky refuse shredder (1000 hp)	1
Bins	3
Feed locks	3
Shaft furnaces	3
Oxygen plant	1
Slag collector and quench tanks	3
Residue removal conveyors	3
Electrostatic precipitators	3
Acid absorbers	3
Condensers	3
Combustion chamber	1
Cooling tower	1

area for raw refuse storage, Union Carbide requires 3 acres. This acreage does not include access roads, truck turn-around space, and overflow storage space; it is only the land required for the building, oxygen plant, storage area, and processing equipment.

OPERATING HISTORY AND EXPERIENCE

The oxygen refuse converter has been developed on a 5-ton-per-day scale (Refs. 67 and 68). The estimated operating experience with this pilot plant covers a period of more than one year. Additional laboratory work extends over a longer period.

On January 11, 1972, the Union Carbide Corporation announced that it had developed the oxygen refuse converter system. At that point, it had been tested for nine months in a 5-ton-per-day pilot plant at the Corporation's Tarrytown, New York technical center. The City of Mount Vernon, near Tarrytown, was selected as a municipal demonstration site. Union Carbide expects that a unit will be on-stream there early in 1974. The capacity of this plant is specified as 150 tons per day.

It is estimated by Union Carbide that the plant would involve funding of about $2-1/2 million, including $300,000 to $400,000 for first-year operating costs. The City of Mount Vernon was hopeful of receiving an Environmental Protection Agency grant for this demonstration. Some of the comments from the Union Carbide Corporation indicate that:

- Full-scale marketing to municipalities will begin after experience is gained from their demonstration plant.

- The hottest temperatures anticipated are in the 2600°F to 3000°F range.

- Major competitors are Torrax Services, Inc.; Melt-Zit; and Monsanto Landgard.

A subsequent communication from the Union Carbide Corporation (Ref. 68) did not mention the demonstration plant proposed to handle 150 tons per day for Mount Vernon, New York, but stated that Union Carbide contemplates operating a 200-ton-per-day demonstration unit by 1974. This unit was planned to be fully funded by Union Carbide and would be built at their West Virginia plant. After successful demonstration, they expect to start making firm commitments for large-scale facilities in the range of 1000 tons per day, needed for major urban areas. The estimates show that it would take two years from initiation of design work to the startup of a 200-ton-per-day facility. The cycle for a 1000-ton-per-day plant would be approximately 2-1/2 years.

At the present time, the pilot 200 ton-per-day plant in West Virginia is operating intermittently while tests are being conducted. The converter was started up about mid-April 1974. The usual minor mechanical design problems were encountered, such as too small a residue conveyor and a small quench pit, which limited the rate of residue flow and hence the input feed rate.

The pilot system in West Virginia consists of the basic refuse converter furnace, a simple cleanup train, and a flare. Oxygen for the plant is obtained from excess oxygen available at the Union Carbide plant. The off-gas from the converter is passed through an electrostatic precipitator and is then flared.

According to recent conversations with Union Carbide personnel, the system has processed a total of about 2000 tons of refuse between April and October 1974. Rates of up to 100 tons per day have been achieved. The output gas has a heating value in the range 260 to 340 Btu per scf. Flow measurements of the volume of gas produced per ton of refuse and gas analysis had not yet been completed at that time.

Union Carbide plans to continue evaluation of their process. One of the next steps in that evaluation process is to try the process on shredded refuse from which the metals have been removed. Successful operation with refuse of this type would allow the process to be used in conjunction with front-end materials separation, thereby increasing the degree of resource recovery possible and minimizing the frit output. Current business plans are to offer the oxygen refuse converter commercially in approximately the middle of 1975.

The Union Carbide UNOX process used in sewage treatment plants can be regarded as the predecessor to their oxygen refuse converter system. In the former process, oxygen is bubbled through the waste water in enclosed tanks, thereby speeding the bacterial action. This, in turn, reduces the time needed to hold the waste water in tanks.

SCALING CONSIDERATIONS

The Union Carbide oxygen refuse converter system concept has been experimented with in the laboratory. Its feasibility was later demonstrated on a five-ton-per-day pilot plant. This latter activity has been carried out for more than one year. The Union Carbide Corporation has been successful in demonstrating recovery of fuel gas under laboratory-type, controlled conditions.

The topics requiring economic as well as technical evaluation when scaling from a five-ton-per-day unit to a 1000-ton-per-day commercial facility include:

- Oxygen generation plant. (No problem exists with the technology; the economics need to be checked.)

- Uniformity of oxygen flow through the refuse bed.

- Need for preprocessing of mixed municipal refuse.

- Thermal gradients in the refractory lining. (A 1000°F/ft vertical gradient exists near the slag line in the pilot plant.)

- Crane and grapple loading versus pit conveyors.

- Shredder for bulky items.

- Advantages and disadvantages of noncombustible separation prior to pyrolysis.

• Size reduction of all combustibles prior to pyrolysis.

The major problems should be met in the 200-ton-per-day facility now under test. A 1000-ton-per-day facility would actually use three trains of 350 tons per day, which would involve only a nominal scale-up from 200 tons per day. In terms of seeing a 1000-ton-per-day facility operational anywhere, it appears that the time frame will be 1977 to 1980, at the earliest.

Residue handling and recovery of metals can be enhanced considerably if the noncombustibles are pulled out, before entering the furnace, and sorted. Marketability of the recovered noncombustibles is also better. Similarly, the cleanliness of fuel gas will be improved considerably if the noncombustibles are not fed into the furnace and if the combustibles are size-reduced ahead of the furnace. This procedure will probably simplify the gas cleaning train now projected for a 1000-ton-per-day facility. The particulate loading and the corrosive effluents mixed with the fuel gas at the furnace discharge will be minimized. In fact, the size of the furnace will be reduced considerably if the noncombustibles do not enter the furnace and if the combustibles are shredded.

The problem with the feed locks is still another item that will show up when equipment is operated continuously. The City of St. Louis demonstration project experience in this respect is an indication of what might happen when mixed refuse is fed through these locks. However, a system of double doors should eliminate most potential problems.

The Union Carbide Corporation has suggested the crane and grapple system for handling raw refuse. This system must be evaluated very thoroughly and carefully. In many modern plants, pit conveyor loading (to feed the bins) is utilized effectively and economically. A small crane can be retained to remove troublesome items.

AIR AND WATER POLLUTION CONSIDERATIONS

The gases discharged to the atmosphere as a result of the operation of the Union Carbide oxygen refuse converter system are nitrogen from the oxygen generation plant and the products of combustion of the fuel gas. The nitrogen stream is basically nonpolluting, because it contains only the constituents of the ambient air. The air compressor itself is nonlubricated; this specifies that oil is not used in the compression cylinder.

The fuel gas produced in the process is cleaned by the gas cleaning train. The sulfur content is low, about 15 ppm. The gas is essentially free of oxides of nitrogen, due to low concentrations of nitrogen in both the refuse and the oxygen fed to the furnace. Also, the reducing conditions within the furnace are not conducive to the formation of nitrogen oxides.

After combustion of the fuel gas, the particulate loading in the gas is estimated to be very low. This loading is primarily due to the gas cleaning equipment used to refine the furnace-generated fuel gas. Pilot plant tests have shown that the measured concentration of the particulates is 0.008 gram per scf, corrected to 12 percent carbon dioxide. This concentration is an order of magnitude better than the Government standards for incinerators and waste disposal plants.

The solid residue from the system is expected to consist of the metal and slag removed from the quench tank. This material is formed from the molten streams discharged from the hearth at a temperature near 3000°F. The Union Carbide Corporation indicates that this material is innocuous, because it is inorganic, noncombustible, sterile, and odorless, but it also includes the salts containing hydrochloric acid and hydrogen sulfide, recycled continuously into the furnace from the acid absorber used in the gas cleaning train. The volume of the residue is 2 to 3 percent of the volume of the refuse processed.

The Union Carbide Corporation feels that this material does not require a sanitary landfill. It makes a good fill material and can be used as subbase in construction or as an aggregate in asphalt paving.

Two water streams are discharged from the converter plant:

- Blowdown from the cooling tower.

- Condensate water from the fuel gas in the condenser.

Cooling tower blowdown will prevent buildup of minerals in the cooling water, due to evaporation of water in the tower. The quality of this water is controllable by the plant designer; it can be matched to the requirements of the site. This water can be discharged into the sanitary sewer.

The estimated amount of water condensed from the fuel gas is small, approximately 80 gallons per ton of refuse or 80,000 gallons per day for the 1000-ton plant. The bulk of the water-soluble organic contaminants is removed from the fuel gas before its entry into the condenser. The organics are removed by the acid absorber.

The Union Carbide Corporation views the condenser discharge water to be satisfactory for sanitary sewer discharge, although it does contain some organics and inorganics, and it is safe to assume that some water purification may be required before its discharge into the sewer.

COST FACTORS*

The projected economics of the Union Carbide oxygen refuse converter system are presented in Table 33. Estimated net disposal costs and investment are shown for 200-, 400-, and 1000-ton-per-day facilities. Both direct use of the fuel gas at 25 psig and electric power generation in an on-site gas turbine-generator are presented.

Table 33

UNION CARBIDE SYSTEM PROJECTED ECONOMICS

Investment/Cost	Facility		
	200 Tons/Day	400 Tons/Day	1000 Tons/Day
Fuel Gas Facility			
Investment ($ million)	3.3	5.7	9.8
Net cost ($/ton)	9.90	6.40	3.75
Fuel Gas/Gas Turbine Facility for Power Generation			
Investment ($ million)	3.9	6.9	12.7
Net cost ($/ton)	10.70	7.00	4.35

*The cost information in this subsection was furnished by the Union Carbide Corporation (Ref. 68).

The facilities include overhead cranes and a grapple system; a refuse pit for two-day storage; a 1000-hp shredder for oversized bulky refuse (approximately 10% for the total); gas cleaning equipment; equipment to recover the energy in the off-gas, where appropriate; residue conveyors; and a building to enclose the storage pit, crane, furnace, residue removal conveyor, switchgear, instrument panels, and employee facilities. Also included are an on-site oxygen supply system (on-site air separation plant and backup), an auxiliary combustion chamber sized to combust all the fuel gas produced at full capacity, and a cooling tower.

All estimates are based on information prepared by the Leonard S. Wegman Company, Inc., a New York-based engineering consulting firm with extensive experience in providing municipal solid waste system.

The net disposal cost presented includes labor, power, materials, maintenance and repair, amortization of capital at 9.63 percent per year, credit for fuel gas at 50 cents per million Btu or electric power at 0.8 cents per kilowatt-hour, and credit for the residue at $3 per ton of residue. Average 1973 construction costs in the United States were assumed, along with a clear and level site. A single converter-gas cleaning train was assumed for the 200-ton-per-day facility, duplicate trains were assumed for the 400-ton-per-day facility, and three somewhat larger trains were assumed for the 1000-ton-per-day facility.

A more recent capital investment estimate of $12 million for a 1000-ton-per-day plant in the New York-Connecticut area has been made by the Union Carbide Corporation.

URBAN RESEARCH AND DEVELOPMENT SYSTEM

COMPONENTS AND PROCESS

The Urban Research and Development Corporation thermal processing system utilizes simplified front-end equipment for preparation of input feed material. The total system includes:

151

- Main reactor

- Fuel gas burner

- Heat exchanger

A schematic diagram of the thermal processing system is shown in Figure 41. Normally, mixed municipal refuse up to 4 feet in one dimension can be fed directly through the top of the reactor. When pyrolyzing bulky waste, a coarse shredder is required to size-reduce it to fit through the feed lock opening.

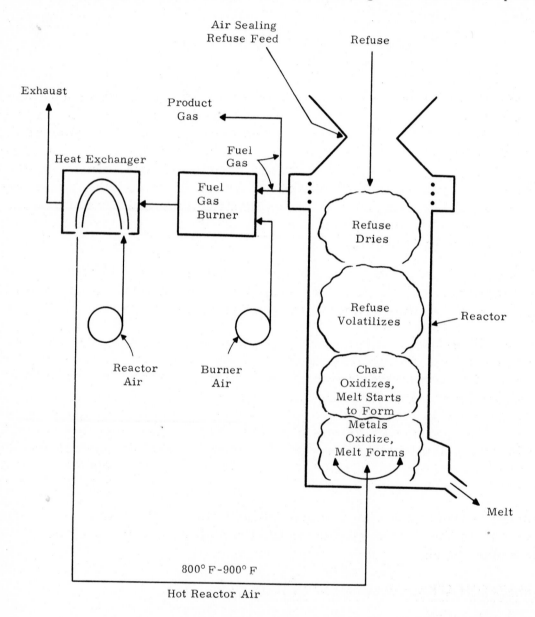

Figure 41. Urban Research and Development Thermal Processing System
(Source: Urban Research and Development Corporation)

Material enters the main reactor or shaft furnace (which has an inside diameter of 9 feet and is 24 feet high) at the top, through the air-sealed feed mechanism. The reactor operates at low temperatures with preheated air entering at 800°F to 900 °F from the bottom of the reactor. Maximum temperatures in the reactor column are 2100°F to 2200°F. The pyrolysis process consists of partial combustion with heated air. It gasifies the organic portion of the refuse to produce a fuel gas, and it oxidized the metals and, together with the inorganic portion, produces a molten slag.

The main product of pyrolysis, the fuel gas, exits the reactor near the top. The fuel gas is relatively smoky. Part of the gas is fed into the gas burner, mixed with burner air, and burned. From the burner, the products of combustion are carried into the heat exchanger and are then exhausted to the atmosphere. The remaining 70 percent of the fuel gas produced is available as a product.

Input and Output Streams

The Urban Research and Development Corporation thermal processing system accepts mixed municipal refuse without any front-end processing. Rugs and crushed 55-gallon drums have been fed along with the municipal refuse.

The system requires no auxiliary fuel. It uses its own fuel gas to preheat the simple metallic reactor (Ref. 64). The output products include low-Btu fuel gas, 30 percent of which is consumed on the site, for reactor preheating. The fuel contains 75 to 85 percent of the refuse heating value (Ref. 64). The gas is best used on the site or nearby, because it should be maintained at 500°F to prevent condensation of energy-bearing tars. The other output product is the slag, which can be used as aggregate in roadbeds. The refuse is said to achieve volume reductions of 96 to 98 percent through pyrolysis.

Physical Requirements

Projecting the space requirements for a 1000-ton-per-day facility, it appears that the Urban Research and Development Corporation thermal processing system would be the smallest among the various processes discussed. The storage area for raw refuse would be similar, regardless of the process. The front-end equipment would be simpler. It would consist of a shredder for size reduction of bulky refuse. Reactor preheat does not call for auxiliary fuel or for a special gas generation plant.

Because this is the simplest configuration of the processes reviewed, it requires a total land area of only 3 acres. Power requirements are also minimum. Including a 1000-hp shredder for bulky items, a total of 2000 to 2500 hp should adequately serve the needs of such a facility.

OPERATING HISTORY AND EXPERIENCE

The Urban Research and Development Corporation demonstration plant located on the property of Roncari Industries, is a joint effort. This plant is designed for 120-ton-per-day capacity. Currently the plant is going through the startup and debugging phase. This size plant is intended to form a basic module to serve any additional capacity requirements. One or more of these requirements will be combined to achieve the capacity desired.

The origin of the Urban Research and Development Corporation dates back to Connecticut Research Commission Research Support Award 67-14, which was granted to Eggen and Powell (then faculty members at the University of Hartford) for a feasibility study of a new solid waste system. This analytical study led to laboratory work under a continuation of the award. The initial work led to the formation of the Urban Research and Development Corporation, using private funding. The Urban Research and Development Corporation carried the development program through several small- and medium-sized pilot plants and to a larger one-ton-per-hour nominal-capacity experimental prototype.

In September 1970, after considerable development effort, this prototype system was brought to the point of continuous operation for runs of one shift or less in duration. Five runs were made in September and October 1970, during which over 36,000 pounds of refuse were processed to produce 9900 pounds of molten residue. These runs demonstrated continuous refuse processing, continuous melt production, self-sustaining operation with essentially no external fuel use, and a clear and odorless exhaust.

The limitations of the original one-ton-per-hour system, especially in the area of refuse handling, made it impossible to run for times much longer than one shift. The plant was shut down, because there was little more that could be learned from operating it.

A new, small bench-scale plant with a nominal capacity of 140 pounds per hour was then constructed. This plant incorporated the development experience gained in operating the large pilot plant. It is being used to both demonstrate the Urban Research and Development Corporation process and to economically acquire detailed design data for a full-scale plant. Its operating characteristics have been quite consistent with the behavior of the one-ton-per-hour system. Thirteen thousand pounds of refuse have been processed to produce 3950 pounds of molten residue. The bench-scale plant has provided the design data for the new, 120-ton-per-day demonstration plant.

The Urban Research and Development Corporation is now prepared to design and construct a full-size installation for a municipality or for a region, preferably with the assistance of a demonstration grant (Ref. 69).

SCALING CONSIDERATIONS

The current demonstration plant is the same size as a full-scale design module. It has a capacity of 120 to 125 tons per day. A basic plant would have two modules, for redundancy, and would be rated at 250 tons per day (Ref. 70).

Experience with the 140-pound-per-hour plant has shown that some iron in the waste is needed to permit proper fluxing for a slag flow at 2100°F. Only two-thirds of the iron in the refuse could be allowed to be removed by front-end processing. Similarly, only 25 percent of the incoming glass could be so removed (Ref. 70). These considerations would have to be evaluated based on the scaled-up, 120-ton-per-day plant. Operating experience, even with the smaller models, has been sporadic. Continuous operating capability, a must for a commercial facility, must be established. The problem of channeling in the refuse bed is expected to be a major problem to be encountered in the 120-tons-per-day unit.

AIR AND WATER POLLUTION CONSIDERATIONS

Air pollution could be a factor, based on the comment by O. Powell, of the Urban Research and Development Corporation, that the fuel gas is quite dirty and heavy in tars (Ref. 64). However, its cleansing through the burner and heat exchanger may make the particulate loading compatible with the established allowables of 0.08 grain per scf, corrected to 12-percent carbon dioxide. Preliminary measurements have been made by the Urban Research and Development Corporation on the particulates, but no data have been released to date (Ref. 64).

Water quenching of the hot slag discharged continuously from the reactor base indicates a potential pollution problem area. The water in the quench tank requires periodic removal and purification before recycling.

COST FACTORS

No written information has been received from the Urban Research and Development Corporation. However, in a meeting on September 26, 1972, O. Powell quoted a capital cost of $10,000 per ton per day and an operating cost of $2.50 per ton for a 250-ton-per-day plant. The capital cost of $2,500,000 would be estimated on a turnkey basis.

TORRAX SYSTEM

COMPONENTS AND PROCESS

The Torrax Services, Inc. solid waste disposal system for high-temperature solid waste disposal uses supplemental energy in the form of very high-temperature preheated combustion air provided by the oil- or natural-gas-fueled super blast heater shown in Figure 42. Combustion air is filtered and then heated in the ceramic-lined super blast heater by passing it through silicon carbide tubes, around which flow the hot combustion products of a fuel, ordinarily natural gas or oil. The air is heated to any desired temperature, up to 2200°F, and the temperature of the air can be increased or decreased rapidly, in order to accommodate changes in demand.

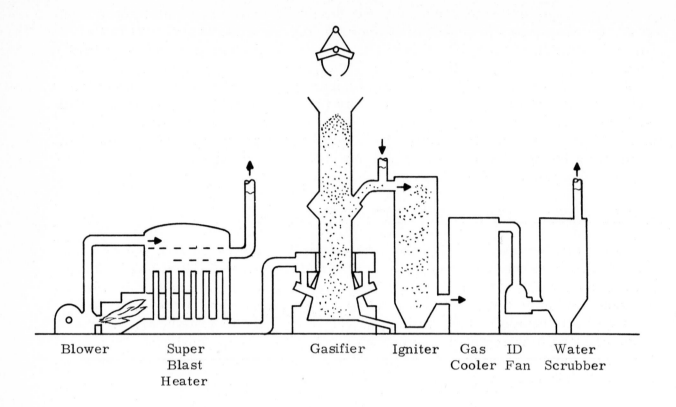

| Blower | Super Blast Heater | Gasifier | Igniter | Gas Cooler | ID Fan | Water Scrubber |

Figure 42. Schematic Diagram of Torrax System

The gasifier itself is a simple steel shaft, tapered slightly toward the top to allow the outside to be cooled by a falling sheet of water. The gasifier is lined with refractory around the bottom, to a height of about 4 feet, with eight tuyeres to supply the air blast. Some gas is fired in the tuyeres, to keep them free of slag, and the slag runs out a tap on one side, which also has a burner in it.

The top of the shaft widens out to a funnel, with a gas offtake at the neck of the funnel. The ID fan maintains a slight negative pressure at this point, and no lock or gate is used. Refuse is hoisted from a pit by a grab bucket crane and is carefully fed a bit at a time into the funnel, to maintain a rough seal between the fire and the atmosphere. When the seal fails, fire and smoke belch up toward the ceiling, and the crane operator dumps a few hundred pounds to reseal it. The refuse is wet by a lawn sprinkler, to improve the plug and suppress fires. As the refuse slowly descends in the gasifier, most of the readily combustible materials never reach the high-temperature zone at the bottom, because the hot gases permeating up through the refuse pyrolyze the organic materials to form combustible gases.

The material that reaches the bottom is comprised of difficult-to-burn objects, pyrolysis char, and noncombustibles. These materials are partially oxidized to create additional combustible gas or are liquefied to form a complex silicate slag and a mixture of molten metals. Temperatures at the base

156

of the gasifier are in the range of 2600°F to 3000°F, depending on the nature of the noncombustibles. A liquid mixture of inorganic and metallic materials flows from the gasifier into a chamber filled with water, where an aggregate-quality frit is formed from the slag, and the metal is frozen into small droplets. The gases flowing from the gasifier contain no free oxygen and consist mainly of carbon monoxide, hydrocarbon gases, and nitrogen. Entrained in the gas stream are particles of carbon and fly ash.

This combustible gas-solid mixture is reacted with ambient air in the igniter. It is estimated that less than 15-percent excess air is required to carry the combustion reactions to completion. The igniter operates above 2000°F, and the entrance gas stream is admitted tangentially, in order to fuse noncumbustible material on the refractory wall of the igniter. Slag flows from the base of the igniter into a water quench tank, in a manner similar in function and operation to that for the tank associated with the gasifier.

The remaining equipment in the process (waste heat boiler, ID fan, and water scrubber) serves to extract heat from the gases issuing from the igniter and to cleanse the gas stream of particulate matter before releasing it to the atmosphere. Originally, the last component in the system was a special fabric filter, which provided for removal of particulate contaminants. However, due to many failures of the filter, it was replaced by a water scrubber.

Input and Output Streams

Mixed municipal refuse is handled without pretreatment or sorting. The diameter of the gasifier imposes a physical size restriction to -4 feet. Average refuse undergoes at least a 95-percent volume reduction. The residue of offers potential as an aggregate and as a source of metal.

The fuel gas produced by the gasifier contains mainly carbon monoxide, hydrocarbon gases, and nitrogen. (An analysis is given in Table 34). A high nitrogen and oxygen content is due to air being entrained at the top of the gasifier shaft.

The Torrax system requires auxiliary fuel to fire the ceramic preheater. This fuel requirement amounts to 20 percent of the heating value of the refuse. Essentially, a premium fuel (natural gas) is being purchased to produce a marginally useful gas (Ref. 64). Torrax personnel have indicated that the fuel gas is not suitable for firing the super blast air heater because of life and maintenance problems in the heater (Ref. 71).

Torrax Services, Inc. also does not favor the idea of eliminating the igniter and using the gasifier output gas directly in a boiler, because the igniter is also a slagging combustor and drops out some of the particulates (Ref. 71). It is felt that the best way to recover energy is with an igniter

Table 34

TORRAX SYSTEM FUEL GAS OUTPUT

$$\left(\begin{array}{c}\text{Sampling Location: Gasifier Discharge -- Crossover Pipe}\\ \text{Between Gasifier and Igniter}\end{array}\right)$$

Time	Volume (% dry gas)*								
	CO_2	N_2	O_2	CO	H_2	CH_4	C_2	C_3	CO/CO_2
8:00	2.9	78.1	14.6	0.0	3.8	0.0	0.0	0.6	0.00
8:	0.8	75.4	19.8	1.8	1.5	0.0	0.0	0.7	2.29
9:30	2.2	75.1	16.0	4.2	2.0	0.4	0.0	0.0	1.89
10:30	4.9	71.6	11.2	9.1	2.0	1.1	0.0	0.1	1.86
11:30	3.2	75.4	15.8	3.9	1.3	0.4	0.0	0.0	1.22
12:30	5.4	73.4	10.6	7.1	2.6	0.9	0.0	0.1	1.31
13:30	4.2	73.5	12.9	6.3	2.2	0.8	0.0	0.1	1.48
14:30	5.4	72.4	11.4	7.1	2.3	1.1	0.2	0.1	1.30
15:10	4.8	74.5	12.6	6.4	0.6	1.0	0.0	0.0	1.33

*Water content was very low, averaging approximately 0.2% and indicating extensive leakage. Calculation of percentages based on wet gas would cause little change in values. The original sample contained about 12% H_2O.

and a water-tube boiler of the type used in the Torrax demonstration plant. This unit produces 3 pounds of saturated steam at 150 psig per pound of refuse. On this basis, the input-output stream becomes, for a 1000-ton-per-day plant:

- Input

Packer truck refuse	1000 tons/day
Natural gas/ton of refuse	2×10^6 Btu

- Output

Saturated steam (150 psig)	250,000 lb/hr
Glassy frit (approximate)	250 tons/day
Gross energy value of steam	6.86×10^9 Btu/day
Energy of natural gas used	2.00×10^9 Btu/day
Estimated electrical power, based on 1500 hp generated at 34% efficiency	$\underline{0.27 \times 10^9}$ Btu/day
Net energy produced	4.59×10^9 Btu/day

An analysis of the glassy frit residue appears in Table 35.

Physical Requirements

A 75-ton-per-day facility in Erie County utilizes a building 113 feet long and 43 feet wide (Ref. 72). This area does not include storage space require-

Table 35

CHEMICAL ANALYSIS OF TORRAX RESIDUE
(December 19, 1972)

Chemical	7:45 a.m.	8:45	9:45	10:45	11:45	12:45 p.m.	2:45	5:00
SiO_2	49.15	34.71	36.42	44.58	47.12	47.10	39.71	32.30
Al_2O_3	9.40	5.70	9.40	9.40	10.90	11.70	10.90	10.90
TiO_2	1.20	0.50	0.80	1.00	1.30	1.30	1.30	1.00
Fe_2O_3	4.33	5.78	22.13	17.33	4.42	4.05	10.06	11.25
FeO	16.25	16.02	16.35	11.14	18.32	16.78	0.35	17.31
Fe	--	19.30	--	--	--	1.74	0.92	10.08
MgO	3.30	3.00	3.00	3.00	3.30	3.30	3.30	2.00
CaO	8.75	5.67	6.52	7.54	8.30	8.18	4.84	4.88
MnO	0.77	1.03	1.03	1.16	1.03	1.03	1.03	1.03
Na_2O	9.27	6.68	6.75	8.50	8.64	7.69	7.81	4.05
K_2O	0.48	0.36	0.36	0.36	0.36	0.36	0.48	0.26
Cr_2O_3	0.32	0.32	1.75	0.55	1.43	1.75	0.55	0.55
CuO	0.22	0.28	0.28	0.28	0.22	0.22	0.22	0.28
ZnO	0.06	0.09	0.08	0.09	0.06	0.06	0.04	0.02
Total	103.50	99.40	104.87	104.93	105.40	105.26	81.51	95.91

Note: Fly ash analysis is probably about the same, except it is much lower in iron and iron oxides.

ments. Extrapolating the equipment space requirements, it appears that 4 acres of land would be needed to house a 1000-ton-per-day facility.

The Torrax system does not require sophisticated front-end processing equipment; however, a bulky refuse shredder is required. The exhaust gas cleaning train would have a high power demand. An estimate of 3000 hp appears reasonable, to meet the power needs of a 1000-ton-per-day commercial facility.

OPERATING HISTORY AND EXPERIENCE

The American Gas Association and American Gas Association research committees have participated in sponsoring the $1.8 million Torrax 75-ton-per-day pilot facility at Tonawanda, New York (Ref. 72). The participants in the pilot project include:

- Erie County
- Solid Waste Management Office of the Environmental Protection Agency
- Torrax Systems, Inc.
- New York State Department of Environmental Conservation

Torrax Systems, Inc. began trial operations in early April 1971. The purpose of the three-phase project has been to:

- Demonstrate the capability of a high-temperature pyrolysis system to convert mixed refuse into an inert residue.

- Develop engineering data and scale up parameters to provide a basis for practical applications of Torrax systems.

The first phase of this project (design, construction, and testing of sub-components) was completed in May 1971. The second phase included testing and installation of emission control equipment and system optimization. The third phase consisted of engineering and economic data, personnel training, and operation of the facility.

The unit has not achieved continuous operational status, the longest run to date being 2-1/2 days (Ref. 71). Total refuse processed to date is 2500 tons. The maximum capacity attained for short periods was 150 tons per day. When P.H. Kydd and H. Bloom of the General Electric Company visited the project on February 22, 1973 (Ref. 71), it was operating at a throughput of 7000 to 10,000 pounds per hour; Kydd and Bloom were favorably impressed with the operation.

Environmental Protection Agency funding support is now ending, because of budget cutbacks, and the political climate in Erie County is not conducive to expanded County funding. Figure 43 shows an article, from a Buffalo, New York newspaper, which summarizes the current situation.

SCALING CONSIDERATIONS

A 75-ton-per-day pilot plant would serve well as a model to project requirements of a 1000-ton-per-day commecial facility. The pilot plant, however, would have to be operational, to produce useful data. The activity has been underway for about two years. Due to problems in the gas cleaning train, the facility has not yet achieved operational status.

Torrax Services, Inc. has shown that the usual problems associated with a blast or a shaft furnace can be overcome (Ref. 64). No major problems have been encountered with furnaces. Moving material through the furnace, without the material hanging up, and the question of whether the gaseous reactant medium can flow through the bed and react with it have been the commonly known drawbacks of a shaft furnace. Torrax seemed to have overcome these drawbacks in their 75-ton-per-day plant. There will still be questions of this nature to be solved in going to larger size units, but a scale-up to 300-ton-per-day furnaces with 9-foot-diameter shafts appears reasonable (Ref. 71).

AIR AND WATER POLLUTION CONSIDERATIONS

The original Torrax design used a system of bag filters to control particulate emissions. These filters did not work out, however, because the bags

160

Experimental Incinerator Torrax May Be Junked by U.S. Fund Cutoff

By PAUL MacCLENNAN
News Environment Reporter

Top decision makers in Erie County government are taking a hard look at the future of Torrax, the experimental high-temperature incinerator developed at a cost of $1.9 million and now undergoing test runs in Orchard Park.

It appears that further federal funding will terminate at the end of the current fiscal year July 1 because the U. S. Environmental Protection Agency's budget has been slashed to $6 million from $30 million.

There was some hope that federal funding would continue on a limited scale to underwrite further testing and experimental work on the project.

Instead, it will be turned over to Erie County for operation and underwriting of costs and this is under review. Some see a future for Torrax, but one top county official said bluntly: "Torrax is dead."

Assuming some of the bugs are ironed out—problems that backers say are normal in any new technology — Torrax still has some major hurdles to overcome. Among those noted at recent county meetings include:

Cost — County Executive Regan has adopted a "users pay" philosophy for county participation and leadership in any solid waste disposal programs.

Estimates of the cost of operation are tentative but the trial runs have produced a cost figure of about $15 a ton to burn refuse.

Landfill runs a quarter of that cost and there is a question of whether—without county subsidies — communities would pay more to dispose of refuse through Torrax than hauling it to landfills.

Edward Rausch, the county's co-ordinator for Torrax, thinks the cost will come down and, at any rate, will be cheaper than conventional incinerators. Regional Engineer William Friedman of the State Department of Environmental Conservation said: "If you get to $10 a ton it will be successful."

Competitive costs of landfill versus incineration are hard to juggle right now because many landfill operations are substandard. Also, if sanitary codes are ever enforced, the cost of landfill will go up.

Air Pollution — Perhaps the most serious and immediate problem Torrax backers — it's financed principally by government but the technology is out of Torrax Systems Inc. — is the need for rather drastic steps to curb air pollution.

The present operation doesn't meet state and federal codes and the Department of Environmental Conservation told Torrax operators this week they are going to have to produce a fourfold reduction in emission of pollutants.

County spokesmen say the company is ready to do this but Mr. Friedman notes that this may drive operating costs up and in turn make it less competitive. Its economic viability is still at issue. Improved air pollution control equipment also involves a costly water treatment system that will effect costs.

Size — The present system is a 75-ton-a-day operation, an optimum scale built for test purposes. It can be expanded to handle 300 tons a day. Whether Erie County would back an expansion and whether it would be eligible for state environmental bond-issue backing is under study.

At its present size it would have limited value. Only Wednesday, it was suggested that the Torrax site might be used for a transfer station — a point where collection trucks transfer their loads to larger, over-the-road trailers for trucking to more remote sites.

Resource Recovery — Torrax produces steam, a valuable byproduct now wasted into the air. Originally it was proposed to build it on the Erie County Community College campus so the steam could be used to heat college buildings, but this plan gave way to community objections.

Some persons question incineration, noting that while it does reduce refuse volume and, therefore, saves space, it does destroy the opportunity to salvage and recycle or reuse byproducts. It is possible that such materials could be sorted out in advance of burning, but the present Torrax site has limited space for expansion.

The trend in Erie County appears to be leaning in the direction of landfill and resource recovery. County officials are weighing whether Torrax has a place in such a system and whether it warrants a piece of the disposal dollar.

Ironically, just this week the Environmental Protection Agency distributed press releases noting the merits of Torrax. One source said it was a desperate attempt to counter the budget cuts by showing what a great job EPA's Bureau of Solid Waste was doing.

About the same time local officials learned that the job of EPA's project co-ordinator for Torrax had been axed in an economy move and transferred to other duties.

Figure 43.   Current Torrax Situation

161

frequenctly caught fire. This system has since been replaced by a water scrubber.

The gas leaving the waste-heat boiler at 600°F contains about 1.2 percent of the weight of the incoming refuse in the form of particulate matter, or about 0.3 grains per scf (Ref. 71). The scrubber reduces the loading by a factor of four, which is nearly legal. Torrax Systems, Inc. feels a better scrubber will accomplish the task.

Tables 36 and 37 summarize analyses of the contents of the waste heat boiler emissions before reaching the scrubber. Table 38 compares these emissions to those of a conventional incinerator.

Table 36

TORRAX SYSTEM FAN DISCHARGE ANALYSES
BEFORE SCRUBBER
(December 19, 1972)

Chemical	Time	Hydrocarbons as CH_4 (ppm)	Hydrocarbons Adjusted to 12% CO_2 (ppm)
Hydrocarbons	9:00	24	48
	9:30	66	113
	10:00	25	34
	10:30	22	25
	11:00	17	20
	11:30	22	28
	12:00	17	21
	12:30	110	140
	13:00	17	23
	13:30	30	42
	14:00	90	114
	14:30	30	35
	15:00	90	114
	15:15	90	164
		NO_x as NO_2 (ppm)	NO_2 (ppm) Adjusted to 12% CO_2
NO_x	8:30	5.8	--
	9:30	1.2	2.0
	10:15	1.2	1.5
	11:00	4.8	5.7
	11:30	3.7	4.6
	12:15	3.4	4.3
	13:00	1.1	1.5
	13:30	1.0	1.4
	14:30	1.6	1.9

SO_x	Sample collected 8:45-10:00: 0.04970g SO_2 in 217,500 cc dry gas or at 16.5% H_2O, 260,479 cc of discharge		Sample collected 13:00-14:30: 0.06288g SO_2 in 261,000 cc dry gas or at 16.5% H_2O, 312,575 cc of discharge
	SO_2 = 66.8 ppm = 80.2 ppm adjusted to 12% CO_2		SO_2 = 70.4 ppm = 84.5 ppm adjusted to 12% CO_2
Chlorides	Sample collected 10:00-13:00: 20.46μg Cl^-/1 of dry gas or at 16.5% H_2O, 17.05μg Cl^-/1 of discharge		
	Cl^- = 10.8 ppm = 13.0 ppm adjusted to 12% CO_2		
Aldehydes and ketones as formaldehyde	Sample collected 13:00-14:30: 0.027μg HCHO/1 of dry gas or at 16.5% H_2O, 0.0230μg HCHO/1 of discharge		
	HCHO = 0.017 ppm = 0.02 ppm adjusted to 12% CO_2		
Cyanide	Sample collected 10:00-13:00: 0.398μg CN^-/1 of dry gas or at 16.5% H_2O, 0.332μg CN^-/1 of discharge		
	CN^- = 0.286 ppm = 0.343 ppm adjusted to 12% CO_2		
Organic acids as acetic acid	Sample collected 10:00-13:00: 0.46μf CH_3COOH/1 of dry gas or at 16.5% H_2O, 0.38μg CH_3COOH/1 of discharge		
	CH_3COOH = 0.14 ppm = 0.17 ppm adjusted to 12% CO_2		

Table 37

TORRAX SYSTEM FAN OUTLET ORSAT ANALYSIS

Date	Time	CO_2 (%)	O_2 (%)	CO (%)	N_2 (%)
12/19/72	9:15 a.m.	6.0	12.6	0.0	81.4
12/19/72	10:30	10.6	7.6	0.0	81.8
12/19/72	11:30	9.6	8.0	0.0	82.4
12/19/72	12:35 p.m.	9.4	6.2	0.6	83.8
12/19/72	1:35	8.6	9.0	0.0	82.4
12/19/72	2:35	10.4	6.2	0.6	83.4
12/19/72	3:30	5.4	13.2	0.0	81.4

Table 38

COMPARISON OF EMISSIONS OF TORRAX
AND CONVENTIONAL INCINERATOR*

Component	Conventional Incinerator (ppm)	Torrax	
		6/20/72	12/19/72
NO_2	53-115	16-33	1.4-5.7
SO_2	55.7-195	19.5	80.85
Cl^-	215-1250	37	13
CN^-	0.02-0.10	3.4	0.34
Organic acids	35.1-178	0.58	0.17
Aldehydes and ketones	2.4-12.6	0.40	0.02
Hydrocarbons	2-30	2.9-5800	21-140

*Reported in Reference 73.

Approximately 75 lb/hr dust ahead of scrubber at 75-ton/day rate:

$$\frac{75 \times 24}{75 \times 2000} = 1.2\% \text{ of original refuse}$$

COST FACTORS

Because of limited operating experience, cost information for the Torrax Services, Inc. system is very sketchy. Environmental Protection Agency estimates place the operating cost in the range of $12 to $14 per ton, when operating the 75-ton-per-day plant on a one-shift basis, and $6 per ton, when operating on a two-shift basis. Four men are required for each shift.

Capital costs are not available for a 1000-ton-per-day plant. It is known, however, that the pilot plant in Tonawanda, New York cost $2 million, including the new water scrubbers (Ref. 71). Of this amount, $500,000 was spent for the building.

CHARLESTON REGIONAL RESOURCE RECOVERY SYSTEM

The City of Charleston, West Virginia, has recently proposed a regional resource recovery system that incorporates a pyrolysis system as the key unit process. The pyrolysis system is an outgrowth of research work conducted by R. Bailie at the University of West Virginia in Morgantown, West Virginia.

COMPONENTS AND PROCESS

Figure 44 presents an overall schematic diagram of the Charleston regional resource recovery system which has an assumed processing capacity of 400 tons per day of municipal refuse. The system utilizes twin fluid beds; the first bed acts as a pyrolyzer, and the second bed acts as a combustor. Shredded and air-classified refuse is fed to the pyrolyzer.

The combustion unit is a fluidized bed that is 3.5 feet in diameter and is 20 feet high. The sand bed height is 4 feet, with a harmonic mean particle diameter of 0.025 inch. The combustor is fed the recycled char, produced in the gasifier unit, at a rate of 31,500 pounds per hour and operates at 1750°F. This bed operates at three times the minimum fluidization velocity, using 2.1 mmscf per day of air.

The off-gas from the combustor passes through two cyclones, to effect gas cleanup. The first cyclone removes large solid particles, while the second removes the smaller particles. The char removed from the cyclones is returned to the combustor for fuel. The combustion products are then passed through a heat exchanger, to preheat the air entering the combustors, for heat conservation.

The fluidized pyrolysis unit is 7.5 feet in diameter, with an overall height of 20 feet. The bed height and sand harmonic mean diameter are the same as the height and diameter of the combustor. The pyrolysis unit will operate at 1500°F and will gasify 400 tons per day of municipal refuse. The gas to fluidize the pyrolysis unit will be supplied by recycling the pyrolysis gas. One third of the gas produced will be recycled, and the bed will operate at three times the minimum fluidization velocity. The pyrolysis stream will pass into a cyclone to remove the product of activated carbon char produced in the pyrolysis reaction. This char is fed to the combustion unit to supply the heat necessary to keep the fluidized sand temperature at 1750°F.

The off-gas may be passed through an optional processing system. The gas could pass through a water gas shift reactor, a carbon dioxide scrubber,

164

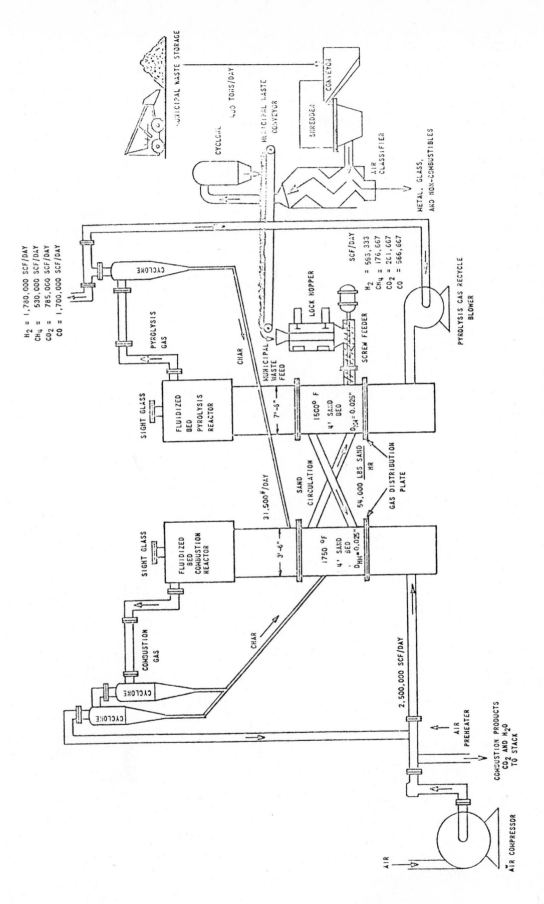

Figure 44. Schematic Diagram of Charleston Regional Resource Recovery System
(Source: Midwest Research Institute)

165

a cleanup washer, and, finally, a methanator, to convert all the pyrolysis gas to methane. This system is purely optional, because the gas coming directly from the pyrolysis unit is an immediately usable energy source. The energy required to maintain the pyrolysis unit at 1500°F is obtained from the sand circulating from the combustion unit at 1750°F. The sand circulation rate is approximately 54,000 pounds of sand per hour.

Most of the refuse feed system has been developed for the Combustion Power Corporation CPU-400 system. The system consists of a refuse storage pit, from which the refuse is removed, as needed, and fed to the conveyor belt by a mechanical lift. The refuse is passed to a sophisticated refuse shredder, where the municipal refuse is reduced in size. The refuse is passed through an air classifier, where 90 percent of the metal, glass, and heavy objects are removed. The classified refuse is then fed by conveyor to a lock hopper/screw feeder apparatus, where the refuse is fed directly into the fluidized bed.

CHARACTERISTICS OF PRODUCTS

Gas, char, and some tar are produced in the pyrolyzer. The gas produced in the pyrolyzer has a heating value of about 400 Btu per scf (gross heating value). Table 39 presents some results of pyrolysis tests on munic-

Table 39

TEST RESULTS OF PYROLYSIS OF MUNICIPAL REFUSE

Component	Proximate Municipal Refuse Analysis	Proximate Activated Char Analysis	Gas Production Solids Fed, Dry Basis (scf/lb)	Dry Gas Composition (vol %)	Dry Gas Composition (CO_2 free)
CO_2	--	--	1.40	16.3	0.0
CO	--	--	3.04	35.5	42.4
CH_4	--	--	0.95	11.1	13.3
H_2	3.56	2.95	3.18	37.1	44.3
Carbon	25.15	60.82	--	--	--
Ash	36.54	11.12	--	--	--
Heating value (Btu/ft^3 dry)	--	--	--	366.0	437.0

ipal refuse. Table 40 indicates the amounts of pyrolysis gas at different steps in the final gas processing on a dry gas basis (400-ton/day plant), while Table 41 gives the composition of the product gas.

The activated char produced by pyrolysis is also a valuable product. This char can be used directly as a solid fuel, or it can be used for general purification and reclamation of liquid and gas streams. In particular, the char could be used to purify sewage sludge to obtain pure water, and then the solids could be used as the energy source for the fluidized bed combustion unit. The

Table 40

PYROLYSIS GAS PRODUCED FROM 400 TONS/DAY
OF MUNICIPAL REFUSE*

Component	Pyrolyzer Exit Dry Basis (scf/day)	CO-Shift Exit, Dry Basis (scf/day)	CO_2 Scrubber Exit, Dry Basis (scf/day)	Methanator Exit Dry Basis (scf/day)
CO_2	785,000	1,610,000	--	--
CO	1,700,000	870,000	870,000	--
CH_4	530,000	530,000	530,000	1,400,000
H_2	1,780,000	2,610,000	2,610,000	--
Total	4,795,000	5,620,000	4,010,000	1,400,000

*Municipal refuse contains, on the average, 30% moisture.

Table 41

COMPOSITION OF PRODUCT GAS FROM TWO-REACTOR SYSTEM

Component	Dry Gas	Dry, CO_2 Free
CO (%)	27.1	31.7
CO_2 (%)	14.7	0.0
H_2 (%)	41.7	48.9
CH_4 (%)	7.7	9.0
C_2 unsaturates (%)	7.1	8.3
C_2H_6 (%)	0.7	0.9
C_3 unsaturates (%)	0.6	0.7
C_3H_8 (%)	0.4	0.5
Total (%)	100.0	100.0
Gross heating value (Btu/scf)	443.0	529.0
Yield of gas (scf/lb dry refuse)	9.3	8.0

activated char could be used to adsorb metallic ions; it could also be used as the fuel for the fluidized bed combustion unit and could be circulated with the sand.

AMERICAN THERMOGEN SYSTEM

The American Thermogen, Inc. Melt-Zit incinerator is described only briefly here, because of its physical resemblance to the Urban Research and Development Corporation and Torrax Services, Inc. systems. The American Thermogen system is basically a slagging incinerator and is not a try pyrolysis system.

A 120-ton-per-day pilot plant is in existence at Whitman, Massachusetts. This plant uses a shaft furnace, which is fed by a conveyor into a side port near the top of the furnace. The refuse forms a bed, as in a pyrolysis furnace, and the noncombustible residue slags at the bottom, running off into a water quench tank where a glassy frit is formed. The residue is inert and has about 4 percent of the volume of the input refuse.

To provide heat to slag the residue, a supplemental fossil fuel is used. Originally this fuel was coke, and all operation from July 1966 through July 1969 used coke as fuel. E. R. Kaiser's study of the process (Ref. 74) was conducted while it was using coke, and some of his conclusions are not valid for oil and gas firing (e. g., limestone and flourspar fluxing is not needed on oil or gas).

All operation in 1970 and 1971 was conducted using oil and gas fuel. American Thermogen, Inc. would not discuss fuel consumption. Their policy is to quote a fixed price per ton to operate a unit, all units being sold on the basis of operation by American Thermogen. From 1967 through 1971, the unit was run virtually every weekday, on an experimental basis, with one 11-day continuous run and several 5-day continuous runs during that period. No breakdowns were experienced during these runs. The furnace is now shut down because of a lack of funds (Ref. 75).

The above unit uses a high-energy scrubber to extract acids at particulates. Stack emissions include:

- 73 ppm NOx

- 6 ppm solids

Primary maintenance is expected to be conveyor preventive maintenance in the air pollution control system, plus a yearly rebricking of the melt zone, which is an eight-hour operation. Rebricking of the entire unit takes three days and is claimed to be required once every ten years.

The Whitman, Massachusetts pilot plant can be easily converted into a heat recovery unit by adding a fire-tube boiler between the shaft furnace and the air pollution control equipment. American Thermogen's design produces 485°F, 625-psig steam. Metal temperatures in the boiler are 500°F, which is conservatively low.

American Thermogen, Inc. has proposals out for a number of different unit configurations. One proposal to Malden, Massachusetts is designed for 2120 tons per day, has a 1650-ton-per-day rating, and has a capital cost of $20 million including air pollution control equipment. It will produce 7600 pounds of 625-psi, 485°F saturated steam per ton of refuse burned. Operating costs at the startup rate of 13,000 tons per month are $9.09 per ton, including coverage of local taxes. Costs go down to $4.16 per ton at a throughput of 50,000 tons per month, but they have an escalator clause. This plant would require a staff of 34 people for a 24-hour day, seven-day-per-week operation (Ref. 76).

American Thermogen, Inc. also offers two varieties of power generating plants, one noncondensing unit producing a refuse heat rate of 42 pounds per kilowatt-hour (equivalent to 81 Btu recovered/pound of refuse) and a more complex, condensing system operating at six pounds per kilowatt-hour (569 Btu/lb of refuse). Because neither of these figures was particularly impressive, the matter was not pursued further.

GENERAL DISCUSSION

For purposes of this discussion, pyrolysis is considered to be a process in which refuse is treated to generate some type of fuel. The baseline resource recovery process against which pyrolysis is compared here is the one in which refuse is burned directly, preferably with some refuse conditioning. A system where prepared refuse serves as a supplementary fuel in a conventional boiler is regarded, for this comparison, as an economic solution. Potential advantages of pyrolysis versus the baseline system, as well as relative advantages within the various pyrolysis processes, are discussed. The City of St. Louis fuel recovery process is regarded as a typical example of a baseline system.

Three fundamental factors are considered for comparative evaluation:

- Refuse conditioning.
- Boiler corrosion and deposit buildup.
- Compatibility of fuel with a wider range of powerplants.

REFUSE CONDITIONING

In a fuel recovery process, the refuse is shredded relatively fine, so it can be burned as supplemental fuel. The City of St. Louis fuel recovery process, for example, shreds 95 percent of the material to less than 1-1/2-inch nominal size. In Section 3, under "St. Louis Fuel Recovery Process, it is

169

pointed out that problems experienced due to the presence of noncombustibles eroding pneumatic ducts and jamming feed locks call for separation and sorting of noncombustibles from the combustible stream. Excessive bottom ash in the boiler also emphasizes this need.

Prelininary Environmental Protection Agency estimates indicate that shredding costs can run from $5 to $12 per ton, with prospects of $2 to $3 per ton (Ref. 77). Added complexity due to air classification and possible second-stage shredding would raise the above costs. If a pyrolysis process requiring reduced refuse conditioning or with no size reduction can be utilized, it could result in a significant cost reduction and high system reliability.

Among the various pyrolysis processes, the three requiring a substanial size reduction are the Garrett, Monsanto, and Hercules systems. In the Garrett flash pyrolysis system the refuse is dried to 2- to 3-percent moisture and is size-reduced to 0.015 inch. Most of the noncombustibles, particularly glass, are thoroughly removed from the feed material to the reactor.

The Hercules system also requires elaborate front-end processing equipment (Ref. 64). It incorporates dual-stage shredding, air classification, and screening. Like the Garrett process, the costs of refuse conditioning associated with the Hercules process make it economically less attractive.

The Landgard system proposed by the Monsanto Company calls for single-stage coarse shredding. A nominal 6-inch size is regarded by Monsanto as the optimum time-temperature profile through a counter-current-flow rotary kiln. The system does not require removal of noncombustibles from the combustible stream. The Monsanto Landgard system therefore can become attractive on that basis.

Among the other pyrolysis systems, the Torrax, Union Carbide, and Urban Research and Development systems can accept material up to a four-foot size. This means only a shredder, for bulky refuse, is needed at any of these facilities. Once a shredder is utilized, these processes become similar to Monsanto's single-stage shredder refuse conditioning facility.

Based on the refuse preparation requirement, it appears that Garrett and Hercules are the least attractive systems, followed by Monsanto, Union Carbide, Torrax, and the Urban Research and Development Corporation systems. In terms of cost, the last four systems might be competitive with a typical fuel recovery process.

CORROSION AND DEPOSIT BUILDUP

In fuel recovery processes where the combustible waste stream is fired into a utility boiler to augment the fuel, one major consideration is the corrosion and deposit of residue or fly ash. This considertaion is discussed in detailed in Section 3, under "Application Considerations." Basically, two corrosion mechanisms exist.

At boiler tube metal temperatures in the 600°F to 900°F range, with a reducing atmosphere, the chlorine in the feed forms volatile $FeCl_3$, which decomposes to iron oxide in the ash layer, releasing chlorine to reattack the tube.

The more serious problem is due to the classical hot corrosion. Simultaneous presence of sulfur and the alkali metals (e.g., sodium and potassium) produce, in combination, low-melting complex sulfates on boiler tubes. These sulfates are extremely corrosive, especially under alternating oxidizing and reducing conditions. Alkali sulfates also act as a binder for ash, forming troublesome deposits. Combined alkali content should be less than 0.4 percent of the dry, ash-free fuel.

Pyrolysis can have a favorable impact in this respect. Most pyrolysis processes can trap the alkali content in the residue (Ref. 64). This situation would reduce the corrosion and deposit problems considerably; thus, it offers a significant gain in flexibility and value of the fuel relative to its use in a fuel recovery process. One estimate shows that the total ash in the gas from a pyrolysis process is reduced to 3×10^{-3}. It can be stated with confidence that raw pyrolysis gas from a shaft furnace (e.g., Union Carbide, Torrax, or Urban Research and Development Corporation systems) can be used directly as boiler fuel without serious corrosion or deposit buildup problems (Ref. 64).

FUEL COMPATIBILITY

The pyrolysis process offers a distinct advantage, compared to shredding and burning refuse in a boiler. Within the broad term "pyrolysis," however, there are some processes offering greater advantage than others.

One potential product of pyrolysis is char, which is like coal and is therefore a compatible and versatile fuel. The Garrett char is dry, with a heating value of 9000 Btu per pound; it burns readily, its sulfur content is 0.3 percent, and chorine at 0.2 percent is acceptable. Its ash content is near 32 percent, with 3-percent alkali, far above the limit at which severe deposits occur. Monsanto char is wet, ash content is 82 percent, containing most of the alkali and sulfur in the original refuse. Its heating value is 2500 Btu per pound. From an ash standpoint, neither char is very appealing as a fuel.

There are two optimum fuels produced from pyrolysis: fuel gas and fuel oil. Some processes yield low-Btu fuel gas, whereas the Union Carbide oxygen refuse converter produces a higher quality gas. The Garrett process, although somewhat flexible (to produce either gas or oil), economically favors production of oil. Its ash content is 0.4 percent, sulfur content is 0.2 percent, and chlorine content is 0.3 percent. The oil is acceptable as is, to fire into a boiler to provide 100 percent of the heat input. It is not compatible with all No. 6 fuel oils, however. It must be burned with a completely separate system of tanks, pipes, and pumps, but can use a common burner. The estimates show that the Garrett oil will be worth 50 to 60 cents per million Btu. It also offers storage and transport advantages over gas fuel.

One conclusion that may be drawn is that pyrolysis is unlikely to produce a fuel that is compatible with hydrocarbon fuels. The major benefit is reduction in ash content, ease of handling, and increased combustion. The pyrolysis system with the lowest cost offers the best alternative fuel recovery process, regardless of the quality of the fuel produced.

One problem in the entire pyrolysis field is the inaccuracy of the information and data. The projections for a commercial facility are made, from small-capacity pilot plants, by the manufacturers. There are no commercial facilities operating, even with small throughput, that would tend to lend credibility to these projections. In approximately two years, larger scale demonstration plants will begin to yield some of these data.

SUMMARY

A review of nine pyrolysis processes is summarized in Table 42. Planned capacity is that for the proposed full-scale operating facility outlined by the respective firms, aimed at resource recovery from solid waste. Input requirements define the quality of refuse to be fed to a furnace or a reactor.

The material presented in Table 42 gives some insight into the relative complexity of a given process, some indication of its proximity to production status, and a comparison of the output products and their relative merits.

Table 43 summarizes the positive and negative features of selected pyrolysis processes. The advantages and disadvantages of each process are given with respect to other processes within the pyrolysis system. The comparison is not between a pyrolysis process and other nonpyrolysis processes.

Table 42

PYROLYSIS SYSTEMS

Process	Planned Capacity	Description	Major Equipment	Input Requirements	Products
Garrett	200 tons/day (1975)	Adiabatic fluid bed reactor. Operates on presorted refuse at 500°C. Produces fuel oil, char, and gas. Gas is used for process heat. The process is known as "flash pyrolysis."	Coarse shredder Air classifier Fine shredder Drier Screen Forth flotation system Pyrolysis reactor Char separator Distillation column Magnetic separator	Dry, fine (0.014 in.) shredder combustible, relatively free of glass	Oil Gas Char Iron Glass Water (polluted)
Monsanto	1000 tons/day (1975)	Rotary kiln with internal heating by counter current flow of deficient air plus supplementary fuel. Requires presorted refuse. The process is known as "Landgard."	Single-stage shredder Rotary kiln Afterburner Boiler Gas scrubber ID fan Plume suppression heater Ash quencher Magnetic separator	Shredded refuse to -6°. Supplementary oil fuel	Low-pressure steam Char Iron Aggregate
Union Carbide	200 tons/day (1975)	Gasification is shaft furnace with oxygen to produce 300°Btu/scf gas and slag.	Shaft furnace Feed lock Gas cleaning train Auxiliary combustion chamber Oxygen generating plant Bulky refuse shredder	Mixed municipal refuse -- 4 ft gaseous oxygen	300 Btu/scf gas slag containing metal
Urban Research and Development	250 tons/day 2 modules (1973)	Gasification with air heated to 900°F in metal heat exchanger fired with gas generated on site.	Air heater Combustor Shaft furnace Feed hopper and lock ID fan Shredder for bulky refuse	Mixed municipal refuse -- 4 ft	Gas 150 Btu/scf Slag aggregate
Torrax	75 tons/day (1973)	Gasification with air heated to 2200°F in ceramic heat exchanger fired with supplementary fuel.	Ceramic heater Shaft furnace Afterburner Gas cooler Venturi scrubber Dust separator ID fan	Mixed municipal refuse -- 4 ft auxiliary fuel	Gas or steam Slag aggregate
City of Charleston	500 tons/day (no date)	Fluid bed reactor with heat addition. 1500°F operating temperature to produce gas.	Shredder Air classifier Drier Pyrolysis reactor Fluid bed combustor Sand transfer, screen, and cleanup Gas recycle cooler and compressor	Dry, finely pulverized, noncombustible, free refuse. Make up sand 30 tons/day	Gas 300 Btu/scf
U.S. Bureau of Mines	Bench Units	Operates on the principle of Bureau of Mines/ American Natural Gas Association apparatus. BM-AGA pyrolysis takes place at 700°C to 900°C.	Electric furnace Cylinderical steel retort Condenser Scrubbing train Electrostatic precipitator	Dry refuse Need not be shredded	High Btu gas 0.500 Btu/scf High-quality tar 1600 Btu/lb Light oil
Hercules	1500 tons/day (1975)	Combination of available components. Pyrolysis of shredded noncompostible organics (20%) in Herreshoff furnace. Compositing of organics proposed for Delaware pilot plant.	Shredder Air classifier -- plastics Air classifier -- paper Magnetic separator Screen Pyrolysis furnace	Shredded material fine enough for two-stage air classification (-1/2 in.)	Gas Compost Iron Glass
Battelle	150 tons/day (1977)	Gasification with air and steam	Shredder Shaft furnace Gas cooler Air heater	Not yet defined	Gas 150 Btu/scf Dry ash

Table 43

ADVANTAGES AND DRAWBACKS OF PYROLYSIS SYSTEMS

System	Advantages	Disadvantages
Garrett flash	Over 2 years operating experience with pilot plant	Requires ultrafine size reduction of refuse -- 0.015 in.
	Does not require auxiliary fuel for preheat	Refuse must be predried -- 2 to 3% moisture
	Produces oil -- storable and transportable (8500-10,500 Btu/lb)	Agglomerating noncombustibles, especially glass, may contaminate fluidized bed
	Potential for recovery of clean glass and iron	Char highly alkaline and corrosive
	Process flexible -- can produce oil (1 barrel/ ton) or gas (6000 scf/ton)	Fuel oil -- low flash point ~130°F; objectionable odor, highly alkaline, and high viscosity
	Produces dry char -- heating value 9000 Btu/lb	Requires extensive water purification
		Entire process -- especially front end equipment -- very expensive
Monsanto	Operating experience with pilot plant since 1969	Shredding of refuse to -6 in.
	Produces gas -- burned to produce low-pressure steam (1 ton/day refuse produces 200 lb/hr steam)	Restricted to normal municipal refuse
		Char is wet, as produced; requires drying
	Potential for recovery of iron	Char -- highly alkaline pH12; low heating value
	Does not require predrying of refuse	Gas scrubber required
		Auxiliary fuel for preheat
Union Carbide	Some operating experience with 5-tons/day pilot plant	Considerable investment in a large O_2 plant
	Generally accepts normal refuse without size reduction	Requires size reduction of bulky refuse
		Requires gas cleaning train -- electrostatic precipitator, acid absorber, and condenser
	Gas produced -- high heating value 300 Btu/scf, relatively clean, and can be upgraded by methanation	Water purification for condenser and cooling tower water
	NO_x production virtually precluded	
	Fuel gas requires less air per scf or per 1000 Btu than natural gas	
	Molten residue -- potential for metals recovery	
	Does not require auxiliary fuel	
Urban Research and Development	Some operating experience with pilot plant	Produces low-grade gas -- 200 Btu/scf
	Generally accepts normal municipal refuse without size reduction	Cleaning of flue gas required
	Uses air with a more conventional degree of preheat	Requires glass and iron in refuse to permit proper fluxing for slag flow
	Gas produced by the process is fired for preheat, no auxiliary fuel needed	Size reduction -- 4 ft required for bulky waste
	Air heater -- a normal boiler	
Torrax	Longest operating experience -- 3 to 4 years, with 75-ton/day pilot plant	Requires separately fired, expensive ceramic preheater
	Accepts normal municipal refuse as received. No shredding	Auxiliary fuel -- about 20% of the heating value of refuse -- required
	Shaft furnace problems -- material hangup and gaseous reactant medium flow through the bed appear to have been resolved	Produces low-Btu fuel gas
		Size reduction of bulky waste required
City of Charleston	Long operating experience, 2-3 years, on bench scale	Requires size reduction of normal municipal refuse
	Fuel gas has high heating value -- half of natural gas	Progress appears slow -- project still in early research phase

Section 5

COMPOSTING PROCESSES

INTRODUCTION

Composting consists of the biochemical degradation of organic materials and can be used as a sanitary process for treating municipal solid waste, sewage sludge, and agricultural and industrial wastes. The end product is a humuslike material, useful as s soil conditioner, as a fuel, or as a base material for more refined products such as fertilizer, wallboard, or building blocks.

GENERAL COMPOSTING PROCESSES

Although composting has been carried out for centuries by farmers and gardeners, the first systematized development of composting as a process took place in India in 1925. (Ref. 78). A. Howard developed the Indore Process, which involved anaerobic degradation of leaves, garbage, animal manures, and night soil for six months in pits or piles. Howard's studies brought out many of the fundamental factors in composting, which still serve as a guideline to compost processors.

The Indian government carried Howard's process further, to include laying down successive layers of refuse and night soil and turning the pile frequently to hasten aerobic action. This system is known as the Bangalore process (Ref. 79) and is now in use in 2500 small installations in India.

Composting has since developed into a multitude of processes, more than 30 being identified by the names of their inventors or by proprietary names. At least 18 types of composting processes have been identified and are listed and briefly described in Table 44 (16 of the processes in Table 44 are from Ref. 80). All processes contain the following steps:

- Preparation. Receiving and sorting of the bulky wastes are required. Many processes grind the refuse (Dano and Geochemical/Eweson processes do not). Some processes magnetically separate and presort and many processes add sewage sludge.

- Digestion. Digestion is either carried out in open windrows or in enclosures. The principal objective of digestion is to create in environment in which microorganisms will rapidly decompose the organic portion of the refuse. Most modern plants use aerobic rather than anaerobic decomposition.

  To furnish the necessary oxygen to the microorganisms, air is introduced into windrows by turning, and into enclosed systems by forced draft and agitation. Heat is generated profusely in aerobic

Table 44

TYPICAL COMPOSTING PROCESSES

Process	General Description	Location
Bangalore (Indore)	Trench in ground, 2 to 3 ft deep. Material placed in alternate layers of refuse, night soil, earth, straw, etc. No grinding. Turned by hand as often as possible. Detention time of 120 to 180 days.	India
Caspari (briquetting)	Ground material is compressed into blocks and stacked for 30 to 40 days. Aeration by natural diffusion and air flow through stacks. Curing follows initial composting. Blocks are later ground.	Schweinfurt Germany
Dano Biostabilizer	Rotating drum, slightly inclined from the horizontal, 9 to 12 ft in diameter, up to 150 ft long. One to 5 days digestion followed by windrowing. No grinding. Forced aeration into drum.	Predominately in Europe
Earp-Thomas	Silo type with eight decks stacked vertically. Ground refuse is moved downward from deck to deck by plows. Air passes upward through the silo. Uses a patented inoculum. Digestion (2 to 3 days) followed by windrowing.	Heidelberg, Germany; Turgi, Switzerland; Verona and Palerma, Italy: and Thessaloniki, Greece
Fairfield-Hardy	Circular tank. Vertical screws, mounted on two rotating radial arms, keep ground material agitated. Forced aeration through tank bottom and holes in screws. Detention time of 5 days.	Altoona, Pennsylvania and San Juan, Puerto Rico
Fermascreen	Hexagonal drum, three sides of which are screens. Refuse is ground and batch-loaded. Screens are sealed for initial composting. Aeration occurs when drum is rotated with screens open. Detention time of 4 days.	Epsom, England
Frazer-Eweson	Ground refuse placed in vertical bin having four or five perforated decks and special arms to force composting material through perforations. Air is forced through bin. Detection time of 4 to 5 days.	None in operation
Jersey (also known as the John Thompson system)	Structure with six floors, each equipped to dump ground refuse onto the next lower floor. Aeration effected by dropping from floor to floor. Detention time of 6 days.	Jersey, Channel Islands, Great Britain, and Bangkok, Thailand
Metrowaste	Open tanks, 20 ft wide, 10 ft deep, 200 ft to 400 ft long. Refuse ground. Equipped to give one or two turnings during digestion period (7 days). Air is forced through performations in bottom of tank.	Houston, Texas and Gainesville, Florida
Naturizer or International	Five 9-ft-wide steel conveyor belts arranged to pass material from belt to belt. Each belt is an insulated cell. Air passes upward through digester. Detention time of 5 days.	St. Petersburg, Florida
Riker	Four-story bins with clam-shell floors. Ground refuse is dropped from floor to floor. Forced air aeration. Detention time of 20 to 28 days.	None in operation
T.A. Crane	Two cells consisting of three horizontal decks. Horizontal ribbon screws extending the length of each deck recirculate ground refuse from deck to deck. Air is introduced in bottom of cells. Composting followed by curing in a bin.	Kobe, Japan
Tollemache	Similar to the Metrowaste digesters.	Spain and Southern Rhodesia
Triga	Towers or silos called "Hygienisators," in sets of four towers. Refuse ground. Forced air aeration. Detention time of 4 days.	Dinard, Plaisir, and Versailles, France: Moscow, U.S.S.R. and Buenos Aires, Argentina
Windrowing (normal aerobic process)	Open windrows, with a haystack cross section. Refuse ground. Aeration by turning windrows. Detention time depends upon number of turnings and other factors.	Mobile, Alabama; Boulder, Colorado; Johnson City, Tennessee; and Israel
van Maanen process	Unground refuse in open piles, 120 to 180 days. Turned once by grab crane for aeration.	Wijster and Mierlo, the Netherlands
Varro	Ground refuse placed in eight-deck digester and moved downward from deck to deck by plows. Each deck pair has own recirculating air supply to control CO_2 level. No inoculum used, no sewage. Output dried, reground, and used as base material for fertilizer, soil conditioner, wallboard, etc. Total digestion time 40 hours.	Brooklyn, New York
Geochemical-Eweson	Unground refuse placed in rotating drum, 11 ft in diameter, 110 ft long, slightly inclined from horizontal. Three compartments in drum, each with own inoculum. Refuse transferred to next compartment every 1-2 days. Total digestion time 3-6 days. Screened output cured in piles 2-3 days.	Big Sandy, Texas and Des Moines, Iowa
Conservation International	Unsorted refuse ground in two stages and then windrowed 12 to 20 days using an inoculum and periodic turning. Composted material then ground and screened. Fine material bagged and sold as soil conditioner/fertilizer. Coarse material recycled.	Jamaica, British West Indies

digestion, reaching temperatures (over 140°F) high enough to destroy pathogenic organisms, weed seeds, and fly ova. No excessive unpleasant odors arise.

If the decomposing mass is not aerated, a different microflora, which obtains oxygen from the waste itself, begins to grow and digestion takes much longer, peak temperatures reach only 100°F to 130°F, foul odors arise, and pathogens may survive.

In general, the aerobic digestion method is much faster (on the order of days) compared to months for anaerobic processes. The fastest processes are aerobic processes in enclosed systems. Digestion is hastened by maintaining an optimum carbon/nitrogen ratio of the incoming waste on the order of 30 to 35 (Ref. 80). Because municipal solid waste can have a carbon/nitrogen ratio on the order of 50 to 80, the addition of nitrogen in some form is desirable.

- Curing. High-rate digesters succeed in decomposing putrescibles such as garbage in two to five days. However, much of the carbon in solid waste is in the form of cellulose and lignin, which do not decompose as readily. Therefore, a curing period is often added after the digestion stage. In this period, additional action takes place on the cellulose and lignin.

- Finishing. Screening and grinding or a combination of similar processes is done to remove plastics, glass, and metals that are not digested. In some processes, such as the Varro process, the output compost is intensively processed to produce fertilizer, wallboard, and other higher value products.

CURRENT COMPOSTING STATUS

Since 1960, the literature has contained reports of about 2600 composting plants operating outside the United States; 2500 are small plants in India (Refs. 80 and 81). Table 45 lists these known installations (Ref. 80) by process type.

In Europe, composting usage ranges from about 1 percent of municipal waste in West Germany to 17 percent in the Netherlands. The high percentage in the Netherlands is based on 1968 data, however, and may be revised downward as the Hague plant switches over to incineration of its refuse. In general, the compost products in Europe are used for the luxury agriculture market. Several composting plants are being kept open to continue to serve this market, but new municipal refuse processing plants use other, lower cost disposal techniques.

Israel makes extensive use of composting. In five of the seven districts that make up the State of Israel, composting processes 43 percent of the total refuse (Ref. 80). The largest windrow plant in the world is at Tel Aviv. Table 46

Table 45

WORLDWIDE DISTRIBUTION AND TYPES OF COMPOSTING PLANTS OUTSIDE THE UNITED STATES
(Operating Between 1960 and 1969)

Process	Argentina	Austria	Belgium	Brazil	Czechoslovakia	Denmark	Ecuador	England	Finland	France	Greece	Iceland	India	Israel	Italy	Jamaica	Japan	Netherlands	New Zealand	Norway	Philippines	Poland	Puerto Rico	Saudi Arabia	Scotland	South Africa	Southern Rhodesia	Spain	Sweden	Switzerland	Thailand	U.S.S.R.	West Germany	Total
Bangalore													2500																					2500
Dano			3	2		12	1	2	2	1		1		2	2		16	4	1	1	1	1		2	6	1			1	6			2	70
Windrow					20									1	2	1												4					3	31
Tollemache																		5									1						1	7
Earp-Thomas								1			1																						1	3
van Maanen																		2																2
Caspari																							1											1
Fermascreen								1																										1
Fairfield-Hardy																														1				1
Jersey	1																																	1
Triga										3																				1		1		5
Becarri										*					*															1				*
Other		1																													1		4	6
Total	1	1	3	2	20	12	1	4	2	?	1	1	2500	3	?	1	16	11	1	1	1	1	1	2	6	1	1	4	1	9	1	1	11	

178

Table 46

EUROPEAN AND MIDDLE EASTERN MUNICIPAL REFUSE COMPOSTING PLANTS

Type and Location of Plant	Population Served	Year Constructed	Year of Observation	Operating Features		Refuse (tons/year)	Compost[1] (tons/year)
				Preshredding	Sewage Sludge		
Windrow							
Arnhem, Netherlands	130,000	1961	1967	Yes		26,500	18,000
Blaubeuren, West Germany	20,000	--	1967	Yes	Yes	--	2,000
Buchs, Switzerland	40,000	--	1967	Yes	Yes	--	--
Lagny, France	75,000	1964	1965	Yes	--	17,500	13,200
St. Georgen, West Germany	14,000	--	1967	Yes	Yes	--	450
Stuttgart, West Germany	75,000	1959	1967	Yes	No	--	--
Tehran, Iran	2,500,000	(2	1969	Yes	No	300,000(3	--
Tel Aviv, Israel	700,000	1963	1965	Yes	--	200,000	74,000
Wyster, Netherlands	800,000	1927	1967	No	No	160,000	55,000
High Rate[4]							
Bad Kreuznach, West Germany	45,000	1958	1967	No	Yes	8,000	2,600
Bristol, England	90,000	1964	1965	--	--	17,000(3	0
Cheadle, England	--	1965	1965	--	--	9,000	--
uisburg, West Germany	90,000	1957	1967	No	Yes	--	10,000
Edinburgh, Scotland	210,000	1958	1965	Yes	--	32,500	5,300
Gladsaxe, Denmark	80,000	1948	1965	--	--	19,000	13,200
Haifa, Israel	170,000	1959/64	1965	--	--	46,500	26,000
Heidelberg, West Germany	30,000	1955/62	1967	Yes	Yes	--	650
Hinwill, Switzerland	100,000	1964	1967	Yes	Yes	19,000	8,800
Jerusalem, Israel	120,000	1968	1969	No	--	--	--
Olten, Switzerland	--	1964	1965	Yes	--	10,000	--
Rome, Italy	700,000	1964	1965	No	--	100,000	66,000
Soest-Baarn, Netherlands	54,000	1958	1965	--	--	10,000	4,500
Soissons, France	27,000	1963	1965	--	--	9,500	5,500
Thessaloniki, Greece	400,000	1966	1968	Yes	No	(5	0
Turgi, Switzerland	70,000	--	1967	Yes	--	--	--
Versailles, France	82,000	1967	1967	--	--	50,000(3	--

1) Volumetric figures converted assuming 300 lb/yd³ (dry weight).
2) Under construction at reported cost of $3,500,000; construction halted prior to installation of compost handling and processing equipment.
3) Estimate, based on 300 days operation at rated capacity.
4) Enclosed systems with mechanical turning, often with forced aeration.
5) Not operating in June 1968 because product could not be disposed of; international loan in default.

list pertinent data on the major composting plants in Europe and the Middle East.

Composting activities in the United States were virtually nonexistent prior to 1950. Since then, a number of plants have been built and operated in the United States (Table 47). Of these, the only ones now continuously operating are located in Altoona, Pennsylvania (Fairfield-Hardy process); Brooklyn, New York (Ecology, Inc. process): and Big Sandy, Texas (Geochemical/Eweson process). These three processes, plus a fourth, the Conservation International process, are analyzed in more detail below.

FAIRFIELD-HARDY PROCESS

COMPONENTS AND PROCESS

Process Flow

The Fairfield-Hardy process is shown in block diagram form in Figure 45. Incoming trash is dumped into a receiving bin and is carried by conveyor to

Table 47

MUNICIPAL SOLID WASTE COMPOSTING PLANTS IN THE UNITED STATES[*]

Location	Company	Process	Capacity (tons/day)	Type Waste	Began Operating	Status
Altoona, Pennsylvania	Altoona FAM, Inc.	Fairfield -Hardy	45	Garbage, paper	1951	Operating
Big Sandy, Texas	Ambassador College	Geochemical/Eweson	40	Mixed refuse, raw sludge	1972	Operating
Boulder, Colorado	Harry Gorby	Windrow	100	Mised refuse	1965	Operating
Des Moines, Iowa		Geochemical/Eweson	40	Mixed refuse, raw sludge	--	Intermittently Under construction
Gainesville, Florida	Gainesville Municipal Waste Conversion Authority	Metrowaste Conversion	150	Mixed refuse, digested sludge	1968	Closed
Houston, Texas	Metropolitan Waste Conversion Corp.	Metrowaste Conversion	360	Mixed refuse, raw sludge	1966	Operating intermittently
Houston, Texas	United Compost Services, Inc.	Snell	300	Mixed refuse	1966	Closed (1966)
Johnson City, Tennessee	Joint USPHS-TVA	Windrow	52	Mixed refuse, raw sludge	1967	Closed
Largo, Florida	Peninsular Organics, Inc.	Metrowaste Conversion	50	Mixed refuse, digested sludge	1963	Closed (1967)
Norman, Oklahoma	International Disposal Corp.	Naturizer	35	Mixed Refuse	1959	Closed (1964)
Mobile, Alabama	City of Mobile	Windrow	300	Mixed refuse, digested sludge	1966	Operating Intermittently
New York, New York	Ecology, Inc.	Varro	150	Mixed refuse	1971	Operating
Phoenix, Arizona	Arizona Biochemical Co.	Dano	300	Mixed refuse	1963	Closed (1965)
Sacramento, California	Dano of American, Inc.	Dano	40	Mixed refuse	1956	Closed (1963)
San Fernando, California	International Disposal Corp.	Naturizer	70	Mixed refuse	1963	Closed (1964)
San Juan, Puerto Rico	Fairfield Engineering Co.	Fairfield-Hardy	150	Mixed refuse	1969	Closed (under negotiation)
Springfield, Massachusetts	Springfield Organic Fertilizer Co.	Frazer-Eweson	20	Garbage	1954 1961	Closed (1962)
St. Petersburg, Florida	Westinghouse Corp.	Naturizer	105	Mixed refuse	1966	Operating intermittently
Williamston, Michigan	City of Williamston	Riker	4	Garbage, raw sludge, corn cobs	1955	Closed (1962)
Wilmington, Ohio	Good Riddance, Inc.	Windrow	20	Mixed refuse	1963	Closed (1965)

[*]References: 2, 13, 22-23, 27-29, 31, 54-55.

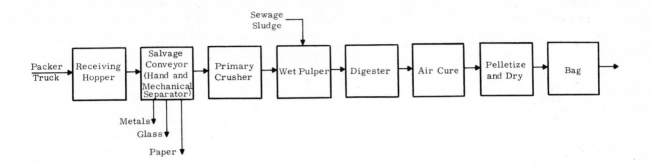

Figure 45. Block Diagram of Fairfield-Hardy Process

the salvage area, where manual and mechanical sorting cull out salvagable materials such as cans and other metals, glass, paper, rags, and cardboard. From the sorting area, the remaining trash is conveyed to the primary crusher which shreds it into a coarse grind. The shredded material is conveyed to the pulpers, where it is mixed with water and is reduced further. At this point, sewage sludge can also be introduced, if desired. The pulped material is then partially dried (to about 50% moisture content) by means of a dewatering press and is conveyed to the digester.

Figure 46 shows three digesters in parallel, as they would be used in a 300-ton-per-day plant. The digester is unique to the Fairfield-Hardy process

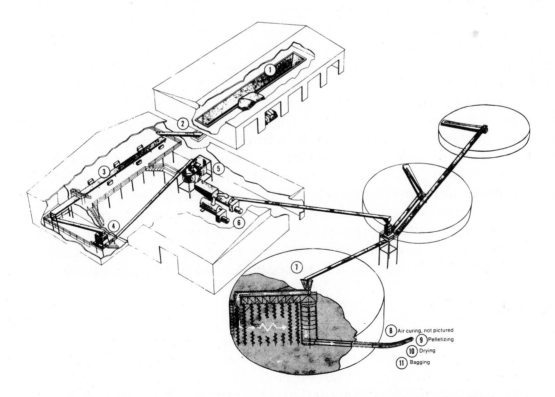

Figure 46. Schematic Diagram of Fairfield-Hardy Process
(Source: Fairfield Engineering Co.)

181

and consists of a circular vessel equipped with a rotating bridge from which are suspended a series of augers. Incoming material is conveyed from the center, out along the top of the bridge to the outer periphery of the tank. As the bridge rotates, the action of the augers agitates the material, providing aeration and gradually working the material toward the center discharge of the digester. Air is forced into the digester by a motor-driven blower and is distributed about the material by pipes. Self-generated temperatures ranging from 140°F to 170°F are produced and maintained by the metabolism of the aerobic-thermophilic mocroorganisms multiplying within the waste material. No starter inoculants or external heat is used.

The speed of the augers, the rate of rotation of the carriage assembly, and the amount of air introduced into the digester are controllable to obtain optimum temperature and correct retention time of material in the digester. Sensors provide the means for automatic control. They measure conditions at a number of locations within the digester, and a control computer calls for automatic adjustment of controls, which maintains desired operating conditions. Total retention time in the digester is three to five days.

The digested material is then conveyed from the center discharge to be air-cured in covered windrows. Following several days of air curing, the material is pelletized, dried, and bagged for sale on the retail market.

Input and Output Streams

The Fairfield-Hardy process is designed to handle packer truck inputs. For each 100 tons of refuse received daily, the process can accommodate 25 tons (dry solids) of sewage sludge. Approximately 25 to 35 tons of dried and pelletized Fairfield Organic Humus Builder is produced from every 100 tons of refuse received at the plant. This pelletized product is sold both on the retail market in 35-pound bags and in bulk, at a price of approximately $25 per ton, to the fertilizer and soil mix industry. Approximately 20 to 25 tons out of each 100 tons of incoming material is presently salvaged and either incinerated or landfilled. This material consists primarily of glass, nonferrous metals, and other inorganic material. Recently, ferrous metal recovery has been added to the process, and baled ferrous scrap is now salvaged and marketed. Tests on the heating value of the compost directly from the digester show a heat content of approximately 4000 Btu per pound, whereas the soil builder (dried and pelletized) has a heat content of 6450 Btu per pound.

Physical Requirements

A 300-ton-per-day Fairfield-Hardy plant, such as the one in San Juan, Puerto Rico, will occupy five acres. Acreage will not decrease drastically for lower capacities. Acreage requirements for larger facilities are not available, nor are any power requirements. However, the considerable number of conveyors and the shredders and pulpers can be expected to consume a fair amount of power.

Manning requirements, as a function of throughput capacity, are listed in Table 48.

Table 48

MANNING REQUIREMENTS VERSUS THROUGHPUT CAPACITY
FOR FAIRFIELD-HARDY PROCESS

Category	Total Input (tons/day)				
	125	250	375	500	1363
Solid waste	100	200	300		1090
Sewage sludge	25	50	75	100	273
Days/wk	5	5	5	5	6
Yearly capacit (tons)	26,000	52,000	78,000	104,000	340,080
Capital costs (without land)					
Total ($K)	1,500	2,500	3,200	4,200	13,000
Capacity/ton/day of solid waste ($K)	15	12.5	10.67	10.5	11.925
Expected life (yr)	20	20	20	20	20
Operating cost*					
Employees	11	18	25	32	71
Cost to operate/ton($)	7.50	6.00	5.25	4.25	5.09

*Real estate, personal property taxes, and amortization and interest are not included.

OPERATING HISTORY AND EXPERIENCE

The first Fairfield-Hardy digester has now been in continuous service at Altoona, Pennsylvania for more than eight years (Ref. 82) and continues to operate at an average daily rate of 25 tons. Another plant, in Puerto Rico, has three 100-ton-per-day digesters, but is temporarily closed because of renegotiations of the contract with the City of San Juan. For approximately six weeks in 1969, the Altoona plant was utilized to successfully demonstrate that sewage sludge could be digested in the Fairfield-Hardy digester without the addition of garbage refuse.

The Altoona operation has processed more than 100,000 tons of refuse since 1960. The current operating rate is about 8000 tons of refuse per year.

SCALING CONSIDERATIONS

The Fairfield Engineering Company has scaled up the 50-ton-per-day Altoona design to a 300-ton-per-day facility in San Juan, Puerto Rico. This scaling has been accomplished by increased shredder and conveyor capacity and by resorting to multiple digesters, each of them twice the size of the original Altoona design. It is expected that further increases in capacity would be obtained by adding additional shredding and pulping lines and by using more digesters of the San Juan capacity.

AIR AND WATER POLLUTION CONSIDERATIONS

The digester exhausts only water vapor and carbon dioxide to the atmosphere and, according to the manufacturer, does not produce any liquid effluent.

COST FACTORS

Cost factors have been supplied by the Fairfield Engineering Company and are summarized in Table 48. The cost per ton of input decreased as the capacity goes up through 400 tons per day, whereas the 1090-ton-per-day unit has a higher cost. Due to this inconsistency, the original Fairfield-Hardy process data are given in Tables 49 and 50. Table 49 covers the 100-, 200-, 300-, and 400-ton-per-day plants; Table 50 describes the 1090-ton-per-day plant and is inferred to be the more complete and recent set of figures.

By letter, the Fairfield Engineering Company indicated that the operating costs should be comparable if the $614,187 interest cost and the $668,534 amortization were to be deducted from the total annual operating cost for the 1090-ton-per-day plant. It was on this basis that the $5.09 per ton was calculated for the 1090-ton-per-day plant in Table 48.

SUMMARY

The Fairfield-Hardy process is a proven process, backed by more than ten years of continuous opertaion. It is the only composting process that has

Table 49

BASIC INFORMATION AND APPROXIMATE COST
FOR FAIRFIELD-HARDY PROCESS
(100-, 200-, 300-, and 400-Ton/Day Capacities)

Plant Design/Size/Cost	Description						
Design	Process separate garbage and trash collection or single unseparated collection.						
	Process sewage sludge.						
	Process combination garbage-refuse and sewage sludge.						
Size	Minimum of 100 tons/day of garbage-refuse						
	No minimum sewage sludge						
	Average of 100 tons/day of garbage-refuse generated for 50,000 population						
	Average of 5 tons/day (dry solids) of sewage waste generated for 50,000 population.						
	Standard 100-tons/day Fairfield digester plant will process:						
	100 tons/day of garbage-refuse						
	25 tons/day (dry solids) of sewage sludge						
	Daily Capacity (tons)	Annual Tons (5-Day Wk)	Employees	Machinery	Construction Cost (Buildings, Plumbing, Electric	Cost to Amortize Plant ($/ton)**	Cost to Operate Plant ($/ton)†
Approximate Cost*	100	26,000	11	$1,100,000	$400,000	$5.50	$7.50
	200	52,000	18	$1,900,000	$600,000	$4.00	$6.00
	300	78,000	25	$2,400,000	$800,000	$3.75	$5.25
	400	104,000	32	$3,000,000	$1,200,000	$3.75	$4.25

*Based on private owner-operator.
**Land not included (5 to 10 acres required).
†Real estate and personal property tax not included.

184

Table 50

FAIRFIELD DIGESTER REFUSE DISPOSAL PLANT
1090 TONS PER DAY, 6 DAYS PER WEEK, 8 HOURS PER DAY

Annual Income

Dumping fee - 1090 tons/day -- 340,080 tons at $9.00			$3,060,720
Organic product - 350 tons/day -- 109,200 tons at $20.00			2,184,000
Metal - 75 tons/day -- 23,400 tons at $10.00			234,000
Total annual income			$5,478,720

Annual Operating Costs

Plant manager	1	$ 24,000	
Plant superintendent	1	18,000	
Salesmen	3	36,000	
Steno-clerks	5	30,000	
Bookkeeper-clerks	4	30,000	
Mobile equipment operators	12	100,000	
Maintenance men	10	84,000	
Laborers and foremen	35	245,000	
Total payroll	71		$ 567,000
Electric power		$170,000	
Water		26,000	
Fuel for drying		150,000	
Supplies		70,000	
Repair and Maintenance		100,000	
Miscellaneous		100,000	
Total			$ 616,000
Salesmen travel			30,000
Payroll tax and workingmens compensation, 15%		$ 85,050	
Real estate and personal property tax (not included)		-0-	
Fire extended coverage and general liability insurance		40,000	
Depreciation building 5%, mobile 20%		668,534	
Interest 20 yr at 9% (4.725%)		614,187	
Total administrative			$1,407,771
Subtotal operation costs			$2,620,771
Contingency and overhead (15%)			393,116
Total annual operation cost			$3,013,887
Profit before tax			$2,464,833
Capital investment without land		$12,998,675	

yet been proven to be economically viable on a long-term basis in the United States. Its advantages and disadvantages are:

- Advantages

  Proven long-term operating and economic viability

  Established control parameters

  No harmful emissions to atmosphere

  No liquid effluents

  Sewage sludge acceptable

- Disadvantages

  Front-end shredding and sorting required

  Moderately high capital cost

GEOCHEMICAL/EWESON PROCESS

COMPONENTS AND PROCESS

Process Flow

Flow of the Geochemical/Eweson process is shown in block diagram form in Figure 47 and is shown in schematic form in Figure 48. The key component of the process is the rotating drum digester, which is 11 feet in diameter and 120 feet long. Externally, the process is similar to the Dano process, but significant differences exist internally.

The drum is divided into three compartments, as shown in Figure 48. Air is circulated axially through the drum at controlled rates, in a flow direction

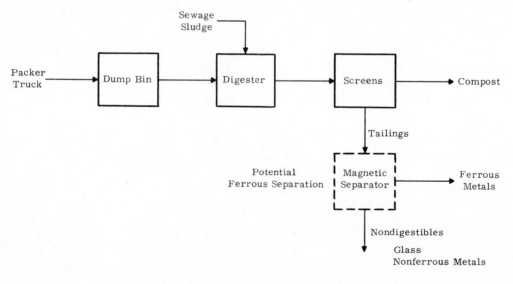

Figure 47. Block Diagram of Geochemical/Eweson Process

186

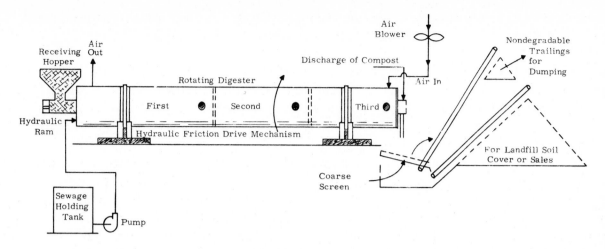

Figure 48.   Schematic Diagram of Geochemical/Eweson Digester

counter to the material flow.   The drum is rotated by hydraulic motors, through a friction drive.   Municipal refuse is dumped into the receiving hopper, and only large metal objects are removed.

The remaining waste is fed into the first compartment by means of a hydraulic ram.   Sewage sludge is added to the first compartment, to bring the moisture content up to about the 50-percent level and to bring the carbon/nitrogen ratio to 35:1 or less.   The charge is maintained in the drum so that a depth of 5 to 6 feet or more prevails.   Internal projections in the drum break up the rubbish bags and other large particles during the tumbling action, thus exposing more surface area to the bacterial action.   The tumbling action produces an effect similar to turning, in windrows, continually exposing new material to the circulating air.   At startup, each section of the drum is charged with a specially prepared inoculum.

The objective is to attain as high a temperature as possible in the first compartment to soften the incoming material.   This objective is achieved by the bacterial action, which is abetted by the circulating air that has already been heated in passing through the second and third stages and is high in moisture and carbon dioxide content.   The carbonic acid from the carbon dioxide dissolves nonwater-soluble microbial nutrients in the waste.   Temperatures up to 160°F are reached in Compartment 1.

After one to two days residency in Compartment 1, a transfer door is opened to Compartment 2.   The drum, which is on a slightly permanent downhill tilt, is sped up, and 85 percent of the material in Compartment 1 is transferred to Compartment 2.   The remaining 15 percent in Compartment 1 serves as an inoculum for the next load.

In Compartment 2, the waste continues to decompose at a high rate, with thermophilic bacterial action.   This compartment is also initially charged with a special inoculum.   After another one to two days in this compartment,

187

the waste is transferred to Compartment 3, again leaving about 15 percent behind in Compartment 2 to maintain its bacterial population.

In Compartment 3, the bacterial activity is less intense and the incoming circulating air has the lowest moisture and carbon dioxide content. A certain amount of drying occurs there, and residence time in the last compartment is again one to two days.

The compost coming from the digester consists of finely digested organic material and undigested organic materials. Plastics and larger pieces of wood are present in the compost, nearly in the form they entered (Ref. 81). Plastic sheet material, such as garbage bags, is torn in the tumbling action, but remains in pieces several inches in size. Cans are bent somewhat, and tin cans and other ferrous metal objects are rusted. Glass size is relatively large, with edges rounded by tumbling and possible microbial action.

This mixture of compost and inorganics is then screened to remove the inorganic material, which is placed in the tailings pile for transfer to a landfill. The remaining organic compost is cured in piles for two to three days.

Additional processing suggested, but not yet implemented, by the Geochemical Corporation is shown in Figure 49.

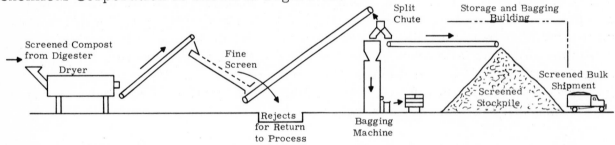

Figure 49. Proposed Secondary Processing
for Geochemical/Eweson Process

Input and Output Streams

Input to the process is packer truck refuse, with large metallic objects such as bicycle frames and large coils of wire being screened out at the dump area. To achieve moisture balance and the correct carbon/nitrogen ratio, 10 tons of sewage sludge is added for every 30 tons of refuse. The process has also operated using chicken manure, dairy lagoon sludge, grass clippings, and slaughterhouse blood and offal. In one test reported by Eweson, 50 pounds of sulfuric acid were thrown into the digester, and bacterial action returned to normal in 20 minutes.

For a total of 30 tons of waste plus 10 tons of sewage sludge input, between 20 and 25 tons of compost will be produced and 2 to 2.5 tons of tailings will result. The wet (40% moisture) compost output has a density of 30 to 35 pounds per cubic foot and will represent roughly in 40 to 45 percent of the volume of the incoming waste. Bomb calorimeter tests performed by the General Electric Company on the compost indicate heat values of about 3000 Btu per pound as taken from the digester, and 4800 Btu per pound when dried.

Physical Requirements

A plant having 150 tons per day of input capacity will require about one acre. Each additional 150 tons per day of capacity will require an additional half acre of land.

Manpower and power requirements are listed in Table 51. Because of the lack of need for front-end processing, power requirements are light. The hydraulic friction drive for one drum consumes only 65 hp, and the blower consumes only 7 hp.

Table 51

MANPOWER AND POWER REQUIREMENTS
FOR GEOCHEMICAL/EWESON PROCESS

Characteristic	Total Daily Input Capacity (tons)				
	80	150	500	1000	2000
Solid waste	60	112.5	375	750	1500
Sewage sludge	20	37.5	125	250	500
Total capital investment ($K)	800	1425	4700	9200	18,000
Ton/day capacity of solid waste ($K)	13.33	12.66	12.53	12.53	12
No. employees	3	3	4	6	10
Man-hr/yr	7488	7488	9984	14,976	24,960
Energy	783	1546	5154	10,309	20,618

Note: Capital costs do not include land or buildings. They are for first-manufactured units and will come down with quantity. Expected life: 20 yr.

OPERATING HISTORY AND EXPERIENCE

One Geochemical/Eweson digester of this design has been operating at Ambassador College in Big Sandy, Texas since mid-June 1972. The capacity is 30 tons per day of refuse and 10 tons per day of sewage sludge. The digester has been running at a rate of about 20 tons per day of refuse, which is due to supply limitations, not operating limitations. The refuse is delivered five days per week, but the digester operates on a seven-day-per-week basis. The refuse sources are the Ambassador College campus and the nearby town of Gladewater, Texas. Retention time is six days in the present operating mode. Since June 1972, between 1000 and 1500 tons of refuse have been processed. The compost is being used experimentally to reclaim farmed-out cotton land on the Ambassador College property.

SCALING CONSIDERATIONS

The digester is considered to be a modular unit. Scale-up of the process would be accomplished by adding more digesters.

AIR AND WATER POLLUTION CONSIDERATIONS

The digester emits only carbon dioxide and water vapor and has no liquid effluent.

COST FACTORS

Cost factors have been supplied by the Geochemical Corporation (Table 51).

SUMMARY

The Geochemical/Eweson process is a new process featuring extreme mechanical simplicity and requiring virtually no front-end processing. Because it does not shred the input, the nonorganic components leave the digester in pieces large enough to separate from the compost by simple sifting techniques.

Because of its simplicity, the process lends itself to portability. The process is still undergoing development, however, and control parameters and operating cycles are still in an early development stage.

The advantages and disadvantages of the Geochemical/Eweson process are:

- Advantages

    Low capital cost

    Portability

    No harmful emissions to the atmosphere

    No liquid effluents

    Accepts sewage sludge

    Low manpower requirement

    Very low power requirement

- Disadvantages

Development in early stage (Some work remains in establishing control parameters and the control mode.)

Sewage sludge or some other nitrogen source is required.

ECOLOGY, INC. (VARRO) PROCESS

COMPONENTS AND PROCESS

Process Flow

The block diagram of the digesting portion of the Ecology, Inc. (Varro) process is shown in Figure 50, and a process flow diagram for the digestion and fertilizer process is shown in Figure 51. Figure 52 shows a schematic diagram of the existing Ecology, Inc. plant in Brooklyn, New York, and Figure 53 shows the plant layout.

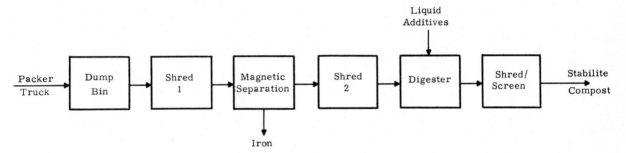

Figure 50. Block Diagram of Ecology, Inc. Process

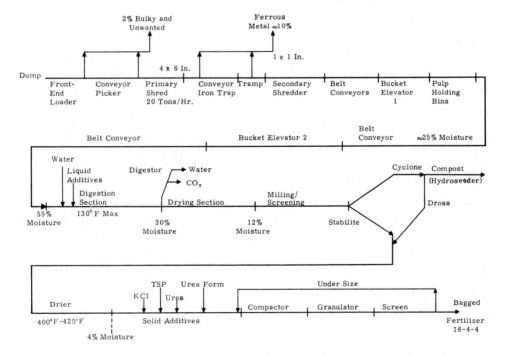

Figure 51. Flow Diagram for Ecology, Inc. Process

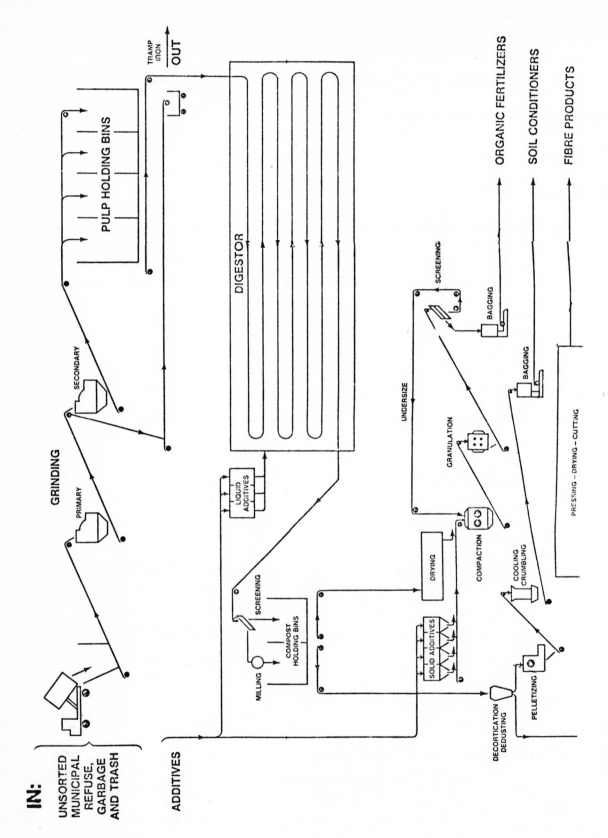

Figure 52. Schematic Diagram of Ecology, Inc. Process (Source: Ecology, Inc.)

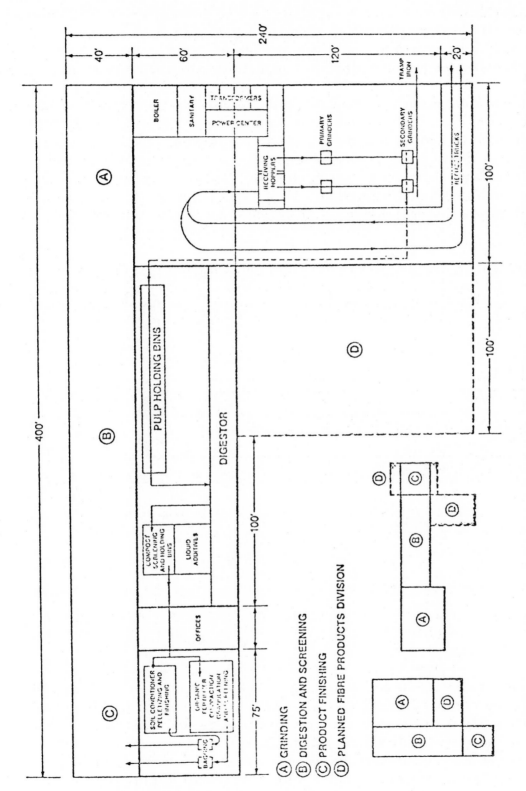

Figure 53. Layout and Modular Features of Ecology, Inc. Plant (Source: Ecology, Inc.)

193

When the plant is operating, city garbage trucks drive over weighing scales and up on a ramp to the roof of the plant where they dump onto a receiving apron. A front-end loader services this area, pushing refuse down either of two chutes serving two parallel primary grinders, each having its own feed conveyor. A hand picker on each conveyor pulls off tires and other items harmful to the system, and the front-end loader operator also separates bulky objects such as rugs.

All shredders in the plant are Pennsylvania Pulverizer Company shredders. Each of the two primary shredders has a 10-ton-per-hour capacity, reducing the size of the refuse to approximately 4 inches by 6 inches. Each shredder has a discharge belt conveyor with tramp-iron trap and magnetic separator, which serves secondary shredders (two in parallel, again). Output of the secondary shredder is about matchbook size. From this point on, the flow paths join. Figure 51 shows the remainder of the process as it now exists in the Brooklyn facility.

The key element in the process is the eight-deck digester, which is about 150 feet long and 50 feet high. Each pair of decks is served by a belt-driven plow conveyor, which periodically drags the waste along the floor (Figure 54).

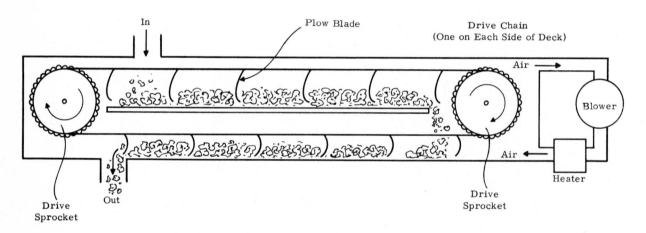

Figure 54. Typical Pair of Decks

Water is added at the first (top) deck to assure a 55-percent moisture of the incoming refuse. This addition is necessary to ensure proper digestion. No water is added elsewhere.

Each deck pair has its own recirculating air blower and heater. Temperature sensors in the compost on each deck control the recirculating air heaters, so circulating air is at the same temperature as the compost bulk temperature. Air is not intentionally circulated either from the outside or between deck pairs, so the atmosphere in them is heavy with carbon dioxide. There is some leakage, however, because a hermetic seal is not attempted.

The top six decks perform the compost function by aerobic digestion, the peak temperature being 130°F, midway through the process. The compost is near room temperature by the time it leaves the sixth deck, where the moisture level is then 30 percent. The two drying decks bring moisture down to 12 percent. After additional conveying and two stages of grinding, the compacted material achieves a form called "Stabilate I," which is fully digested, but does contain small shards of glass, plastic, and nonferrous metals.

Currently, about one-third of the Stabilate I is further processed to produce a fine compost (Hydroseeder) for spraying in water suspension on such applications as golf courses and sides of highway embankments. The fine compost is obtained simply by an air cyclone separator. The remaining dross goes to the fertilizer line, where the Satbilate I is dried to a 4-percent moisture, solid chemicals are added, and the result is then processed to form a fertilizer similar in physical appearance to Scott Turfbuilder.

Input and Output Streams

The Ecology, Inc. process uses city packer truck refuse as its input. In addition, water and liquid additives (probably a nitrogen source) are added to achieve the necessary composting conditions in the digester. For 1 ton of input refuse, 1000 pounds of Stabilate I are produced.

The process is quite flexible relative to the products to be made, starting with Stabilate I as the feedstock. Alternate outputs, for 1 ton of municipal refuse input (Ref. 83) are:

- 1000 pounds of Stabilate I

- 2000 pounds of Ecology, Inc. fertilizer (No. 12)(involves adding 1000 lb chemicals, including K, Cl, TSP, urea, and urea form)

- 800 pounds of Ecopeat/Hydrosealer

- 640 pounds of Stabilate fuel, plus 360 pounds of aggregate for building blocks

- 780 pounds of heavy wallboard

Tests of the Stabilate I fuel show heating values ranging from 6620 Btu per pound to 8500 Btu per pound. The lower number was obtained at the General Electric Materials and Processes Laboratory, and the higher number was obtained using identical tests (but different batches of fuel) at New York University.

Samples of wallboard have been made from Stabilate I feedstock by equipment manufacturers using production-size machinery. A 1/2-inch fiberboard that passes ASTM specifications has been produced.

Physical Requirements

Figure 53 shows the layout of the existing 150-ton-per-day plant in Brooklyn, New York. Approximately 2 acres are occupied by the plant. A

300-ton-per-day plant requires 2 acres, and a 600-ton-per-day plant requires 3 acres, excluding storage for finished products.

For a 300-ton-per-day system to produce Stabilate I, 3500 hp, installed, are required. Manpower requirements for various modes of the Ecology, Inc. process are given in Table 53.

OPERATING HISTORY AND EXPERIENCE

The Ecology, Inc. process was first demonstrated on a pilot scale (1 ton/day) at Manhattan College. This plant operated for six years, from 1962 to 1969. In September 1969, ground was broken on a 150-ton-per-day plant in Brooklyn, New York, which started operating on a reduced scale in May 1971. Production on a regular basis started in mid-1972, and the plant is now operating at the input rate of 40 tons per day and will be operating at 60 tons per day by December 1972 (Ref. 83). Four thousand tons of refuse have been processed to date.

Most operating problems encountered to date have pertained to the materials handling systems and include:

- Bucket Elevator 1 can only handle 10 tons per hour, due to light mass density. Actual value is 9 to 10 pounds per cubic feet versus a design value of 20 to 50 pounds per cubic feet. The material bridges and jams, requiring about one hour for clearing.

- Hammermill wear is uneven. The chute conveyor concentrates the load in the center, wearing the center hammers. The takeup adjustment cannot compensate, as a result. Hammers are repaired approximately every 150 operating hours, which requires 2-1/2 men one full day's work.

- The screw conveyor in the fertilizer line jams from fine dust; it clears by reversing. The net result is that it runs forward 2-1/2 minutes and is reversed, to clear, for 1-1/2 minutes.

- The Hydroseeder cyclone now only recovers fines.

- Fires in the primary shredder are a problem because of inaccessibility.

- Automobile coil springs in the shredder are difficult to remove, but do not always cause a shutdown. Some problems with throwback exist.

These problems are not associated with the digester itself and are cited here to show the problems that can be encountered in any solid waste processing plant. Similar problems are known to occur in equipment for other processes.

196

SCALING CONSIDERATIONS

The Ecology, Inc. process does not afford any advantages in scaling over a 600-ton-per-day capacity, according to S. Varro, President of Ecology, Inc.

AIR AND WATER POLLUTION CONSIDERATIONS

The process emits only water vapor and carbon dioxide to the atmosphere. There are no liquid effluents.

COST FACTORS

Cost factors have been supplied by Ecology, Inc. and are summarized in Tables 52 and 53. Table 52 lists capital costs for various stages of the Ecology, Inc. process. In the table:

Disposal (D)	=	Cost to produce Stabilate I compost, which contains glass shards and other dross. Stabilate I can be the basis of the other finishing processes or the product can be disposed of by landfill (1000 lb/ton/hr).
Fertilizer (F)	=	Added costs, beyond disposal, to produce Ecology fertilizer.
Wallboard (W)	=	Added costs, beyond disposal, to produce wallboard.
Blocks (B)	=	Added costs, beyond disposal, to produce building blocks.
Ecopeat (E)	=	Added costs, beyond disposal, to produce Ecopeat (which can also serve as a fuel).

Table 53 lists operating costs using the same code.

Ecology, Inc. states that no further cost reduction will result for capacities over 600 tons per day. The process does not use sewage sludge, and all capacities are in terms of solid waste. The expected life of the plant is 20 years.

SUMMARY

The Ecology, Inc. system converts municipal waste to the highest value products of any of the composting processes. At the same time, it is the most complex of the processes, mechanically, and requires the highest initial investment of capital among the composting processes considered here. To date, most of the operating problems can be traced to material handling systems, and it is expected that these will be worked out as more operating experience is gained.

Table 52

PLANT COSTS FOR ECOLOGY, INC. PROCESS

Plant Capacity (tons of refuse) Daily	Yearly	Disposal Total	Disposal Per Ton	Fertilizer Total	Fertilizer Per Ton	Wallboard Total	Wallboard Per Ton	Blocks Total	Blocks Per Ton	Ecopeat Total	Ecopeat Per Ton
100	30,000	$ 4,000	$11.70	$ 800	$2.40	$1,000	$2.90	$ 600	$1.80	$ 200	$ 0.60
150	45,000	5,000	9.70	1,000	2.00	1,400	2.70	800	1.60	200	0.40
200	60,000	5,800	8.50	1,200	1.80	1,800	2.60	1,000	1.50	300	0.45
250	75,000	6,500	7.60	1,400	1.70	2,200	2.60	1,200	1.40	300	0.35
300	90,000	7,000	6.80	1,600	1.60	2,600	2.50	1,400	1.40	300	0.30
450	135,000	10,000	6.50	2,000	1.30	3,600	2.40	2,000	1.30	400	0.25
600	180,000	12,000	5.80	2,500	1.20	4,500	2.20	2,500	1.20	500	0.25

Table 53

ECOLOGY, INC. PROJECTION OF OPERATING COSTS

Category ($ thousand)	100-150 Tons/Day D	F	W	B	E	200-250-300 Tons/Day D	F	W	B	E	450 Tons/Day D	F	W	B	E	600 Tons/Day D	F	W	B	E
Per Month																				
Men	16	6	10	6	4	30	10	16	10	6	40	16	20	10	10	50	16	20	16	10
Plant labor	$ 13	$ 5	$ 8	$ 5	$ 3	$ 24	$ 8	$ 13	$ 8	$ 5	$ 32	$ 13	$ 16	$ 8	$ 8	$ 40	$ 13	$ 16	$ 13	$ 8
Plant supervision	3	1	3	2	1	4	1	2	2	1	4	1	2	2	1	5	2	4	3	2
Utilities	5	2	3	3	1	10	4	6	6	2	14	6	10	6	4	20	8	14	8	5
Maintenance	2	2	2	2	1	3	3	3	3	2	4	4	4	4	2	6	4	6	6	2
Services	2	1	2	2	1	2	2	2	2	1	2	2	2	2	1	2	2	2	2	1
Insurance	2	2	2	1	1	3	3	3	2	1	4	2	4	3	1	5	3	6	3	1
Office	2	2	2	1	1	2	2	2	1	1	2	2	2	1	1	2	2	2	1	1
Total monthly	$ 29	$ 15	$ 22	$ 16	$ 9	$ 48	$ 22	$ 31	$ 24	$ 13	$ 62	$ 30	$ 40	$ 26	$ 18	$ 80	$ 34	$ 50	$ 36	$ 20
Total yearly	$ 348	$ 180	$ 264	$ 192	$ 108	$ 576	$ 264	$ 372	$ 288	$ 156	$ 744	$ 360	$ 480	$ 312	$ 216	$ 960	$ 408	$ 600	$ 412	$ 240

Per Ton of Refuse

Tons/Day	Tons/Year	D	F	W	B	E
100	30,000	$11.60	$6.60	$8.80	$6.40	$3.60
150	45,000	7.72	4.00	5.87	4.27	2.40
200	60,000	$9.60	$4.40	$6.20	$4.80	$2.60
250	75,000	7.72	3.53	4.96	3.84	2.08
300	90,000	6.40	2.94	4.14	3.20	1.74
450	135,000	$5.51	$2.67	$3.56	$2.31	$1.60
600	180,000	$5.34	$2.27	$3.34	$2.29	$1.34

D = Disposal   B = Blocks
F = Fertilizer   E = Ecopeat
W = Wallboard

The advantages and disadvantages of the Ecology, Inc. process are:

- Advantages

    High value of products

    Sewage sludge not required

    Harmful emissions to atmosphere not present

    Liquid effluents not present

    Possible fuel source

- Disadvantages

    High capital cost

    Complex front-end and back-end processes

    Not portable

    High power requirement

CONSERVATION INTERNATIONAL PROCESS

COMPONENTS AND PROCESS

This Conservation International, Inc. process is shown schematically in Figure 55. It is basically a windrow plant using grinding of both the incoming and outgoing waste.

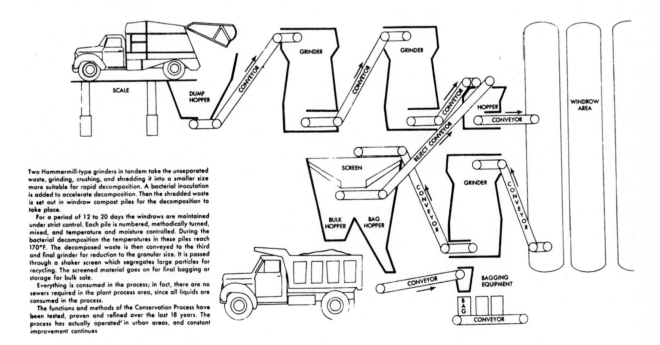

Two Hammermill-type grinders in tandem take the unseparated waste, grinding, crushing, and shredding it into a smaller size more suitable for rapid decomposition. A bacterial inoculation is added to accelerate decomposition. Then the shredded waste is set out in windrow compost piles for the decomposition to take place.

For a period of 12 to 20 days the windrows are maintained under strict control. Each pile is numbered, methodically turned, mixed, and temperature and moisture controlled. During the bacterial decomposition the temperatures in these piles reach 170°F. The decomposed waste is then conveyed to the third and final grinder for reduction to the granular size. It is passed through a shaker screen which segregates large particles for recycling. The screened material goes on for final bagging or storage for bulk sale.

Everything is consumed in the process; in fact, there are no sewers required in the plant process area, since all liquids are consumed in the process.

The functions and methods of the Conservation Process have been tested, proven and refined over the last 18 years. The process has actually operated' in urban areas, and constant improvement continues

Figure 55. Schematic Diagram of Conservation International Process (Source: Conservation International, Inc.)

199

Process Flow

Two Hammermill-type grinders (200-hp Gruendlers) in tandem accept the unseparated waste -- grinding, crushing, and shredding it into a smaller size more suitable for rapid decomposition. A bacterial inoculation is added to accelerate decomposition. Then the shredded waste is set out in windrow compost piles, for the decomposition to take place.

For a period of 12 to 20 days, the windrows are maintained under strict control. Each pile is numbered, methodically turned, mixed, and temperature- and moisture-controlled. During the bacterial decomposition, the temperatures in these piles reach 170°F. The decomposed waste is then conveyed to the third and final grinder for reduction to the granular size. The waste is passed through a shaker screen, which segregates large particles for recycling. The screened material goes on for final bagging or for storage for bulk sale.

Everything is consumed in the process; in fact, there are no sewers required in the plant process area, because all liquids are consumed.

Input and Output Streams

Input to the process is unsorted packer truck refuse, all of which is shredded and composted. No attempt is made to separate metallic or inorganic wastes, other than very heavy metal objects. After composting, the material is pulverized and screened. About 85 percent passes the screen and is sold in bag or bulk form. The remaining 15 percent is fed back into the incoming waste. The only limitation on incoming waste is that it cannot contain chemical concentrates from industrial plants and toxic waste from hospitals.

The end product is a dried, finely ground product representing 60 percent of the input tonnage. General Electric Company tests showed its higher heating value (HHV) to be 1355 Btu per pound. Test conducted by Environmental Engineering, Inc. (Ref. 84) on the product showed a very low carbon/nitrogen ratio of 1.44/1, which indicates the compost is well decomposed. The sample tested by Environmental Engineering, Inc. contained 6 percent nitrogen, which is unusually high for compost, and places the product in a low-grade fertilizer category. Tests were also conducted to determine the load carrying capability of the product when used as a fill. The sample failed at 170 psi. A typical Florida soil will fail at 70 psi (Ref. 84).

The analysis for other nutritional elements, reported on a dry weight basis, is:

Element	Amount
Sulfur	0.36%
Chlorine	0.78%
Boron	0.0042%

Element	Amount
Iron	0.134%
Calcium	3.68%
Magnesium	1.15%
Copper	16 ppm
Manganese	69 ppm
Zinc	0.003%
Molybdenum	1.0 ppm

The content of magnesium is high. All the other elements, except iron, fall within the range that can be expected from compost samples. The content of iron is low.

Tests were also performed on insecticide residues. Insecticide residues were determined by the method given in the Food and Drug Administration pesticide analytical manual (Ref. 85). The results are:

Residue	Amount
DDE dichloro phenyl dichloro ethane	0.27 ppm
DDD 1, 1 - dichloro - 2, 2 - bis (p-chloro-phenyl) ethane	0.11 ppm
DDT 1, 1, 1 - trichloro - 2, 2 - bis (p-chlorophenyl) ethane	0.20 ppm
Dieldrin	0.18 ppm

All of these values are reasonably low. A concentration of any one of these residues in soil or compost of less than 1 ppm is good (Ref. 84).

One word of caution is in order regarding uniformity of product quality. A sample received by the General Electric Company was tested for nitrogen by the Kjeldahl method and was found to contain 0.4- to 0.5-percent nitrogen, compared to the 6 percent found in the 1970 tests conducted by Environmental Engineering, Inc. Most compost products fall into the lower category. Because Conservation International, Inc. can offer no explanation as to why their product should contain more nitrogen (Ref. 86), further investigation is in order before it is accepted as a low-grade fertilizer.

Physical Requirements

Land requirements for the 350-ton-per-day Jamaica plant is 5.7 acres. An 850-ton-per-day plant (8-hr day) will occupy 20 acres, with a building area of 160,000 square feet.

Water requirements for the 350-ton-per-day plant are 100 gpm, intermittently. Power for the 350-ton-per-day plant is concentrated in the 600-hp total for the three shredders (200 hp each). For an 850-ton-per-day plant, 400-hp shredders will be used, bringing the total to 1200 hp.

OPERATING HISTORY AND EXPERIENCE

A plant was operated successfully in McKeesport, Pennsylvania, and then moved to Jamaica, British West Indies, in 1957 after a disagreement with the City of McKeesport. The plant has been operating on a 350-ton-per-day, 5-day-week basis and claims to have processed several hundred thousand tons of waste in that time (Ref. 86). It currently is being expanded to an 850-ton-per-day capacity. The main difference between this process and the Fairfield-Hardy process is twofold:

- It processes all waste received, without separation.

- It uses controlled windrowing, instead of high-rate digestion.

SCALING CONSIDERATIONS

Scale-up appears to present no problems. Shredder capacity must be increased, and additional windrow area must be provided.

AIR AND WATER POLLUTION CONSIDERATIONS

The process emits only carbon dioxide and water vapor and has no liquid effluent.

COST FACTORS

Capital investment figures supplied by Conservation International, Inc. (exclusive of land) are listed in Table 54.

Table 54

CAPITAL INVESTMENT
FOR CONSERVATION INTERNATIONAL PROCESS

Input Capacity (tons/day)	Capital Cost	Capital Cost (tons/day)
150	$ 1,200,000	$8000
500	3,200,000	6400
1000	6,000,000	6000
2000	10,000,000	5000
3000	12,518,000*	4172

*1968 data.

Operating costs submitted may not be typical, due to lower wage scales in Jamaica. For the 350-ton-per-day Jamaica plant operating expenses are those listed in Table 55. The plant is operated for 8 hours per day and for seven days per week.

Table 55

OPERATING EXPENSES FOR CONSERVATION INTERNATIONAL PROCESS
(350 Tons/Day)

Expense	Monthly	Input Ton (9000 tons/yr)
Wages (based on 8 to 10 men/shift plus one checker and one extra tractor man)	$2730	$0.3033
Manager's expenses	560	0.0622
Lease of premises	28	0.0031
Fuel (vehicles)	98	0.0109
Electricity	504	0.0560
Water	168	0.0187
Maintenance materials and repairs	952	0.1058
Total	$5040	$0.56

Note: Cost is for bulk output. If bagging costs are added, the cost goes up to $3.36/input ton when bagged in 56-lb bags.

When expressed in terms of product, the costs are $.93 per ton of bulk product and $5.60 per ton of bagged product. The product is marketed in Nassau at $2.75 per bag, retail. The Kingston plant charges a tipping fee of $1.40 per load, which averages out to $.56 per ton.

In a separate estimate prepared in 1968 for a 3000-ton-per-day (750,000 tons/yr) plant proposed for Pittsburgh, Pennsylvania, an operating cost of $2.50 per ton was quoted. A dumping fee of $6 per ton was proposed, and $10 per ton produced was assumed, giving a net income of $9.50 per input ton before depreciation and taxes.

SUMMARY

The Conservation International process is a proven process with a record of handling several hundred thousand tons of refuse. It has been successfully applied in the economic climate of Jamaica. Further analysis is needed to determine its applicability to the State of Connecticut situation. When questioned about marketability of the product, H. W. Burr, General Manager of Conservation International, suggested the Caribbean Islands and the Middle East as areas desperately needing topsoil. Adaptations of the process, which he suggested, involved shredding in Connecticut and composting at sea, while in transit to the market. In the case of the Caribbean market, barges would be used, whereas the holds of tankers could be used when deadheading empty back to the Middle East. Kuwait has already shown interest in this possibility.

Although there may be serious technical problems in providing proper turning and aeration during transit, these problems could be solved if the economics appear attractive.

For the existing process, the advantages and disadvantages are:

- Advantages

    Proven long-term operating viability

    No harmful emissions to the atmosphere

    No liquid effluents

    High-quality product

- Disadvantages

    Requirement of front-end and back-end shredding and pulverizing

GENERAL DISCUSSION

COMPOST AS SOIL IMPROVER AND COVER MATERIAL

Compost was originally developed as a soil improver, replenishing the organic matter in the soil as it is depleted by the growth of crops. The process closely parallels nature's own method for recycling the soil. Neither composting nor natural soil renewal are well understood processes and offer opportunities for research.

Compost, however, cannot be considered a fertilizer. With regard to supplying plant nutrients, compost neither performs as well as chemical fertilizers nor meets the legal requirements established by several states for designation as a fertilizer. A typical compost contains approximately 1 percent nitrogen, 1/4 percent phosphorus, and 1/4 percent potassium. The slightly higher values that result when sewage sludge and municipal refuse are composted together are derived from the sludge.

Studies by C. Tietjen and S. S. Hart (Ref. 87) have shown that compost, when used with chemical fertilizer, does produce an increase in yields. Figure 56 shows the effect of such treatment on three different crops.

The major benefits of adding compost to the soil are: improved workability, better structure with related resistance to compaction and erosion, and increased water-holding capacity. Better structure and improved water-holding capacity are particularly important for erosion control on steep slopes. These factors will be especially useful in reclaiming old landfills. Work conducted in Bad Kreuznach, Germany by Dr. Banse (Ref. 88) showed compost to be very effective in reducing erosion and water runoff (Figure 57).

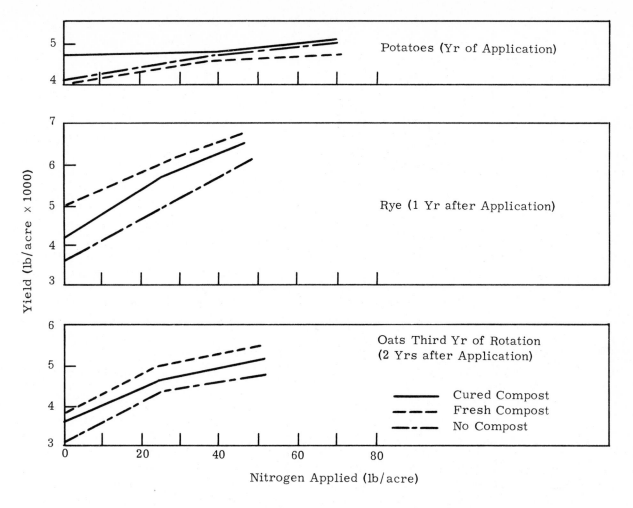

Figure 56. Data from a Nine-Year Experiment by Tietjen and Hart
(Source: Ref. 80)

These data were obtained on 30. 1-degree vineyard slopes. The use of
compost on vineyard slopes is an accepted practice by the grapegrowers of
Germany and Switzerland. About 60 percent of the compost produced from
municipal solid waste in Germany is used for this purpose. The 79-ton-per-
acre application represents about a 1/2-inch-deep layer, and the 158 tons
per acre represents about a 1-inch-deep layer. The 158-ton-per-acre application virtually eliminates erosion and cuts water runoff drastically.

It could be argued that compost, applied over an old sanitary landfill,
will reduce erosion of the cover and, by reason of its moisture-holding capacity, will reduce the rate of leaching of rain water down through the landfill.
The compost will also permit quick planting of a cover crop to further reduce
erosion and leaching. If the area is well drained, up to 1 foot of compost
cover can be applied; if not, there is a danger of it becoming anaerobic. If
the general soil conditions are alkaline, the bacteria will continue to grow,
and methane will form. If the soil is acidic, the compost will be stable.

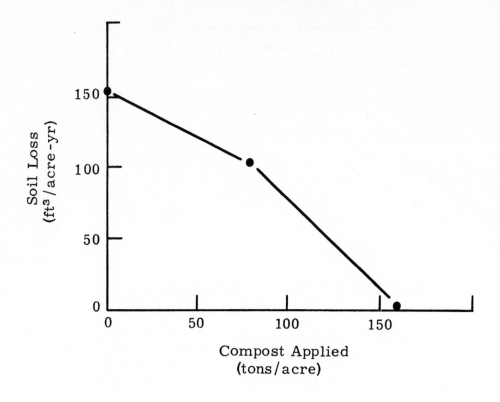

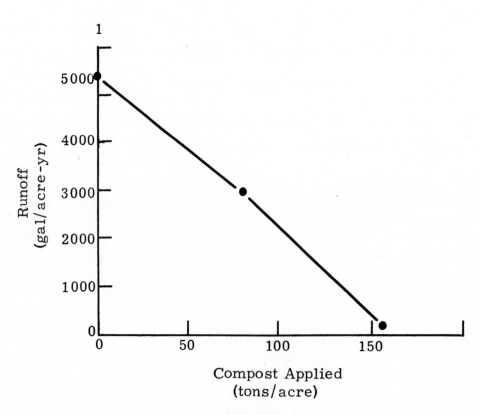

Figure 57. Soil Loss as a Function of Compost Application
(30. 1-Degree Slope)

When considering using compost as a landfill material or as a landfill cover, some measure of biological activity is desired, to attest to the degree of inertness (Ref. 89). No such coefficient of biological activity (CBA) has yet been developed. The rate of oxygen consumption under a specific set of conditions could be one such criterion (Ref. 89).

Another related property is the flammability of the compost. Tests discussed below under "Comparison of Processes" indicate that considerable variation in flammability does exist among the products of the various processes, although even the most flammable one still falls within the highly non-flammable range.

COMPARISON OF PROCESSES

Compost can also be a feedstock for other processes, such as the various alternatives offered by Ecology, Inc., which make wallboard, building block aggregate, fertilizer, and fuel. The fuel capability is common to all three of the high-rate digestion processes described above in this section, and a comparison of results is shown in Table 56.

Table 56

COMPARISON OF FUEL CAPABILITY

Process	Fuel Yield/ Input Ton (lb)	HHV of Fuel (Btu/lb)	Heating Value/ Input Ton (Btu)
Conservation International	1200	1355	1.6×10^6
Ecology, Inc.	640	6622	4.2×10^6
Eweson/Geochemical	1330	3000	4.0×10^6
Fairfield-Hardy	600	6450	4.0×10^6

Table 56 shows the energy yield per input ton of the three high-rate digester processes tested to be close enough that differences are statistically insignificant. The Conservation International energy yield is much lower. In general, the fuel energy yield is poor compared to that of other processes. However, the nature of the composting is such that the undesirable fuel contents, such as metals and plastics, are not decomposed in the composting process. The Eweson process has no front-end shredding, so that the inorganic materials will leave the digester in virtually the same form they entered. As a result, metallic objects and plastics are easily separated from the compost by simple sifting.

The processes can also be compared in other ways. Figure 58 shows a comparison of the four processes in terms of manpower requirements as a function of capacity. Figures furnished by Eweson/Geochemical show much lower manning requirements than either Ecology, Inc. or Fairfield-Hardy,

207

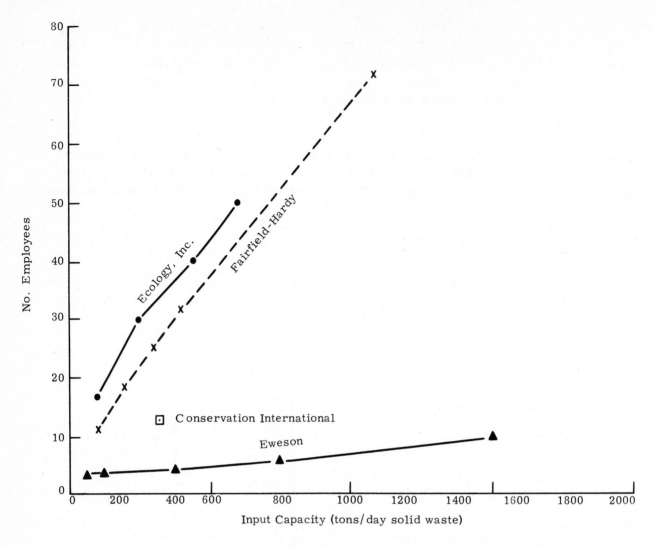

Figure 58. Number of Employees as a Function of Input Capacity

and figures that are somewhat lower than those of Conservation International, Inc. Although probably on the optimistic side, the figures do reflect the relative simplicity of the Eweson/Geochemical process.

Figure 59 shows a similar comparison of the processes on the basis of capital investment. Here, as expected, the Ecology, Inc. process is more expensive than the other three, because it uses more equipment. However, the costs of the Fairfield-Hardy process and the Eweson/Geochemical process are comparable. The costs for Conservation International, Inc. are much lower, which is surprising because that process involves as much front-end shredding as the Ecology, Inc. process and also has one output shredding stage.

At this point, the capital cost comparisons of Figure 59 are to be considered strictly tentative, because it is not known whether each estimate supplied is directly comparable to the other estimates.

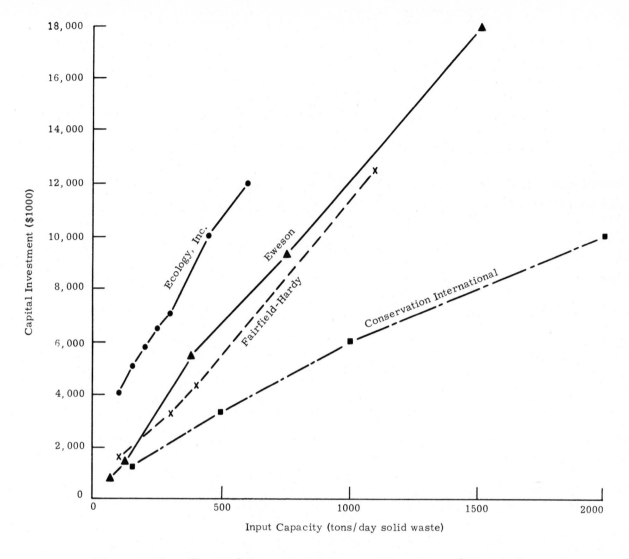

Figure 59. Capital Investment as a Function of Input Capacity

All of the processes described in this section are felt to be effective in digesting the putrescible content of municipal solid waste. However, further testing is desirable before reaching firm conclusions as to the applicibility of the products for covering open dumps, for instance. Very crude odor tests have been run to determine the activity of the products when immersed in water. Tests were run with 150-milliliter samples of compost immersed in 500 milliliters of water. Results are:

Process	Result
Fairfield-Hardy	Very strong, unpleasant odor developed within one day and persisted
Geochemical/Eweson	Less strong, not-unpleasant, swamplike odor developed within one day and persisted

Process	Result
Ecology, Inc.	Barely perceptible odor
Conservation International	Barely perceptible odor

Note that the two composts with strong odor contain sewage sludge. The fact that the Fairfield-Hardy product has been successfully marketed for years as a soil improver may indicate that the odors that develop in an open-air landfill may not be objectionable.

Flammability tests have also been run using the ASTM-D-2863-70 procedure developed to determine flammability of plastics. In this test, the sample is burned in an oxygen/nitrogen atmosphere, and the oxygen concentration is reduced until the sample no longer supports combustion. The percentage of oxygen concentration at this point is referred to as the "oxygen index." R sults are:

Sample	Oxygen Index (%)
Fairfield-Hardy	32.25
Ecology, Inc.	38.67
Geochemical/Eweson	79.9
Conservation International	Over 90

No standards exist to relate the oxygen index to flammability of the compost. However, for plastics, the following definitions are used:

Oxygen Index (%)	Flammability
0-20	Highly flammable
21-30	Fire resistant
Over 31	Highly fire resistant

By these standards, the Conservation International and Geochemical/Eweson products are particularly attractive. As a matter of fact, the Conservation International sample never did catch fire, even when a propane flame was applied to it in an atmosphere over 90 percent pure oxygen. Minute sparks did shoot from the sample as the very fine iron particles in the sample ignited, but no combustion persisted.

The above test results are encouraging; however, before reaching firm conclusions, more samples should be tested, and a small, pilot-scale test of the use of the composts for closing dumps should be conducted.

Section 6

COMPARISON OF FRONT-END SYSTEMS

Although the processes studied under the technology assessment task differed considerably in both the end product and in key processing steps, several patterns did emerge. One observation is that a few processes required virtually no front-end preparation (e.g., Geochemical/Eweson, Urban Research and Development, Union Carbide, and Torrax), while most of the other processes did. Those processes requiring no front-end preparation either do not fit the fuel recovery strategy or are in early stages of development as fuel processes.

On the other hand, many of the fuel processes (including pyrolysis) incorporate front-end shredding and various degrees of material separation. Because fuel preparation by shredding or wet pulping is the most promising near-term strategy, this approach raises the possibility of using shredding and materials separation for the near term and later incorporating the same equipment for front-end processing as pyrolysis processes are perfected. The equipment could also be used for front-end processing for several compost processes.

Four existing processes were considered for an in-depth study, based on a 1000-ton-per-day plant:

- City of St. Louis process (scaled-up Horner-Shifrin process, with air classifier added)

- Garrett pyrolysis process (front-end system only)

- Combustion Power CPU-400 process (front-end system only)

- A.M. Kinney/Black-Clawson process

Of these, the St. Louis and the A.M. Kinney/Black-Clawson processes were originally designed to produce utility boiler fuel. The front-end system for the Garrett process produces a finely shredded product from which most of the inorganic material has been separated. This product will make a very good boiler fuel as well as a feedstock for pyrolysis systems. The front-end system for the Combustion Power CPU-400 is designed to produce a specialized gas turbine fuel that should also be a satisfactory boiler fuel.

In reviewing the above processes, the concept of a hybrid system emerged. This system is believed to maximize the advantages and minimize the disadvantages of the above four candidate systems.

INVESTIGATIVE PROCEDURE

The investigative procedure entailed analysis and comparison of information and previous appraisals of the individual systems. The sources included:

- Midwest Research Institute reports

- General Electric Research and Development Center direct inquiry responses

- Environmental Protection Agency project records and reports

- Experts in waste management (ASME Incinerator Committee, universities, EPA, equipment vendors, private and public collectors, disposal and salvage officials, and corporate-level facility managers)

- On-site visits

- General Electric Industrial Heating Business Department waste management files

- General Electric Real Estate and Construction Operation facility cost estimating experts

As an example of the type of activity involved in the investigation, the Combustion Power Corporation system was studied as follows:

1. J. Mirabal (General Electric Research and Development Center) requested information concerning the application of the Combustion Power Corporation process to the Connecticut project (August 16, 1972).

2. F. Walton (Combustion Power Corporation) responded with detailed proposals for a 686-ton-per-day system and a 477-ton-per-day system, as previously quoted to Grand Rapids, Michigan and Palo Alto, California (August 23, 1972). This information was studied and subsequent questions were answered by the Combustion Power Corporation.

3. The Midwest Research Institute reviewed the Combustion Power Corporation process at a presentation to the General Electric solid waste management team September 28, 1972.

4. The Midwest Research Institute questionnaire completed by the Combustion Power Corporation on May 11, 1972 for Midweat Research Institute Project 3634-D was reviewed, and questions were listed.

5. S. Ali (General Electric Industrial Heating Business Department) was interviewed concerning his recent visits to the Combustion Power Corporation and regarding proposals he has received from them for items of waste pretreatment equipment.

6. A comparison of all inputs to date revealed discrepancies in such items as cost per square foot of floor space (in different proposals) anticipated downtime on the equipment, and cost of installation and startup.

7. Costs of individual items of equipment were checked with vendors. Also, vendor recommendations concerning the application of the

equipment were compared to the Combustion Power Corporation proposal.

8. The overall system was broken into standard subsystems, to permit comparison with other front-end processes.

9. A modified 1000-ton-per-day system evolved that was based on the Combustion Power Corporation process, but included sufficient redundancy for desired system reliability. Also, realistic installation, startup, and contingency costs were included in the capital equipment estimate.

10. A report of the system study was prepared (included in Section 3, "Refuse Fuel Recovery Processes," of this report).

HYBRID SYSTEM

After investigating the four existing candidate systems, the results were compared with regard to project status, history, physical requirements, scale-up considerations, environmental impact, and costs. By comparing similar subsystems of the various processes, it was determined that an optimum system might be a hybrid system, utilizing a combination of components to maximize the advantages and minimize the disadvantages of the individual candidate systems.

A report on the hybrid system, prepared by the General Electric technology assessment team, is included as Appendix II, "Hybrid Facility Description and Capital Cost Estimate."

SUMMARY

Each candidate front-end system has certain advantages and disadvantages (Table 57). The investment and operating costs are based upon assumptions regarding the capacity and reliability of the equipment, as determined by the process developer. Because these assumptions may vary considerably, it is possible to compare apples with oranges unknowingly.

A hybrid process evolved by choosing a combination of compatible subsystems having desirable characteristics. The ground rules for capacities, redundancy, contingencies, and other items of judgment are defined in Appendix II, "Hybrid Facility Description and Capital Cost Estimate." After applying these same ground rules to the various candidate systems, the capital costs for a 1000-ton-per-day (input) waste processing facility compare as follows:

Process	Capital Cost ($ million)
St. Louis process with air classifier (Ref. 13)	11.7×10^6
A. M. Kinney/Black-Clawson process (Ref. 90)	13.7×10^6

Process	Capital Cost ($ million)
Combustion Power process	11.6×10^6
Garrett process (Ref. 91)	15.3
Hybrid system (Ref. 92)	13.8

The above costs do not include site development or land costs. The A.M. Kinney/Black-Clawson figure is based on a scale-down of the Hempstead proposal minus the powerplant, but including an added fuel drier to produce 25 percent moist fuel. Costs for the A.M. Kinney/Black-Clawson process, based on equipment costs supplied by the A.M. Kinney, Inc. without a detailed

Table 57

COMPARISON OF FRONT-END SYSTEMS

Candidate System	Advantages	Disadvantages
St. Louis process	Low capital cost Low operating cost Saves fossil fuel Existing technology Operating experience Materials recovery feasible	Boiler effects still unknown Air pollution effects unknown No oil-fired boiler experience Bottom ash provision required on boiler
Black-Clawson/A.M. Kinney process	Low operating cost Operating experience on process Few apparent problems Saves fossil fuel	Higher capital costs Requires presorting Operating costs questionable Dryer required for fuel Additional pollution control equipment required Bottom ash provision required on boiler No boiler experience -- coal or oil
Combustion Power front-end process	Low capital cost Many components off-the-shelf Simple process Saves fossil fuel	Operating experience lacking Single-stage shredding unsatisfactory Slow progress record in development Cost estimates questionable Bottom ash provision required on boiler Air pollution effects uncertain No boiler experience -- coal or oil
Garrett front-end process	Allows subsequent addition of flash pyrolysis unit with the following potential advantages: Fuel storage feasible Low ash content in fuel (no bottom ash provision required) Less air pollution Fuel burns in conventional oil-fired boiler -- saves fossil fuel at higher rates than others here EPA-sponsored demonstration plant in San Diego will provide some answers in 1974-1975	Extensive pretreatment required New technology Undesirable fuel characteristics Highest capital cost Requires presorting
Hybrid process	Tailored to Connecticut application as now known Existing technology Materials recovery feasible Saves fossil fuel Simple process Redundancy provided Cost estimates have sound basis	Capital costs high Operating experience lacking Boiler effects unknown Air pollution effects unknown Bottom ash provision required on boiler

breakdown of equipment items, come to $9.6 million. The complete Garrett process, including the pyrolysis system, will cost $25.4 million when estimated on the same basis.

The operating costs are still questionable on all systems because of a lack of operating experience with a full-scale plant and because of the developmental nature of many of the existing operations.

Although the list of disadvantages for the hybrid system appears long (Table 57), some of these disadvantages could become advantages with little effort. For instance, the operating experience is closely related to the City of St. Louis experience, and the time logged on the equipment at St. Louis will complement the experience on the first hybrid system application. Further, the boiler and air pollution effects of the hybrid system should be resolved at the St. Louis project. If the results are satisfactory, these items will appear as advantages for the hybrid system, because the fuel characteristics are the same.

Section 7

SANITARY LANDFILLS

Many estimates indicate that in the United States approximately 90 percent of collected trash is disposed of in some type of landfill (Ref. 93). Over 200 million tons of solid waste are collected annually. Of this quantity, 180 million tons go to land disposal at 12,000 sites throughout the United States. Only 6 percent of these sites were labeled "sanitary" (Ref. 94).

Approximately 8 percent of the collected waste is processed by incineration at 300 municipal incinerators. About 2 percent of the waste is disposed of by composting, recycling, and hog feeding. Only 25 percent of the incinerators were considered "adequate" by 1968 standards.

This gross inadequacy in waste disposal means is the result of a lack of planning and financing, and of public apathy to the environment. Urban areas are rapidly being depleted of adequate disposal sites. This fact has a direct impact on disposal costs, due to hauling expenses associated with remote landfill sites.

Open and burning dumps contribute to water and air pollution. These dumps also provide food and breeding grounds for insects, birds, and other carriers of disease. In addition, these dumps are unsightly and very adversely influence the value of nearby land and propetry. Recently, legislation has been passed by local, state, and Federal agencies to aid the development of satisfactory disposal practices and to plan for all aspects of solid waste management. The development and implementation of such plans, however, requires the cooperation of all citizens, industry, and government.

Sanitary landfill is an acceptable alternative to current damaging practices of waste disposal. This practice involves planning and application of sound engineering principles and construction techniques. This method of disposing of waste consists of spreading the waste in thin layers, compacting its volume, and providing a daily cover to protect the environment. No burning is permitted. A sanitary landfill offers an acceptable and economic solution to disposal of solid waste and recovery of otherwise unsuitable or marginal land.

Site selection, design, operation, and completed use call for proper planning and application of sound engineering practices. An understanding of the variables affecting the decomposition rate, decomposition products, and their impact on the environment is very essential. Physical stability of the fill, gas generation and movement, and water pollution aspects must be thoroughly assessed.

Sanitary landfill generally offers an economically attractive method of waste disposal to the taxpayers of a community. In most cases, it appears to be the least expensive means of disposal. However, shrinking acreage available for landfill and the type of operations required to manage a sanitary landfill are rapidly closing the gap between this and the next best economical alternate.

It would be worthwhile to consider briefly the significant Federal regulations that, in effect, influence site selection, design, and operation. Local regulatory agencies and criteria usually govern site approval. However, more states are passing environmental protection legislation, which parallels Federal requirements.

Following is a summary of Federal legislation:

- Refuse Act. This act is part of the River and Harbor Act of March 3, 1899, which originally was interpreted to be "applicable only to that refuse considered dangerous to navigation." The courts recently expanded its intent to include all refuse. Therefore, it is unlawful to discharge refuse of any kind from any source except that flowing in a liquid state from streets and sewers into a navigable water of the United States.

- Executive Order 11575, December 23, 1970. This order directs the implementation of the permit program under the authority of the Refuse Act.

- Federal Water Pollution Control Act. This act, passed in 1956, has been amended three times. It requires states to prepare the standards and submit them to the Federal Government for approval. The Environmental Protection Agency administrator has the approval authority.

- Water Quality Improvement Act of 1970. This act strengthens the Federal Water Pollution Control Act. It requires a Federal license, and a certificate of assurance is called for indicating that the activity will not violate the applicable standards. Additional legislation includes the Fish and Wildlife Act of 1956, the Migratory Marine Game-Fish Act, and the Fish and Wildlife Coordination Act. These acts state that the U. S. Fish and Wildlife Service must be consulted whenever a Federal or state agency impounds a body of water or otherwise controls it.

- National Environmental Policy of 1969. This policy established that Congress is interested in restoring and maintaining environmental quality. It directs all Federal agencies to provide appropriate environmental considerations in decision making, along with economic and technical considerations. It calls for inclusion of: the environmental impact of the proposed action, adverse environmental effects that are unavoidable, alternatives to the proposed action, short-term use and enhancement of long-term productivity, and any irreversible or irretrievable commitment of resources if the proposed action is implemented. The act requires an annual report by the President to the Congress on the environmental quality of the nation. It also establishes the Council on Environmental Quality.

- Executive Order 11514. This order basically outlines the role of the heads of the various Federal agencies in implementing the National Environmental Policy of 1969.

- Council on Environmental Quality Guidelines for Statements on Federal Actions Affecting the Environment. These guidelines require that the impact on ecological systems be thoroughly evaluated. Sufficient analysis of the alternatives and their cost impact on the environment are called for.

- Solid Waste Disposal Act of 1965. This act encourages resource recycling. A system of grants was established to encourage research. The act carries no enforcement powers.

- Amended Resources Act of 1970. Of particular interest among the Federal legislation is this amendment, which reqiures that guidelines for sanitary landfill sites be established by the Office of Solid Waste Management. The act carries no enforcement powers.

In response to this last act, a preliminary draft of Guidelines for the Land Disposal of Solid Wastes has been prepared by the Environmental Protection Agency. The draft, dated January 26, 1973, will appear in the Federal Register when finalized.

TYPES OF LANDFILL

The landfill method offers the most versatile means for utilizing a variety of land topography for the disposal of solid waste. This method can be made effective for disposal above or below ground level, or partially above and below the existing site elevation.

Practically any type of site is suitable for landfill. The four common types of landfill methods, depending on the site, are (Ref. 93):

- Area method
- Slope/ramp method
- Trench method
- Pit/canyon/quarry method

The site is assumed to be usually flat, for the area method. The solid waste, after being deposited, acquires a slope. Additional waste can be spread and compacted against this slope. Cover material can then be applied over the compacted layers of waste.

In the slope/ramp method, cover material is excavated from the toe of the slope, opening up a cavity. The solid waste is inserted into this opening and is compacted. After the first lift is completed, the procedure then becomes similar to the area method. A lift is defined as the depth of compacted solid waste plus the cover material.

The trench method utilizes a large ditch especially excavated for landfill purposes. The solid waste is placed in the trench and is compacted. A layer of cover material is placed on top.

In the pit/canyon/quarry method, an existing depression is utilized to store and compact solid waste. Layers are compacted, beginning at the bottom. The procedure after the first lift is similar to that for the area method, until the final elevation is reached. Figure 60 illustrates these four methods.

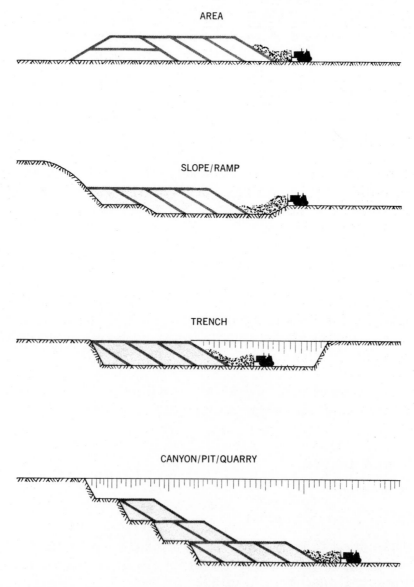

Figure 60. Types of Landfill (Source: Ref. 93)

Another attempt at classifying the landfill into various types has been made by the Caterpillar Tractor Company (Ref. 95).   This attempt utilizes the following grouping:

- Trench method

    Progressive slope method

    Separate trench method

- Area ramp method

- Area fill method

Schematic diagrams of each of the above methods are presented in Figures 61 through 65.

In the progressive slope method, the earth cover material is obtained from the area just ahead of the working face of the landfill. The cover material is therefore obtained, and the trench excavation is performed in one continuous operation. The dirt is handled only once. Figure 61 illustrates this technique.

The second variation of the trench method is the separate disposal and excavation trench method. In this procedure, the first trench is excavated, and refuse is emptied and compacted on the face of the trench. The earth cover for this trench is obtained by excavating an adjacent trench. This arrangement is shown in Figure 62.

The word "ramp" means to rear up. In the ramp method, the cells containing solid waste are formed, compacted, and covered above the natural ground level. If cover material is obtained ahead of the ramp, then some of the deposited solid waste is below the ground level. If the cover material is obtained elsewhere, then all the refuse is above the ground level. This mode of operation is shown in Figure 63.

In the area fill method, low areas such as eroded areas, ravines, and swampy places are used to landfill solid waste. The waste is deposited, compacted, and covered with earth cover. Depressions and pockets in the ground are thus filled to the ground level of the surrounding land. This type of application is shown in Figures 64 and 65.

The Environmental Protection Agency suggests that there are two basic types of landfill (Ref. 96):

- Trench method

- Area method

In general, the trench method is used when the ground water is low and the soil is more than 6 feet deep. It is best suited for flat or gently rolling land. The area method can be followed on most topographies and is often used if large

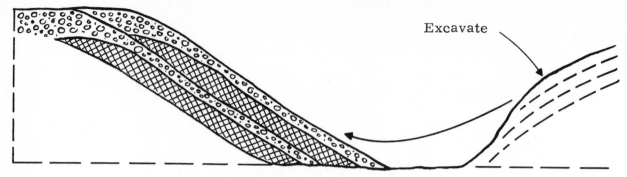

Excavate

Longitudinal Cross Section

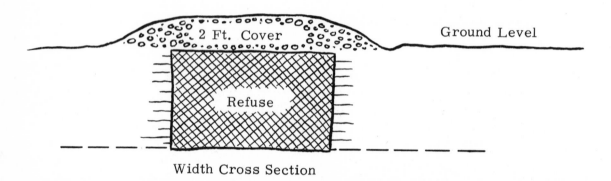

2 Ft. Cover

Ground Level

Refuse

Width Cross Section

Figure 61. Progressive Slope Method (Source: Ref. 95)

Top View

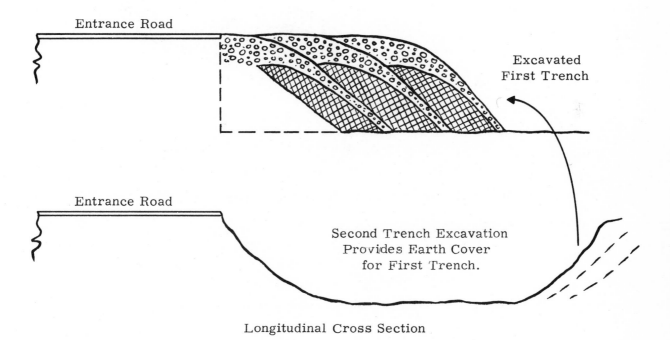

Entrance Road

Excavated
First Trench

Entrance Road

Second Trench Excavation
Provides Earth Cover
for First Trench.

Longitudinal Cross Section

Figure 62. Trench Method (Disposal in Prepared Trench with Earth Cover
Secured by Excavating Adjacent Trench) (Source: Ref. 95)

223

(a)   Sloping Terrain Made Level
        by Ramp Method

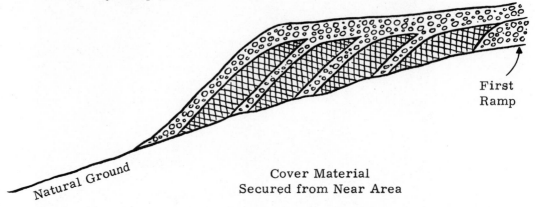

First
Ramp

Natural Ground

Cover Material
Secured from Near Area

(b)   Area Ramp Method Integrated with
        Progressive Slope Trench Method

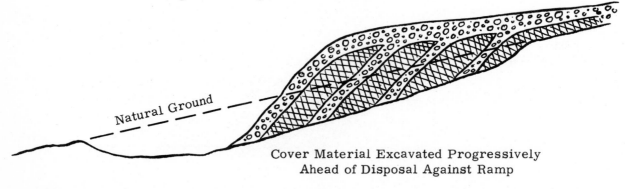

Natural Ground

Cover Material Excavated Progressively
Ahead of Disposal Against Ramp

Figure 63. Area Ramp Method (Source: Ref. 95)

(a)   Longitudinal Cross Section of
        First Fill, Partially Completed

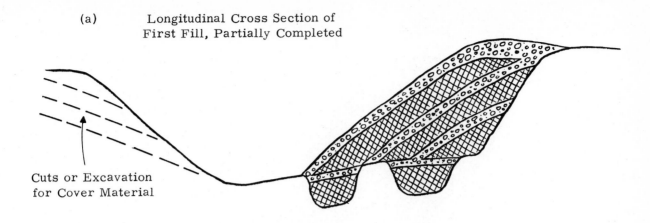

Cuts or Excavation
for Cover Material

(b)   Longitudinal Cross Section
                                        of Second Fill

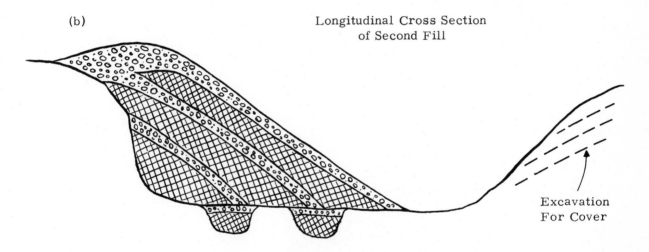

Excavation
For Cover

Excavation Is Done on the Opposite Side to
Keep Both Sides of the Ravine Uniform.

Figure 64. Area Fill Method (Source: Ref. 95)

(a)   Two Separate Lifts -- Upper and Lower

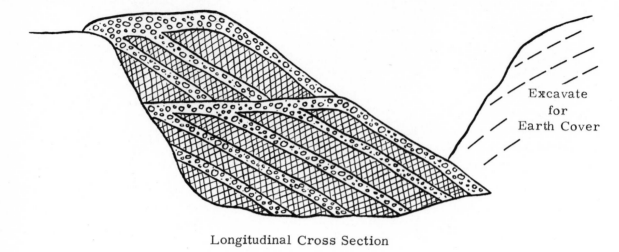

Excavate
for
Earth Cover

Longitudinal Cross Section

(b)   One Lift -- Progressive Stages

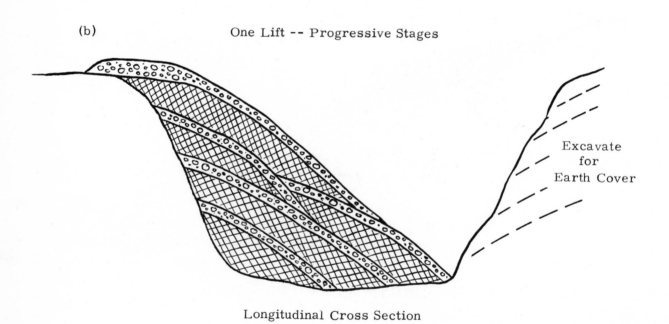

Excavate
for
Earth Cover

Longitudinal Cross Section

Figure 65. Area Fill Methods in Deep Ravines (Source: Ref. 95)

quantities of solid waste must be disposed. Other approaches are modifications of these basic methods.

The building block common to both methods is the cell. All the solid waste is spread and compacted in layers within a confined area. It is then covered completely with a thin, continuous layer of soil.

In the trench method, waste is spread and compacted in an excavated trench. Cover material, taken from adjacent excavation, is spread and compacted over the waste to form the basic cell structure. Any excess of excavated material not used as the cover may be stockpiled for later use as cover for the area fill operation. Figure 66 illustrates this method.

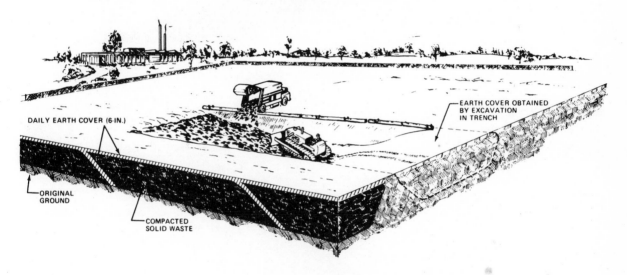

Figure 66. Trench Method (Source: Ref. 96)

The area method utilizes the principle of building cells on the natural surface of the ground. The waste is spread, compacted, and covered. This method is used on flat or gently sloping land and also in quarries, strip mines, ravines, valleys, or other land despressions. A typical schematic of the area method is shown in Figure 67.

TECHNICAL CONSIDERATIONS

The success of a sanitary landfill is intrinsic to proper planning of the entire project. It also requires a thorough understanding of the respective contribution and the interaction among the following factors:

- Site selection
- Landfill design
- Operation
- Equipment requirement

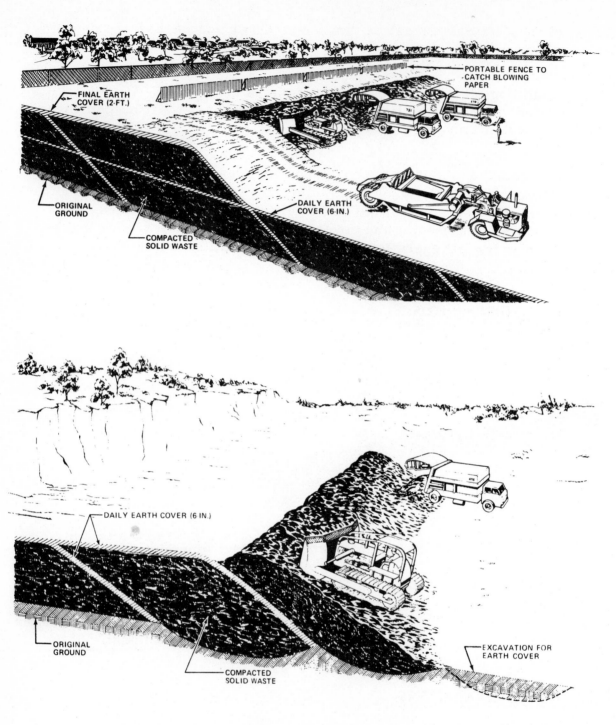

Figure 67. Area Method Examples (Source: Ref. 96)

228

SITE SELECTION

Selection of the site is a very critical parameter for successful operation of a sanitary landfill. The type of site selected has a direct impact on landfill design and operation and on the equipment required.

One significant factor controlling the economics of a sanitary landfill is the location of the landfill with respect to an area of highly concentrated solid waste generation. Every effort should be made to keep the waste hauling distance to a minimum. Hauling distances greater than 20 miles generally make the landfill site economically unattractive.

In addition to proximity to the centroids of solid waste production, a landfill site should offer easy accessibility. In selecting a site, preference should be given to those sites having access via major highways. Sites near or adjacent to railroads may also be desirable. Any location requiring access across main line railroad tracks should be avoided.

Entrance to the site should be shielded from the surrounding residential areas to minimize public impact and for aesthetic reasons. Natural separation provided by trees or topography is also highly desirable. If natural separation is not available, the site should be suitable for screening with transplanted trees, fencing, or earthen dikes. Potential sites are more acceptable to the public if located in industrial, agricultural, or open space areas. The sites should be compatible with present and future land use planning (Ref. 97).

The available acre feet of the landfill site should be adequate for a minimum five-year operating life, based on the region it serves. Because of the equipment and building investment and in order to operate a landfill as a commercial facility, charging a reasonable tipping fee, the site should serve the community for as long a time period as possible.

Geological characteristics of a proposed site should be another major consideration. Sites that have thick deposits of relatively impermeable material, such as glacial till, lake silts, and clays; residual soils; and wind-blown silts (loess) offer more natural protection to ground-water pollution than deposits of sands and gravels (Ref. 98). Major aquifers below a landfill site can pose a problem unless the overburden soils are particularly thick and impermeable. Aquicludes directly under the site are generally quite desirable. Bedrock that is shallow, relatively sound, and intact acts as an aquiclude and is preferable to jointed, fractured, and weathered rock, which will permit rapid transmission of ground water and leachate.

A sound geotechnical assessment of the proposed landfill site would help determine:

- Cover Material. Sufficient material should be available at a nearby site. The material should be compactable, to provide a good seal without cracking excessively when dry. The material should be relatively clean and free of putrescible materials.

- Water Pollution. Ground water levels should be low, and the site should not produce either ground water or surface water pollution from percolation or runoff.

- Drainage. The site should allow fill limits to extend to the upper end of a drainage area, for control of surface water flow. The site should not require excessive grading or storm sewers for intercepting tributary surface runoff.

LANDFILL DESIGN

The design of the sanitary landfill should include topographical maps of the proposed site and the adjacent area, 1000 feet beyond the site. Information on the maps should include as a minimum (Refs. 96, 99, and 100):

- Proposed fill area.

- Roads on and off the site.

- Grades for proper drainage of each lift.

- Borrow areas and volume of material available.

- Typical cross section of a lift.

- Surface drainage and ground water.

- Fire protection facilities.

- Profiles of soil and bedrock.

- Fencing.

- Utilities above and below ground.

- Location of public and private water supplies, wells, springs, streams, swamps, and other bodies of water within 1 miles of the site property lines.

- Buildings within 1000 feet of the property lines (residential, commercial, and agricultural).

- Gas control devices.

- Direction of prevailing winds.

- Equipment shelter and employee facilities.

- Areas to be landfilled, including special waste areas, and limitations on types of waste that may be disposed.

- Sequence of filling.

- Entrance to facility.

- Leachate collection and treatment facility.

- Landscaping and completed use.

The scale of the maps should not be over 200 feet to an inch, and contour intervals should not be greater than 5 feet.

Capacity

If the daily volume of solid waste collected and the capacity of the landfill site are known, the useful life can then be estimated. The ratio of solid waste to cover material volume usually ranges between 4:1 and 3:1. The thickness of the cover and the cell configuration influence the ratio. If the cover material is not excavated from the fill site, this ratio may be compared, with the volume of compacted soil waste and the capacity of a site already determined. Figure 68 can be used, for example, to determine the acre feet per year used by a population of 10,000, for various compaction densities, and the daily rate of solid waste (Ref. 96).

The total tonnage to be disposed of at a proposed sanitary landfill can be obtained from the incoming waste to disposal sites. The daily volume of compacted solid waste can be determined from Figures 69 and 70 (Ref. 96). Figure 69 is applicable to large communities, and Figure 70 applies to small communities.

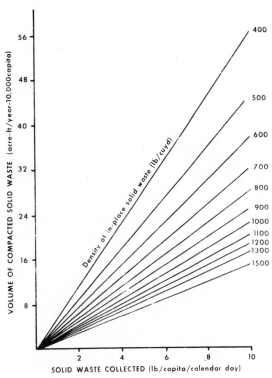

Figure 68. Yearly Volume of Compacted Solid Waste Generated by Community of 10,000 (Source: Ref. 96)

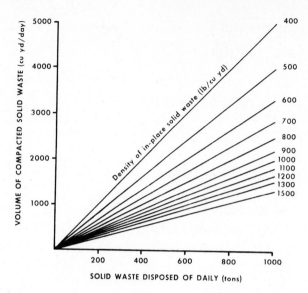

Figure 69. Daily Volume of Compacted Solid Waste
Generated by Large Communities
(Source: Ref. 96)

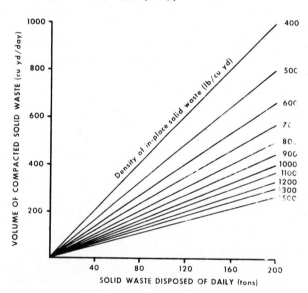

Figure 70. Daily Volume of Compacted Solid Waste
Generated by Small Communities
(Source: Ref. 96)

Field density or solid waste density at landfills is the weight of a unit of volume of solid waste in place. A well compacted landfill will have a field density of 800 to 1000 pounds per cubic yard. Landfill density is the weight of a unit of volume of in-place solid waste divided by the volume of solid waste and its cover material. Both methods include moisture at the time of the test, unless otherwise stated.

Land Preparation

The proposed plan should define how the site will be improved to provide proper operation. This plan may include land clearing and construction of

232

buildings, roads, and utilities. Trees and brush that hinder equipment movement must be removed. The ground should be free of stumps that interface with compaction or that obstruct vehicles.

Clearing of a large site should be accomplished in increments, to avoid erosion. Natural windbreaks and trees in strategic areas should be saved to improve appearance of the site.

Roads

Permanent roads should be provided from the public streets to the site. A large site may require extension of permanent roads to the vicinity of the working area. Minimum width of these roads should be 24 feet (two lanes) for two-way traffic. Uphill grades should be less than 7 percent, and downhill grades should be less than 10 percent. The initial cost of permanent roads is higher than for temporary roads, but the savings in equipment repair and maintenance more than offset these additional costs. Temporary roads a are used from the permanent road to the working face. These roads may be changing constantly and are constructed by compacting the natural soil. Binders may make such roads more serviceable, and calcium chloride may be used as a dust inhibitor, when more than 50 daily round trips are anticipated.

Scales

The selection of available scales is extensive. Depending on whether the site is small or large, anything from portable scales to an elaborate system with load cells, electronic relays, and printed output could be used.

Solid waste delivered to the site must be properly weighed to help regulate and control the landfill operation. The scales should be capable of weighing the largest vehicle using the landfills; 50-ton capacity is usually adequate. The platform should be long enough to weigh all axles simultaneously, to avoid errors and to speed operation. A platform 10 by 34 feet would be adequate for most vehicles without trailers. However, a 50-foot platform would be needed to handle all vehicles, including tractor-trailer transfer vehicles and dump trucks.

Generally, a scale accuracy of ±1.0 percent is acceptable. This accuracy would enable weighing most of the applied loads, between 8 and 14 tons, to the nearest tenth of a ton. Periodic inspection (quarterly) of the scales should include:

- Check of the indicated weight as a heavy load is moved from the front to the back of the scale.

- Observation of the dial action for irregularity or pauses in its motion.

- Use of test weights.

Buildings

A building to serve as an office and for employee facilities is needed for each landfill. For small operations, the building can also serve as a scale house. An equipment storage and repair shed and a separate scale house may be required as the site becomes larger.

Because the site will probably be used for ten years or less, the buildings constructed should be temporary and, preferably, movable. The design and location should consider gas movement and differential settling of the land from waste decomposition.

Utilities

All sites should have electric, water, and sanitary services. A remote site may require acceptable substitutes. Water should be available for drinking, fire fighting, dust control, and employee sanitation. At a large site, a sewer may be necessary where the effluent may be combined with leachate for further treatment. Telephone and radio communication may also be required.

Fencing

Two types of fencing are generally utilized: the peripheral fence and the litter fence. The former is used to control access, to keep out people and stray animals, to screen the site, and to define the property line. A litter fence controls blowing paper.

Trench operation requires less litter fencing, and blowing paper is a problem in an area operation. Depending on the type of operation, fencing from a 4-foot snow fence to a 6- to 8-foot litter fence could be used.

Surface and Water Control

It is essential to divert any surface water to pipes in gullies, ravines, and canyons. Open channels may be used to divert runoff from surrounding areas. Figure 71 shows a proposed technique for diverting runoff water and for transferring drainage around a landfill site.

The use of mechanical equipment, such as sump pumps, for water control is highly undesirable, because of frequent operating and maintenance requirements. Low-cost, portable drainage channels can be constructed by bolting together half sections of corrugated steel pipes. These half sections can be constructed to intercept and remove runoff water. Surface water running off stockpiled cover material should be ponded, to remove suspended solids, before entering water courses.

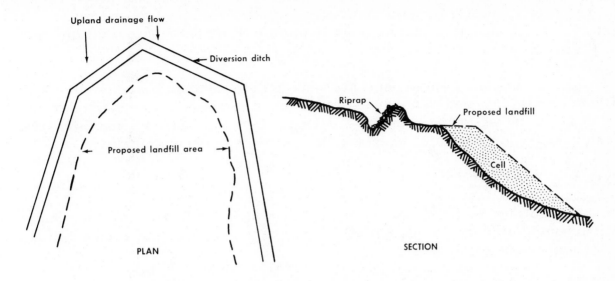

Upland drainage flow

Diversion ditch

Proposed landfill area

Riprap

Proposed landfill

Cell

PLAN

SECTION

Figure 71. Plan and Section View of Use of Diversion Ditch
to Transmit Upland Drainage Around Sanitary
Landfill (Source: Ref. 96)

Ground Water Protection

Interaction between ground water and deposited solid waste should be
avoided. It is unwise to assume that a leachate will be diluted in ground
water, because very little mixing occurs in an aquifer and the ground water
flow there is usually laminar.

In general, a 5-foot separation will remove enough readily decomposed
organics and coliform bacteria and make the liquid bacteriologically safe.
Mineral pollutants, on the other hand, can be conveyed long distances through
soil or rock formations. The important design considerations are:

- Current and projected use of area water resources

- Effect of leachate on ground water quality.

- Flow direction of ground water.

- Surface water and ground water interrelationship.

Some landfills have experienced ground water mounds (Ref. 96). These
mounds are rises in the piezometric level of an aquifer. The mounds are
above the surrounding ground water level and intersect with deposited waste.
One theory (Ref. 101) is that the permeability of the landfill boundary decreased due to excavation and rework or that additional water entered the
area through the cover material and the solid waste.

An impermeable liner, synthetic or from compacted clay, may be used
to control the movement of fluids. The use of such a barrier requires some

235

provision for removal of the contained fluid. This removal can be accomplished using gravity outlets, drains, and ditches, to intercept the ground water. If clay is used, the liner must be installed in a manner to ensure that the clay will always stay moist (to avoid cracking).

One means of detecting leachate, and to help establish whether the land-fill is creating ground water and surface water pollution, is to use a series of observation wells and sampling stations. Data on upstream and downstream water quality would enable proper assessment of any contamination.

Gas Movement Control

Controlling the movement of decomposition gases is an important design requirement of a sanitary landfill. Mainly carbon dioxide and methane (CH_4) evolve as waste decomposes.

Methane is highly explosive in concentrations of 5 to 15 percent. In a few cases, methane has moved from a landfill into a sewer line or into buildings in explosive concentrations. Methane kills nearby vegetation by excluding oxygen from the root zone. The effect of carbon dioxide is to increase the hardness of the water. Carbon dioxide is soluble in water and forms carbonic acid, which in turn dissolves minerals, particularly carbonates, in the waste. Water in wells located near a lanfdill becomes harder.

Two methods are proposed to control movement of gas in landfills. In the permeable method, gravel vents or gravel-filled trenches are employed. The trenches are deeper than the fill, to make sure that these trenches intercept all lateral gas flow. In another version of the permeable method, vent pipes are inserted through a relatively impermeable top cover. Collecting laterals can be tied to the vertical riser. It is important that pipe vents be avoided near buildings. Figures 72 and 73 illustrate the two permeable methods.

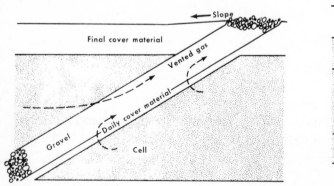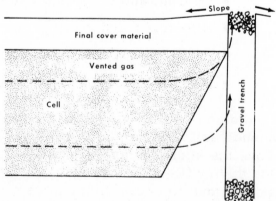

Figure 72. Gravel Vents or Gravel-Filled Trenches Used to Control Lateral Gas Movement in Sanitary Landfill (Source: Ref. 96)

236

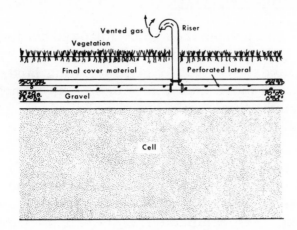

Figure 73.   Gases Vented from Sanitary Landfill via Pipes
Inserted Through Relatively Impermeable Top
Cover (Source: Ref. 96)

Impermeable materials can also serve to control gas movement through
soils.   An impermeable carrier can be used to contain the gas and vent it
through the top cover or simply to block the gas flow.   The most common
material used is compacted clay.   Clay can be used as a liner to block underground gas flow.   A clay layer of 18 to 48 inches is probably adequate.   Figure 74 shows the use of clay to form an impermeable layer.

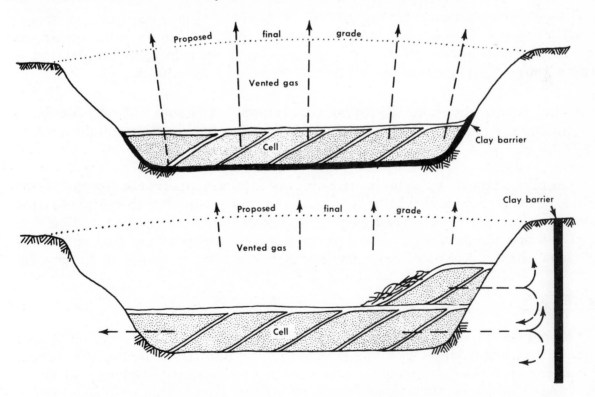

Figure 74.   Clay Used as Liner in Excavation or Installed as Curtain
Wall to Block Underground Gas Flow (Source: Ref. 96)

Cell Construction and Cover Material

The design of cells and procedures to be followed in disposing of waste must be properly outlined for a given site. The two basic methods are the trench and the area methods. On flat or gently rolling land, where the ground water is low and the soil is more than 6 feet deep, the trench method is recommended. The area method can be followed on most topographies, especially when large quantities of waste are disposed of at the site.

The building block of a landfill is the cell. All the solid waste is spread and compacted in layers within a confined area. At the end of each working day, the waste is covered completely with a thin, continuous layer of soil, which is also compacted. The compacted waste and soil cover are defined as a cell. A series of adjoining cells makes up a lift.

All dimensions are generally determined by the volume of the compacted waste and the density of the in-place waste. The field density of most compacted waste within the cell should be at least 800 pounds per yard. If the land and cover material are readily available, an 8-foot height for a cell might be appropriate. Heights up to 30 feet have been used in large operations. The key consideration for the designer is to keep cover material volume at a minimum, while disposing of as much waste as possible.

The cell should generally be square, with sides sloped as steep as operation will permit. The slope should be about 30 degrees, to keep the cover material volume at a minimum. This angle will also provide good compaction of waste, particularly when spread in layers about 2 feet thick.

A narrow working face should be maintained. It should be wide enough not to hinder the waste dumping operation from the trucks, but should not exceed 150 feet.

Cohesive soils (e. g. , glacial till or clay silt) are desirable for the trench method. The trenches should be aligned to be perpendicular to the prevailing winds, to minimize blowing of litter. The bottom of the trench should be slightly sloped for drainage. The trench can be as deep as the soil and water conditions allow. A typical cell construction (Ref. 93) is shown in Figure 75.

OPERATION

An operations and maintenance plan is basically the specifications to run the landfill in accordance with the ground rules established during the design and construction phases. Routine procedures and anticipated abnormal situations are defined. Instructions should be adequate to enable operation without discontinuity, even with a new crew. An up-to-date site status log should be maintained.

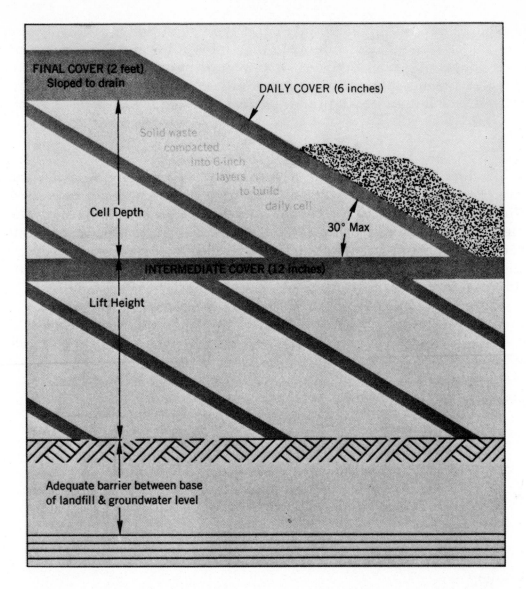

Figure 75. Typical Cell Construction (Source: Ref. 93)

Operation Hours

Hours of operation depend on the collection and delivery schedules. The landfill should usually operate eight to ten hours per day, six days per week. In large cities, the collection systems operate 24 hours a day. In these cases, the landfill can remain open longer, unless it is located in a residential area.

At the site entrance, the operational hours, information on acceptable waste, and the tipping fee should be posted. Fees are usually levied on a per-ton basis for large loads and on a flat-rate basis for small loads.

Weighing

Weighing of the incoming waste is necessary to determine the rate at which a landfill is being used. This information is also useful for estimating the amount of settling that will occur.

239

For landfills accepting 1000 tons per day of waste or more, two or more scales are usually required. Adequate provision for weighing and simplification of this operation would include a two-gate system, one-way exit barricades, signal lights, curbing, alarms, and automatic recording devices. Dirt, water, snow, and ice should not be allowed to accumulate on the deck of the scale.

Traffic Flow

The efficiency of the operation depends on the smooth flow and unloading of trucks. Proper barricades, rails, and signs should be posted to direct the traffic to the exact landfill location in use. Waste is delivered to the site in vehicles equipped with fast mechanical unloading apparatus and also in vehicles requiring manual unloading. To avoid major delays and tie-ups, it may be necessary to direct these two types of vehicles to separate locations or to minimize manual unloading of vehicles during busy hours.

Scavenging should not be permitted. Waste should be deposited at the toe of the working face, to enable better compaction. Sufficient warning markers and poles should be provided if dumping and unloading is at the top of a slope or into a trench. The unloading area should be level to facilitate trucks with a high center of gravity in the raised position.

Waste Handling

The waste to be disposed of at a landfill can be:

- Residential, commercial, and industrial
- Bulky
- Pathological
- Sludges
- Volatile
- Fly ash and residue
- Manure
- Radioactive
- Explosive

Waste in the first category is usually highly compactible. Cushioning and bridging can be minimized using this waste, and greater volume reduction achieved if it is spread in 2-foot-deep layers and is compacted by tracked, rubber-tired, or steel-wheeled vehicles. A proposed method of spreading and compaction is shown in Figure 76.

Bulky items do not lend themselves to sufficient volume reduction through normal landfill practices. These items should be crushed on solid ground and

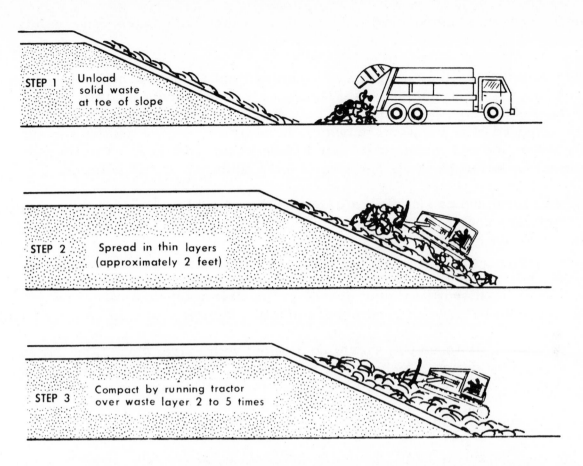

STEP 1 Unload solid waste at toe of slope

STEP 2 Spread in thin layers (approximately 2 feet)

STEP 3 Compact by running tractor over waste layer 2 to 5 times

Figure 76. Proposed Method of Spreading and Compaction
(Source: Ref. 96)

then pushed onto the working face, near the bottom of the cell. A greater volume advantage can be achieved by shredding bulky items prior to landfilling. Demolition waste can be used, at the site, for roads.

Institutional waste and dead animals should be spread immediately, compacted and enclosed by a layer of other waste, and then covered completely with cover material. Pathological waste should be buried immediately, under 1 foot of cover material. Large dead animals are placed in a pit and are covered with 2 feet of compacted soil. State laws have restrictions on the manner in which such waste should be handled.

Industrial process waste varies in physical, chemical, and biological characteristics. Sufficient knowledge of the waste should be obtained beforehand, to evaluate its influence on the environment and on the health and safety of landfill personnel. The effect of the waste on surface water or on ground water should be evaluated, and appropriate disposal measures should be used for each type of waste from a given process.

Dewatered sludges from water treatment and waste water plants can be disposed of at a landfill. In most cases, these sludges can be placed normally in the fill, but must be covered immediately. Other solid waste may be mixed if the moisture content of the sludges is very high.

Some waste, such as paints, dry cleaning fluids, and magnesium shavings, are volatile. Any flammable waste should be excluded from landfilling. If these wastes are not highly volatile, they can be mixed with other waste and can be disposed of in the usual manner. Incinerator residue and fly ash should usually be mixed with municipal waste and deposited in a cell. The fly ash should not be left uncovered, because of the possibility of air pollution.

Animal manure can be placed in the fill but should be covered immediately. If the moisture content is too high, the manure should first be mixed with dry waste before covering.

Any toxic materials and pesticide containers should normally be excluded from a landfill. Landfills do not accept radioactive waste. Any such waste detected should be isolated, and proper health authorities should be contacted.

Explosives, if accepted, should be handled with extreme caution. The exact location of this waste should be recorded on the final plan of the complete site. Appropriate security fencing and warning signs should be erected.

Cover Material Placement

The type of soil to be used, where to obtain it, and how to place it over the compacted waste should be specified in the operations plan. Cover materials are grouped as daily, intermediate, and final; the classification depends on the thickness of the soil used. A guide for each type of cover material is given in Table 58.

Table 58

APPLICATION OF COVER MATERIAL

Cover Material	Minimum Thickness	Exposure Time
Daily	6 in.	0 to 7 days
Intermediate	1 ft	7 days to 1 yr
Final	2 ft	>1 yr

All cover material should be well compacted. Coarse-grained soils can be compacted to 100 to 135 pounds per cubic foot; fine-grained soils can be compacted to 70 to 120 pounds per cubic foot.

Daily covering controls vector, litter, fire, and moisture. A minimum of 6 inches of cover is necessary. The cover is applied to the compacted waste at the end of each operating day. It should be compacted on the top and sides as the cell construction progresses, leaving only the working surface exposed until the end of the day.

The intermediate cover provides gas control and serves as a road base, in addition to the functions of a daily cover. The minimum compacted depth is 1 foot, with periodic grading to repair erosion damage.

Final cover must be capable of supporting vegetation growth, in addition to functioning as the intermediate cover. A minimum of 2 feet in compacted 6-inch layers should be used. Proper grading is important. Water should not be allowed to pond. Grades should not exceed 4 percent, and side slopes should be less than one in three. A final layer of topsoil may be used.

Maintenance

Proper maintenance during construction and after completion will distinguish a landfill from an open dump. Dust is a problem, especially in dry climates. It causes equipment damage, is a health hazard to site personnel, and can be a nuisance to nearby homes and businesses. Wetting roads is an effective means of controlling dust. Calcium chloride, waste oils, and water are used for this purpose. Litter control is another important parameter. Blowing litter can be minimized when a small-size working face is maintained. Fences also are helpful for controlling the spread of litter.

An equipment maintenance program should be established as a part of the operations plan. Periodic maintenance and the ability to correct most of the unscheduled maintenance actions will determine how successfully a landfill can be operated.

To overcome infestation of a landfill by rats, an anticoagulant poison should be used periodically. An insecticide should also be considered for use during the summer, to control flies. Compacted waste should be covered immediately. Birds can also be discouraged by proper compaction of daily waste and cover material.

Weather

Operations plans should specify how the landfill can be operated during inclement weather, as well as in good weather. Two major problems due to weather are frozen ground and excess water. In freezing weather, sufficient cover material should be kept on hand and kept dry, to prevent freezing. Clay soils may be unsuitable for cover material during freezing weather.

Rain causes operational problems: roads can become unusable. Gravel, crushed stone, and demolition waste should be applied to road surfaces. Trucks picking up mud on the site should be cleaned before leaving the site. This effort is justifiable for both sanitary and aesthetic reasons.

Fires

Fires occur at the site as a result of careless handling of open flames or hot waste. Daily cover should keep a fire contained in a cell.

If the fire is too large, waste in the burning area must be spread so water can be sprayed on it. Equipment operators should carry fire extinguishers on their machines. Adequate fire-fighting procedures should be stablished, and site personnel should be made thoroughly familiar with these procedures.

Salvage and Scavenging

Unless salvaging and recovery measures were planned and incorporated into the design, none should be permitted. It is estimated that salvaging at the site is expensive (Ref. 96). Scavenging should be strictly prohibited at all times.

EQUIPMENT SELECTION

Sanitary landfill machines can be categorized as machinery used to:

- Directly handle waste
- Handle cover material
- Perform support functions

The primary objective of a sanitary landfill is to accomplish the practical and safe disposal of solid waste. Handling of waste resembles handling of earth. Waste is less dense, more compactible, and more heterogeneous, requiring less energy to spread a given volume than is required for spreading of soil. Compaction is carried out by the compressive forces.

Some of the problems peculiar to machines spreading and compacting solid waste are: overheating due to clogged radiators, broken fuel and hydraulic lines, punctured tires, and damage to drives.

The excavating, hauling, spreading, and compacting of cover material is analogous to an earth moving operation in highway construction. However, rigorous control of moisture content to achieve maximum soil density is not usually required in landfill operations. Depending on the specific type of soil available for cover material, a crawler equipped with a rock ripper may be required. Soils may vary from glacial till to shale, cemented siltstone, and granite rocks. Seasonal variations may also require a rock ripper to remove the frost layer, during freezing weather.

Support equipment may be required at a landfill to perform road construction, maintenance, dust control, and fire control and to provide assistance in unloading. Roads must be provided to reach the working face in all weather conditions. Dust control may require a water wagon and sprinkler. Mobile fire fighting equipment should be available.

Equipment used for a landfill operation should include:

- Crawler dozer

- Crawler loader
- Rubber-tired dozer
- Rubber-tired loader
- Compactor
- Scraper
- Dragline
- Special purpose equipment (e.g., road grader and water wagon)

All of these machines, except the special purpose ones, are illustrated in Figure 77.

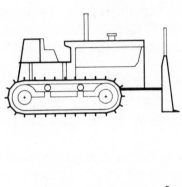

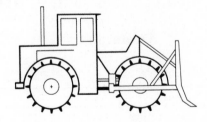

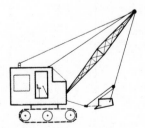

Figure 77.   Types of Landfill Equipment (Source: Ref. 93)

Crawler machines have good flotation and traction capabilities. A crawler is excellent for excavation, but it can operate only up to 8 mph. Unlike a dozer, a crawler loader can lift materials off the ground. It can carry soil as far as 300 feet.

Rubber-tired machines, both dozers and loaders, are faster than crawlers (up to 29 mph). However, these machines are not as good for excavation; they are effective for compaction and are mobile. These machines can be used for carrying cover material economically over distances of up to 600 feet.

Compactors are equipped with large trash blades. The power train is similar to that of rubber-tired machines. Their major asset is their steel wheels, which are effective in crushing and compaction. A compactor works well for spreading and compacting on flat or level surfaces and on moderate slopes. Its maximum speed is 23 mph. A compactor is not suited for clay soil, because the soil becomes lodged between the load concentrators.

Scrapers are available as towed models or as self-propelled models. They can haul cover material economically over large distances. The prime function of scrapers is to excavate, haul, and spread cover material.

Large excavations can be accomplished economically using a drag line. A drag line can dig moderately hard soil and deposit it away from the excavation; it can also spread cover material over compacted surfaces and is especially valuable in wet lands. A drag line is common at large landfills, where the trench method is used. As a rule of thumb, the boom length of a drag line should be twice the width of a trench. Characteristics of these machines are listed in Table 59 (Ref. 96).

Some of the recommended and optional accessories for commonly used landfill equipment are tabulated in Table 60 (Ref. 96).

Table 59

PERFORMANCE CHARACTERISTICS OF LANDFILL EQUIPMENT

Equipment	Solid Waste		Cover Material			
	Spreading	Compacting	Excavating	Spreading	Compacting	Hauling
Crawler dozer	E	G	E	E	G	NA
Crawler loader	G	G	E	G	G	NA
Rubber-tired dozer	E	G	F	G	G	NA
Rubber-tired loader	G	G	F	G	G	NA
Landfill compactor	E	E	P	G	E	NA
Scraper	NA	NA	G	E	NA	E
Dragline	NA	NA	E	F	NA	NA

E = Excellent   G = Good   F = Fair   P = Poor   NA = Not applicable

*Basis of evaluation: Easily workable soil and cover material haul distance greater than 1000 ft.

Table 60

RECOMMENDED AND OPTIONAL ACCESSORIES FOR LANDFILL EQUIPMENT

Accessory	Dozers (Crawler, Rubber-Tired)		Loaders (Crawler, Rubber-Tired)		Landfill Compactor
Dozer blade	O	O	--	--	O
U-blade	O	O	--	--	O
Landfill blade	R	R	O	O	R
Hydraulic controls	R	R	R	R	R
Rippers	O	--	O	--	--
Engine screens	R	R	R	R	R
Radiator guards, hinged	R	R	R	R	R
Cab or helmet air conditioning	O	O	O	O	O
Ballast weights	O	O	R	R	R
Multiple-purpose bucket	--	--	R	R	--
General-purpose bucket	--	--	O	O	--
Reversible fan	R	R	R	R	R
Steel-guarded tires	--	R	--	R	--
Life arm extensions	--	--	O	O	--
Cleaner bars	--	--	--	--	R
Roll bars	R	R	R	R	R
Backing warning system	R	R	R	R	R

O = Optional   R = Recommended

The functions and evaluation of equipment performance must be compatible with the size of a landfill. One machine is not capable of performing all the required functions effectively at one site. Therefore, it is particularly difficult to select a single type of machine for a site. A machine must be capable of spreading and compacting waste and cover material, and its dependability must be high.

For municipalities with an operation of less than 10 tons per day, the cost of a dozer or loader may be prohibitive. If contract work can be used for excavation and stockpiling of cover material, and if a large fill area is available, then a farm tractor equipped with a blade or bucket may be adequate.

Large landfill sites are well suited for multiple machine operation, scrapers and compactors become more economical to use, and dependability requirements are not as severe. In case of one machine breakdown, other equipment can serve on an interim basis. Replacement machines should be available through a lease, a contract, or a borrowing arrangement. Table 61 shows the type of equipment that is compatible with the site capacity.

The equipment selected for a landfill must also be economically attractive for a given site application. A crawler machine weighing 29,000 pounds

Table 61

LANDFILL EQUIPMENT NEEDS*

Solid Waste Handled (tons/8 hr)	Crawler Loader		Crawler Dozer		Rubber-Tired Loader	
	Flywheel (hp)	Weight** (lb)	Flywheel (hp)	Weight** (lb)	Flywheel (hp)	Weight** (lb)
0-20	<70	<20,000	<80	<15,000	<100	<20,000
20-50	70 to 100	20,000 to 25,000	80 to 110	15,000 to 20,000	100 to 120	20,000 to 22,500
50-130	100 to 130	25,000 to 32,500	110 to 130	20,000 to 25,000	120 to 150	22,500 to 27,500
130-250	150 to 190	32,500 to 45,000	150 to 180	30,000 to 35,000	150 to 190	27,500 to 35,000
250-500	†	†	250 to 280	47,500 to 52,000	†	†
500-plus†						

*Compiled from assorted promotional material from equipment manufacturers and based on the ability of one machine in a stated class to spread, compact, and cover within 300 ft of the working face.
**Basic weight without bucket, blade, or other accessories
†Combination of machines

without accessories costs about $29,000. With accessories, the base price goes to $32,000. A new drag line costs between $75,000 and $110,000, depending on boom and cable lenghts. The expected life of landfill equipment is five years or 10,000 operating hours. In some applications, a used machine might be the most desirable economic alternative.

Operating costs, excluding labor, run about $3 per hour. These costs include fuel, filters, lubricants, tires, oil, and any material required to provide routine maintenance. Some dealers offer lease agreements and a maintenance contract, which may be very attractive -- especially in industrialized urban areas.

Actual operating and maintenance expenses should be determined during site operation by use of a cost accounting system. The range of equipment costs for the more common machines is given in Table 62.

ECOLOGICAL IMPACT

Sanitary landfilling is regarded as one of the best methods available for solid waste disposal. It is generally neat, safe, and inexpensive. Despite the apparent simplicity, there are certain problems that can only be avoided by proper landfill design, construction, and operation. Protecting ground water from leachate is of primary importance. Finely suspended solid matter, microbial waste products, and dissolved materials are produced when ground water or surface water is infiltrated through the waste.

Table 62

MACHINE CAPITAL COSTS

Machine Type	Equipped Machine				
	Flywheel (hp)	Weight (lb)	Approximate Weight* (lb)	Approximate Cost** ($)	Comment
Crawler dozer	<80	<15,000	19,000	21,000	Landfill blade
	110-130	20,000-25,000	32,000	38,000	Landfill blade
	250-280	47,500-52,000	67,000	70,000	Landfill blade
Crawler loader	<70	<20,000	23,000	21,000	GPB-1 yd^3
	100-130	25,000-32,500	31,000	30,000	GPB-2 yd^3
	100-130	25,000-32,500	32,000	32,000	MPB-1-3/4 yd^3
	150-190	32,500-45,000	45,000	46,000	GPB-3 yd^3
	150-190	32,500-45,000	47,000	49,000	MPB-2-1/2 yd^3
Rubber-tired loader	<100	<20,000	17,000	21,000	GPB-1-3/4 yd^3
	<100	<20,000	18,000	23,000	MPB-1-1/2 yd^3
	120-150	22,500-27,500	23,000	33,000	GPB-4 yd^3
	120-150	22,500-27,500	26,000	36,000	MPB-2-1/4 yd^3

GPB = General-purpose bucket   MPB = Multiple-purpose bucket

*Basic machine plus engine sidescreens, radiator guards, reversible fan, roll bar, and a landfill blade, general-purpose bucket, or multiple-purpose bucket, as noted.

**June 1970

Decomposition of solid waste produces various gases. These gases include methane, carbon dioxide, nitrogen, oxygen, and hydrogen sulfide. Some of these gases can have serious effects on the environment.

LAND POLLUTION

Decompostion of landfills depends on permeability of cover material, depth of burial, rainfall, moisture content, putrescibility of the waste, and the degree of compaction. Solid waste is composed primarily of carbohydrates, fats, and proteins, which decompose to form humus and the gaseous end products shown in Figure 78 (Ref. 93).

Anaerobic decomposition occurs at elevated temperatures, usually 100°F to 120°F and sometimes up to 160°F. Over 90 percent of the gas produced in large landfills is carbon dioxide and methane. Both of these gases were found in one study (Ref. 102) in concentrations up to 40 percent, at distances up to 400 feet from the edge of the fill. Gas production is noted to decrease as moisture decreases. When sea water is allowed to enter a landfill, hydrogen sulfide generation increases.

Waste decomposition is a slow process and may require several generations. One investigation (Ref. 103) estimates total decomposition times, in California, of more than 950 years.

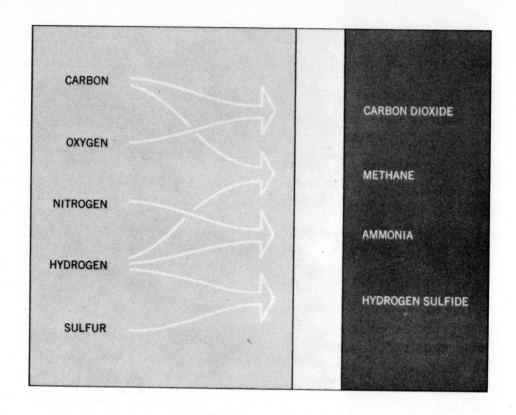

Figure 78.   Gaseous End Products (Source: Ref. 93)

The chief constituents of waste are carbon, oxygen, and hydrogen, with some nitrogen and sulfur. Microbiological fermentation changes these constituents to produce four gases: mainly carbon dioxide, and methane, with some ammonia and hydrogen sulfide. Theoretically, a pound of waste can produce 2.7 cubic feet of carbon dioxide and 3.9 cubic feet of methane (Ref. 103).

The gas movement mechanism is primarily convection, and the basic rate equation is Darcy's Law (Ref. 103). This law assumes that large pressure differences will not occur and that the gases may be treated as incompressible fluids. Based on these assumptions, the results of both the theoretical and actual calculations for carbon dioxide movement are:

Direction	Time (days)	Observed Diffusion (lb/acre/yr)	Theoretical Diffusion (lb/acre/yr)	Density Convection (lb/acre/yr)
Vertically downward	550	24,400	19,100	12,800
Horizontally	550	22,700*	25,400*	--
Upward through cover	718	1.9×10^5	9.2×10^5	--
	809	3.0×10^5	9.2×10^5	--

*Acre of vertical interface.

250

It is generally believed that in a landfill, anaerobic conditions prevail approximately one month following deposition of the waste. Air in the fill is consumed by aerobic decomposition. Following are typical reactions of carbohydrates and stearic acid:

$$C_6H_{12}O_6 + \rightarrow 6CO_2 + 6H_2O$$

$$C_{18}H_{36}O_2 + 26O_2 \rightarrow 18CO_2 + 18H_2O$$

When the oxygen is used, the anaerobic decomposition results in an increased quantity of gas, which creates increased pressure and higher rates of gas diffusion:

$$C_6H_{12}O_6 \rightarrow 3CO_2 + CH_4$$

$$C_{18}H_{36}O_2 + 8H_2O \rightarrow 5CO_2 + CH_4$$

Aerobic action is produced by microbes that thrive on free oxygen; anaerobic action is produced by microbes thriving in an oxygen-deficient environment. The former produces carbon dioxide and the latter produces methane. Carbon dioxide prouuction is highest during the first months of landfill operation. As carbon dioxide production declines, methane rises; then both level off. A graphic representation of this phenomenon for carbon dioxide, methane, and nitrogen is shown in Figure 79.

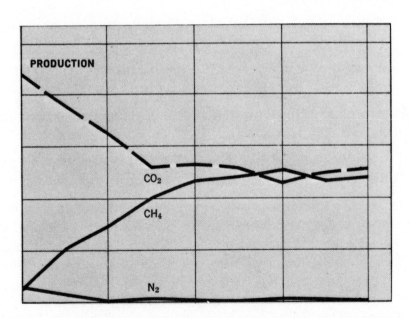

Figure 79. Gas Production and Composition from Experimental Landfill (Source: Ref. 93)

The results of a study conducted over a 2-1/2-year period define the time phase evolution of various gases in a landfill (Ref. 104). The significant conclusions are:

251

- Hydrogen, oxygen, nitrogen, carbon dioxide, and methane constitute the major gases. Hydrogen sulfide appears in trace form during the first two years. No ammonia appears, unless high-pH conditions exist.

- Hydrogen in great quantities appears during the first three weeks. The masimum amount is 20.6 percent, during the first and second weeks; 'the maximum is down to 1 percent by the end of the first month and is 0.1 percent after one year. Molecular hydrogen is present, indicative of a reducing system.

- Carbon dioxide reaches 85 percent after two weeks, but is down to 40 percent after two months. It stabilizes at this lower level.

- Methane appears in trace amounts during the second week. In one study, it reached 2.7 percent after two months, 6 percent in six months, 13 percent after one year, and 20 percent after two years.

- Stratification of gases occurred. Carbon dioxide, methane and hydrogen were higher in the lower portion, and nitrogen was higher in the upper portion.

- Composition of gas was dependent on compaction densities. Higher compaction densities yielded more gas per unit of volume.

- The most pronounced changes in the organic materials occurred within the first 60 days. At the end of two years, however, landfills were still far from stabilized.

It has been indicated (Ref. 103) that gas production quantities vary directly with temperature, moisture, and aeration. Some of the test results show that:

- Dry refuse and saturated refuse produce 0.035 and 0.21 cubic feet of gas per pound of refuse, respectively, on a dry basis.

- Gas movement was 0.22 to 0.8 foot per day, vertically, and 0.24 to 1.4 foot per day, horizontally.

- Carbon dioxide increases hardness and bicarbonates in ground water. Depending on the pH, the water may become corrosive.

Carbon dioxide movement velocities through landfills have been investigated (Ref. 105). The test cell contained 4290 tons of waste, placed in through 6- to 7-foot layers in an abandoned gravel pit. The mean gross vertical and horizontal velocities were 0.22 foot per day and 0.24 foot per day, respectively.

The net carbon dioxide passing through a certain area was claculated. The amount of carbon dioxide entering the soil from solid waste was 18,700 pounds per year (vertically) and 3300 pounds per year (horizontally). These quantities are equivalent to 5.1 pounds per year for each ton of waste.

Based on the field test work in California (Ref. 106), some procedures were developed to control gas migration. Placing wells 300 to 400 feet apart, with a withdrawal rate of 200 cfm per well, was effective in providing control of gas migration. It was found that the withdrawal rate is a function of well spacing and that well depth is not a significant factor.

In the test area, two-foot-diameter ventilation wells were drilled in natural soil and held a six-inch-diameter perforated PVC pipe packed in gravel. The wells were connected to a vitrified clay pipe header that ran 3/4 mile to a blower. Gases from the blower were fed to a fume incinerator. Gases, especially methane, are highly explosive in concentrations from 5 to 15 percent. Therefore, adequate provisions should be made to vent a landfill properly.

In addition to the timely and proper control of gas migration from a landfill site, erosion control, vector control, and prevention of littering and blowing of waste are important considerations for controlling land pollution.

The type of soil selected for cover material should be able to withstand local weather conditions. After compaction, the cover material should not crack or become dusty. Both the wind erosion and the surface water erosion should be minimized. Regardless of whether the trench method or the area method is used, appropriate grading and sloping should be used to avoid ponding of surface water. Area roads should also provide a minimum of erosion damage.

Scattering of solid waste due to winds can be reduced by providing daily coverage and by the use of wind breaks and fencing. Initially this type of nuisance has more of an aesthetic impact than a practical impact. However, if permitted to continue unchecked, this problem could lead to a degradation of vector control and gradually to deterioration of site sanitary conditions.

Land pollution, except for the underground gas evolution, is more visible than water pollution. Therefore, it is highly important that adequate practices be incorporated into the daily operations plan, to enable proper checks on elements causing land pollution.

WATER POLLUTION

Probably the most significant problem is pollution of ground water by leachate. Leachate develops at sanitary landfills by ground water or surface water filtering through the solid waste. The contaminants in a leachate are a function of the composition of the waste material and the combined physical, chemical, and biological activities occurring in the fill. The significant pollutants in leachate are reported to be BOD, COD, chloride, and nitrate. Bacteriological contaminants usually are filtered from the leachate after several feet of travel through most soils. Suspended solids, however, travel greater distances, creating ground water pollution.

253

The chemical and biological characteristics of leachate in two studies are presented in Table 63. The data exhibit a significant range of values (e.g., the pH variation is between 6.0 and 6.5 in Study A and between 3.7 and 8.5 in Study B). Such differences further indicate the variability of leachate composition with time and between sites.

Figure 80 is a schematic diagram showing how leachate forms. As water passes through the cover material and down through the compacted waste, it picks up solids and dissolves some into the solution. With this burden, the liquid is a potential contaminant. As it percolates through the strata of the

Table 63

LEACHATE COMPOSITION

(Mg/ℓ/ft³ of compacted, representative, solid waste)

Component	Study A		Study B	
	Low	High	Low	High
pH	6.0	6.5	3.7	8.5
Hardness, CaCO₃	890	7,600	200	550
Alkalinity, CaCO₃	730	9,500	--	--
Ca	240	2,330	--	--
Mg	64	410	--	--
Na	85	1,700	127	3,800
K	28	1,700	--	--
Fe (total)	6.5	220	0.12	1,640
Ferrous iron	8.7	8.7	--	--
Chloride	96	2,350	47	2,340
Sulfate	84	730	20	375
Phosphate	0.3	29	2.0	130
Organic-N	2.4	465	8.0	482
NH₄-N	0.22	480	2.1	177
BOD	21,700	30,300	--	--
COD	--	--	809	50,715
Zn	--	--	0.03	129
Ni	--	--	0.15	0.81
Suspended solids	--	--	13	26,500

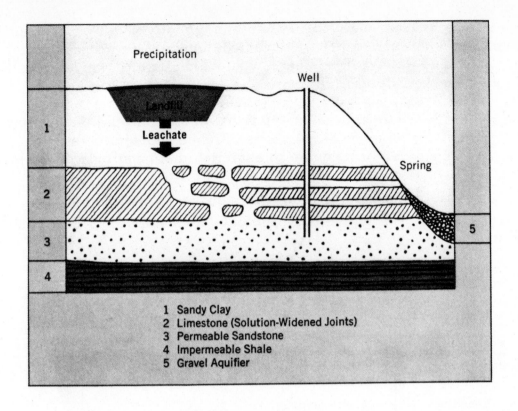

1 Sandy Clay
2 Limestone (Solution-Widened Joints)
3 Permeable Sandstone
4 Impermeable Shale
5 Gravel Aquifer

Figure 80. Leachate and Infiltration Movement
(Source: Ref. 93)

permeable material containing ground water, the leachate can become mixed
with the ground water, causing a deterioration of its quality. An impermeable
barrier, possibly clay or a synthetic liner between the ground water and the
source of the leachate will protect the ground water supply.

The impact of leachate on the ground water quality is shown in Table 64
which lists three wells and the quality of samples drawn from each. The first

Table 64

GROUND WATER QUALITY

Ground Water Characteristic	Background (mg/l)	Fill (mg/l)	Monitor Well (mg/l)
Total dissolved solids	636	6712	1506
pH	7.2	6.7	7.3
COD	20	1863	71
Total hardness	570	4960	820
Sodium	30	806	316
Chloride	18	1710	248

Source: Ref. 101.

well was upstream from the landfill, one was directly beneath the landfill, and the third was downstream. As expected, the deteriorating effect of leachate on ground water can be seen by comparing the background well data with the monitor well data.

The time of leachate appearance may be offset from the initial time of the waste deposit by as much as 20 years. Hence short-term studies on leachate may be entirely inadequate to establish the magnitude of the problem (Ref. 106).

Table 65 contains information pertaining to the relative distances of contamination and their respective time periods, for various types of landfill pollutants.

Table 65

RELATIVE CONTAMINATION DISTANCES

Nature of Pollution	Pollutant	Observed Distance of Travel	Time of Travel
Industrial wastes	Tar residues Picric acid	197 ft several mi	
Garbage leachings	Misc leachings	1476 ft	
Industrial wastes	Picric acid	3 mi	4-6 yr
Industrial wastes in cooling ponds	Mn, Fe, hardness	2000 ft	
Garbage reduction plant	Ca, Mg, CO_2	500 ft	
Chemical wastes	Misc chemicals	3-5 mi	
Industrial wastes	Chromate Phenol Phenol	1000 ft 1800 ft 150 ft	3 yr
Salt	Chlorides	200 ft	24 hr
Gasoline	Gasoline	2 mi	
Weed killer waste	Chemical	20 mi	6 mo
Radioactive rubidium chloride	Radioactivity	--	5 days

Preliminary laboratory scale results (Ref. 107) show that biological treatment of leachate is efficient for removing a substantial amount of the organic pollutants. Anaerobic treatment of leachate concentrate was the most effective, reducing more than 90 percent of BOD for detention times greater than ten days at temperatures of 23°C to 30°C. Aerobic refinement of anaerobic effluent produced BOD values comparable to surface water discharge. At loadings of less than 0.03 pound BOD per cubic foot per day, aerobic treatment also proved promising, removing in excess of 90 percent BOD. Foaming and poor solid-liquid separation, however, occurred in the bench models.

These studies indicated that mixing leachate (5% by volume, COD 10,000 mg/ℓ) with domestic waste water in an extended aeration, activated sludge plant did not seriously impair effluent quality. Waste water plant effluent degraded rapidly when the added leachate increased more than 5 percent.

Tests were conducted to treat leachate chemically. Several chemicals (e.g., lime, Na_2S, Cl_2, aluminum, $FeCl_3$, and $KMnO_4$) were used. The results indicated that chemicals alone will not provide efficient removal of COD; however, effective removal of multivalent cations and color may be expected with lime, the oxidents ($KMnO_4$), and the coagulants ($FeCl_3$ and aluminum). Undesirable aspects of chemical treatment include high dose requirements and large amounts of sludge produced. Chemical treatment seems to be particularly useful for treating biological process effluent. Lime precipitation was effective for removing color and iron from both anaerobic and anaerobic-aerobic polished effluent and did provide some incremental removal of COD.

After reviewing the literature on landfills, one point becomes clear almost immediately: there are very few case histories of serious or even troublesome contamination of ground water directly attributable to leachate. Considering the number of landfills and dumps and the amount and variety of waste generated, this situation is remarkable. Most soils are able to attenuate the leachates. The soil provides for microbial degradation of the organics. The inorganics are absorbed to the soil surface, and the extremely low velocity of underground water provides the necessary time for stabilization and confines most of the degradation to the immediate vicinity of the landfills.

PUBLIC RELATIONS

The possible disturbance of the ecological balance in the general region of a site is being seriously studied on many projects. This possible imbalance becomes more significant as marginal land (e.g., wet lands, steep hillsides, and remote virgin forests) must be used. Even minor environmental changes can have significant impact if the delicately balanced ecological systems are upset by landfill operations.

More obvious pollution problems, which are nuisances rather than true pollution concerns, include noise from earth moving operations, hauling equipment, and compactors. Odors from anaerobic decomposition could also affect nearby residences.

Visual aesthetics have often been overlooked in the past. This area alone has caused much of the general public opposition to opening new sanitary landfills near residential areas. Proper operating procedures are vital to overcome such opposition. Also, the use of the completed landfill for recreational purposes must be properly emphasized.

To enhance public cooperation for either adding a new sanitary landfill or for ameliorating an existing one, proper planning and execution of a community education program is a vital prerequisite. These sites represent an extremely emotional issue, particularly to those in the immediate vicinity. The program should begin early in the planning stages and should continue after operations begin. Public information should stress that the waste is covered daily, access is restricted, insects and rodents are adequately controlled, water quality is unaffected, and open burning is prohibited. Examples of properly operated landfills near residential areas should be pointed to. Help provided by community organizations can do much to increase public support.

An elected or appointed official who firmly believes that a waste disposal system is acceptable and is needed can be a very important factor in winning public support.

A comprehensive management plan should be developed, preferably on a regional basis. Land suitable for sanitary landfilling is scarce within a large city. Smaller communities nearby may be able to provide land and dispose of their own waste in a very economical manner.

A key aspect of public relations is a procedure for handling citizen complaints. Deficiencies in operating methods or employee courtesy should be investigated and acted upon promptly. If this practice is followed, citizens are apt to be less hostile toward the operation, and employees will become more conscientious.

A sanitary landfill is a positive and relatively inexpensive step for a community to provide a safe and attractive environment. Proper balance among design, operation, and management can result in an acceptable method of solid waste disposal by a community.

LANDFILL ECONOMICS

Project costs for a sanitary landfill include:

- Land.
- Capital investment, excluding land.
- Operation and maintenance.
- Hauling and transportation, including a transfer station (if needed).

These latter costs may be omitted, becuase they are not directly associated with the disposal of solid waste.

ASSUMPTIONS

To develop total costs of a landfill operation, the following groundrules were observed:

- The site will operate six days per week and will remain open to receive waste from 8 a.m. to 5 p.m. daily.

- Site selection, design, and operation will comply with the 1973 Environmental Protection Agency guidelines.

- The site will be capable of disposal of 1000 tons per day. An alternate disposal rate of 500 tons per day will also be examined.

- The completed site will serve as a recreational area, supporting no buildings.

- Waste will be compacted to a density of 800 to 1000 pounds per cubic yard. Generally shredded or baled waste is not included.

- No waste from an industrial process, especially chemical plants, will be disposed of at the site. Only residential, commercial, and normal industrial waste will be disposed.

- No reclamation, recovery, or salvage operation will operate at the site.

- Land costs and hauling and transfer station costs will not be considered.

- The site operating life will be approximately ten years.

- Amortization and debt service charges will not be considered.

TOTAL COST ESTIMATES

Preliminary estimates showing total costs based on 1972 dollars, for a 1000-ton-per-day sanitary landfill site, exclusive of land, are shown in Table 66.

Table 66

PRELIMINARY COST BREAKDOWN FOR SANITARY LANDFILL WITH TEN-YEAR LIFE

Cost	1000 Tons/Day	500 Tons/Day
Capital costs		
Administrative	$ 317,600	$ 257,600
A/E and consultants	191,200	162,700
Construction	3,699,000	1,849,500
Equipment (including replacement every 3 yr)	1,833,000	1,200,000
Total	$6,040,800	$3,469,800
Annual operating costs		
Administrative	$ 120,000	$ 95,000
Operation	230,800	142,900
Maintenance	55,000	36,000
Consultants	15,000	14,000
Total	$ 420,800	$ 287,900
Disposal cost/ton*	3.31	4.10

*Disposal costs quoted here do not include interest charges.

259

Management costs include expenses for administrative and engineering consultants: costs of surveys for proper site selection and specifications, site development supervision, and site monitoring during the operation phase.

Construction costs include such items as site preparation, soil borings, ground water wells, buildings, drainage, road work, gas monitoring system, leachate drain, and fencing.

Equipment purchase costs are a part of the initial capital investment. If the operation of a sanitary landfill is contracted out, then the equipment costs would show up in the operating cost category (Ref. 108). Other options, such as leasing equipment from a dealer or leasing with intent to buy can be considered instead of an outright purchase. From an economic standpoint, leasing with intent to buy might be a more attractive option. However, in this preliminary estimating, equipment costs are considered as initial investment costs.

Operation costs are associated with the labor and material costs to operate the landfill on a daily basis (disposal of incoming waste and proper maintenance of the site and equipment in a clean operable condition).

Total costs are simply an addition of the capital investment and management costs equally divided over a ten-year life of the landfill, plus the annual operation and maintenance costs. It is assumed that while there may be periodic variation in the quantity of incoming waste on a daily basis, it does, for the most part, stabilize at 1000 tons per day, or 500 tons per day for the alternate cost estimate.

The cost per ton is given at the bottom of Table 68. These costs are $3.31 per ton for a 1000-ton-per-day facility and $4.10 per ton for a 500-ton-per-day facility. The economics of operating a larger facility show up in terms of 79 cents per ton. This amount should not be interpreted, however, as a scaling factor when smaller or larger facilities are considered.

Table 66 contains the details in each of the two major cost categories (capital investment and operation and maintenance). Management costs are included in each category and consist of the costs for administrative and architectural engineers and consultants. Kansas City and Omaha-Council Bluffs sites (Refs. 108 and 109) were used as guides and were modified to generate cost estimates in all categories.

Table 66 also shows that management cost components do not vary significantly between a 500-ton-per-day and 1000-ton-per-day facility, because the additional costs for these services, beyond a minimum site, are incurred on a few cost elements within a subcategory, and most of the major elements remain unchanged.

The construction of the site and equipment cost categories are influenced directly by the size of the operation. This situation is reflected more in the construction category. The equipment category does not show a linear cost increase, because instead of doubling the number of machines as the site capacity is doubled, larger capacity machines have been selected. Equipment costs includes replacement every 10,000 operating hours (3 yr) to obtain the total cost.

Under operation and maintenance costs, operating costs include personnel expenses as well as equipment operation costs. Personnel expenses cover equipment operators and service functions (e.g., scale man and foreman). Maintenance costs are assumed to be approximately 10 percent of the initial equipment costs. To operate the site as a commercial facility, equipment availability must be high. This availability can be achieved either through redundance or through effective maintainability. If a redundancy option is exercised, maintenance costs can be reduced; however, the equipment cost category would correspondingly increase to a new number. In this analysis, "better maintenance" rather than "more equipment" was selected.

CLOSING DUMPS

A dump is a waste disposal site that has been used without regard to pollution control and aesthetic impact. Dumps are generally referred to as "open" because no material covers the deposited waste. Very little supervision has been involved in operating and maintaining the dumps. Practically all types of topography have been used for dumping, and the waste deposited ranges from abandoned tires and automobiles to demolition waste, agricultural waste, and municipal waste.

An open dump also serves as a burning dump. The fire may be due to spontaneous combustion, smoldering waste, or may be intentional. It serves the purpose of some volume reduction, at the expense of air pollution.

A dump is a major health hazard. In addition to its adverse impact on air, it pollutes surface and ground water. Insects, rodents, and birds attracted to the dumps in search of food can carry disease and biological and chemical contaminants to man and his domestic animals. Uncontrolled scavenging at the dumps also presents a major health hazard from pathogenic organisms, toxic chemicals, and fires.

The aesthetic degradation of the landscape from an open dump is also very real. The fear in the minds of people regarding any waste processing and disposal activity near their neighborhood is due to their association of such activity with a dump. Approximately 90 percent of solid waste in 1970-1971 was disposed of by this means (Ref. 110). Besides the aesthetic impact, the economic costs of an open dump should be fully appreciated. A dump will probably prove to be more expensive than a sanitary landfill after all factors are considered.

The effects of smoke on laundry and paint and the health hazard from water pollution are constant sources of concern and expense to the community. A sanitary landfill, on the otherhand, is more acceptable during operation and becomes a useful site when complete.

In developing a plan to close a dump, governmental agencies, industry, citizens, and environmental effects -- and their interactions -- should be considered; an ecceptable substitute should be planned, funding requirements should be outlined, and anticipated use of the closed site should be identified. Conversion of a dump to a sanitary landfill operation should also be considered.

COMMUNITY INVOLVEMENT

Appropriate information must be disseminated to the public and to industry. The following types of information should be highlighted:

- Reasons for closing
- Plans to accomplish the task
- Alternate disposal means
- Costs

Closing a dump may require several months. Extermination of rodents and rats normally requires about two weeks. Compacting and covering may take more than two months. An alternate waste depositing site must be made available first, before further dumping can be stopped, to proceed with the closing program. Complete information about the new facility should be displayed at the entrance. Disposal procedures should be thoroughly explained to all parties concerned.

RAT AND FIRE ELIMINATION

The closing operation must be conducted properly to avoid compounding the rat problem. If the rats are not destroyed at a dump, they migrate and pose a more serious problem. Trained personnel and pest control specialists should conduct this operation. Guidance may be obtained from State health officials, the Fish and Wildlife Service, the Public Health Service, and the Environmental Protection Agency.

It may be difficult and expensive to extinguish fires. Heavy earth moving equipment may be required to spread and expose burning waste. Water can then be applied to the smoldering remains. The fire extinguishing program can be conducted concurrently with the rat extermination effort.

COVERUP OPERATION

Following the operations involving rat and fire control, the dump surface should be graded, compacted, and covered with at least 2 feet of soil. Large crawler dozers may be required to perform these functions. The trench or the area method may be used.

Figure 81 shows the trench method procedure. Waste is spread in thin layers in an excavation, is compacted, and is covered with the excavated soil. This process provides maximum density and minimum settlement. The cover material should be graded to prevent ponding. The bottom of the trench should be kept above the ground water level.

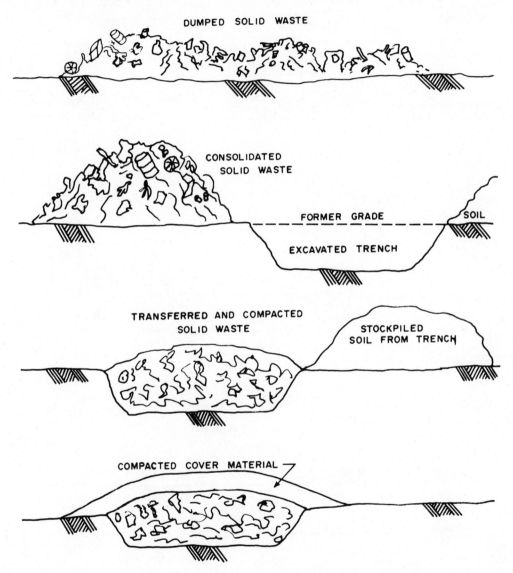

Figure 81. Trench Method of Covering Dump (Source: Ref. 110)

The recommended procedure using the area method is shown in Figure 82. The waste is spread in thin layers, is compacted, and is covered with a minimum of 2 feet of compacted soil. Grading of the cover material is essential to prevent ponding. A variation of this method, used to close bank-type dumps, is shown in Figure 83.

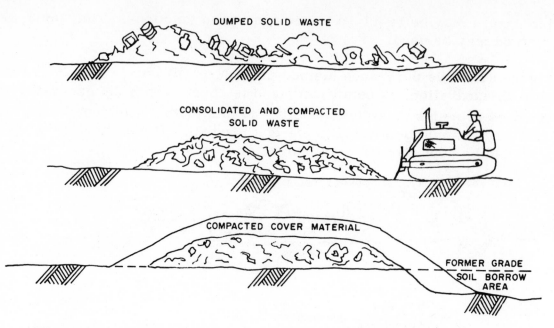

Figure 82. Area Method of Covering Dump (Source: Ref. 110)

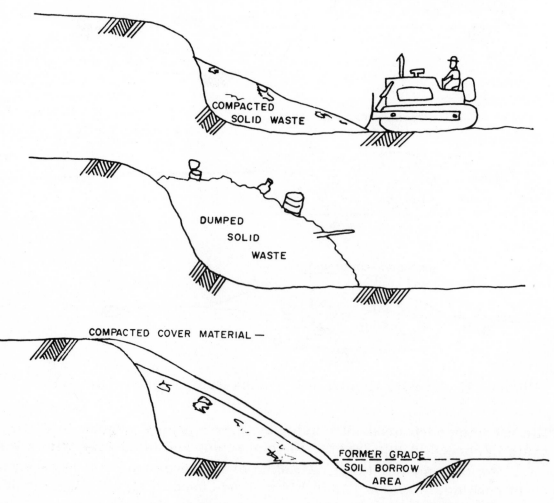

Figure 83. Modified Area Method (Source: Ref. 110)

Cover material should be able to:

- Limit the access of vermin to the solid waste

- Control moisture

- Control gas movement from decomposing waste

- Provide a neat appearance

- Support vegetation

Usually 2 feet of earth is sufficient cover when grass is grown, but more depth is required for shrubs and trees. For dumps along a lake front or at the edge of a stream, riprap (Figure 71) is often required to prevent erosion.

WATER QUALITY PROTECTION

If the dump is a marshland, or if ground water or surface water has been contaminated by it, then the solid waste should be removed from the water. Treating the water may not be feasible, because of the difficulty in collecting contaminated water. Diverting the flow or removing solid waste from the water course offers some means of separating the water for treatment.

Removed waste should not be allowed to create new problems. An impervious berm should be constructed around the perimeter of the new site. The berm should be keyed to the underlying impervious silt layer and should be constructed higher than the outside water level. A mat to serve as an operating platform for a drag line may become the foundation for the excavated waste. Inert materials such as rocks and demolition waste can be used for this purpose.

ULTIMATE USE

A closed dump with vegetation cover becomes a recreational area. Uneven settlement of the site should be recognized in planning this ultimate use. Construction of any buildings should be avoided because of the inevitable settling and the possibility of gas formation from the decomposing waste. The gas can reach explosive concentrations in dumps, because these operations were not properly planned.

SUMMARY

As an ecological alternative to open dumping, Federal and state officials are recommending and promoting sanitary landfills. A landfill is an accepted method of solid waste disposal. The waste is spread into layers, is compacted, and is covered daily with suitable material. This technique seals off odors; controls rats, vectors, and other disease carriers; and enhances the physical appearance. Sanitary landfills are regarded by many as the most inexpensive method of waste disposal. An additional plus is that they reclaim otherwise useless land. However, proper planning, design, operation, and maintenance call for a multidisciplinary contribution -- from hydrology and geology to civil and mechanical engineering.

Two methods are used in sanitary landfill operations: the trench and the area methods. In the former, a trench is cut in the ground. Waste is spread, compacted, covered, and compacted again in the trench. The cycle is repeated daily. When the trench is filled, a thicker layer of cover material is placed and compacted for sealing. This method is common in areas with low water table and with sufficient depth of soil. It is well suited to flat or gently sloping land.

The area method utilizes the same operational technique: spreading of waste in layers, compaction, covering with earth, and further compaction. The cycle is repeated daily. When the site is used, a thicker layer of final cover is provided. The area landfill is ideal for ravines, canyons, valleys, and other sites with existing depressions. Suitability of various soil types as cover material is listed in Table 67.

Table 67

SUITABILITY OF GENERAL SOIL TYPES FOR COVER MATERIAL

Function	General Soil Type					
	Clean Gravel	Clayey-Silty Gravel	Clean Sand	Clayey-Silty Sand	Silt	Clay
Prevent rodents from burrowing or tunneling	G	F-G	G	P	P	P
Keep flies from emerging	P	F	P	G	G	E*
Minimize moisture entering fill	P	F-G	P	G-E	G-E	E*
Minimize landfill gas venting through cover	P	F-G	P	G-E	G-E	E*
Provide pleasing appearance and control blowing paper	E	E	E	E	E	E
Support vegetation	P	G	P-F	E	G-E	F-G
Be permeable for venting decomposition gas**	E	P	G	P	P	P

E = Excellent   G = Good   F = Fair   P = Poor

*Except when cracks extend through the entire cover
**Only if well drained

In both methods of sanitary landfill, the compacted waste and cover material used in daily operation are identified as a cell. A series of adjoining cells, all the same height, make up a lift. A completed fill is one or more lifts on top of each other. Generally, a cell is approximately square. The width, or the working face, is made compatible with the site equipment and the rate of incoming waste, but never exceeds 150 feet. A compaction density of 800 to 1000 pounds per cubic yard is regarded as good.

Protecting ground water from leachate is of primary importance. Ground water or infiltrating surface water, moving through the solid waste, can develop a solution (leachate) containing dissolved and finely suspended solids and microbial waste products. Contamination of either ground water or surface water with leachate can make the water unfit for domestic or irregational use.

The three approaches to controlling leachate are:

- Location of landfills at safe distances from streams, wells, and other water sources.

- Avoiding site location above a subsurface stratification that could provide an easy flow path for leachate to water sources (e.g., fractured limestone).

- Proper operation of the site, using cover material nearly impervious after compaction, grading to prevent ponding, and providing adequate drainage to carry the surface water away from the site.

A critical review (Ref. 111) of ground water pollution from landfill sites and dumps was completed in 1971. In one statement, the author indicates that while there have been some fears that the current waste disposal techniques may seriously contaminate ground water, no strong evidence seems to exist. Refuse dumps, operated for long periods, have not often been proven at fault in this respect. This lack of contamination is primarily due to leachate attenuation by most types of soils, except highly permeable sandy and gravel soils or limestone with fissures where the protective mechanisms (active microbial degradation and absorption) seem to break down.

The most extensive work in the area of water pollution from landfills has been confined to three states: California, Illinois, and South Dakota. Maryland, Pennsylvania, and Wisconsin have become active lately. The policy on landfill-to-water-well distance appears to be a tenuous one. Eight states do not specify any distance. For the states that give a value, the distances vary from 50 to 1000 feet, with most of them in the 100- to 500-foot range. However, most states have become active within the last year in the area of rules and regulations for solid waste disposal. More emphasis is being placed on the ground water pollution problem, and there appears to be a trend toward more rigid state control than in the past (Ref. 111). The Environmental Protection Agency guidelines recommend a minimum of 1/4 mile between a site boundary and a water source (Ref. 112).

Solid waste decomposition, producing gases, is a very real problem for landfills. Gases produced include methane, carbon dioxide, nitrogen, and hydrogen sulfide. Methane and carbon dioxide pose a threat. Methane is hazardous when accumulated in enclosed areas; it reaches explosive limits when present in air at a concentration between 5 and 15 percent. Carbon dioxide causes mineralization of ground water as it dissolves and forms carbonic acid. Permeable vents and impermeable barriers can make a landfill safe from decomposing gases.

Waste decomposition and erosion of fine particles cause settling in landfills. Studies have shown that 90 percent of the ultimate settlement occurs in the first five years after a landfill is completed. Measurements in the Los Angeles area have shown settling of 2.5 to 5 percent in three years.

Public acceptance is the most important factor in deciding whether a selected site will be used. Active public relations and education, with emphasis on the completed use of a site, are vital. Another major problem is getting communities to coordinate disposal needs with the sites available. One community may have a shortage of available land, while an adjoining community may have land that could be reclaimed through a properly planned sanitary landfill.

Initial proper planning of a land disposal site is important. A designer must consider the topography, geology, hydrology, and environmental impact. The ultimate use of a landfill should be decided in these early planning stages. Monitoring is essential to avoid water pollution or undesirable gas movement. Peripheral wells in strategic locations are used to monitor any ground water contamination. Collecting and treating leachate also prevents water pollution. Two sites in Pennsylvania use this approach and, in Seattle, Washington, leachate is treated in an aerated lagoon.

To operate a sanitary landfill properly, ground disposal of industrial chemical waste should be avoided. Chemical wastes can be handled in concrete-lined vaults or can be treated in other ways, depending on their composition. This is usually a complex problem beyond the capability of any landfill operation. In areas with a high water table, or where leachate could become a potential problem, appropriate liners (clay or synthetic materials), together with a leachate collecting system, should be incorporated in the construction of the site.

The Solid Waste Research Division of the Environmental Protection Agency has recently been active in this area. Field studies are conducted in a test facility in Boone County, Kentucky and are compared to laboratory tests for correlation. Various funded projects have also been initiated by this Division:

- Recycling of leachate over the fill.
- Developing treatment methods for leachate.
- Leachate attenuation.
- Settlement characteristics.
- Shredding or baling waste prior to disposal.
- Slurry injection into landfills.

Sanitary landfills are a step function improvement over open dumps. The disposal costs also seem to be considerably lower (about one-third to one-half) than other alternative means of disposal, especially if recovery is not a criterion.

Section 8

BALING AND BALEFILL

Baling of solid waste originally was investigated as an outgrowth of a study of the potential benefits of rail hauling in solid waste disposal systems (Ref. 113). A short series of spot tests during that study showed that high-pressure compaction of municipal solid wastes was feasible and could contribute substantially to the idea of rail hauling of solid wastes away from the urban environment (Ref. 114).

Because of this promising start, a much more exhaustive study of the baling of municipal solid waste was conducted by the American Public Works Association for the City of Chicago, under a Federal demonstration grant from the Bureau of Solid Waste Management (then part of the Department of Health, Education, and Welfare). Results of this study are also reported in Reference 114.

BALING AND BALEFILL TECHNOLOGY

An objective of the baling study was to develop bales that would remain stable for nine days and that would be suitable for rail transport up to 150 miles during this period. This objective was amply exceeded.

HIGH-PRESSURE COMPACTION OF SOLID WASTE MATERIALS

The American Public Works Association compaction experiments were carried out using a 17-year-old, three-stroke scrap baler donated by the General Motors Corporation. This baler was reconditioned and modified. A simplified diagram of the baler is shown in Figure 84. The first two rams operate sequentially in a horizontal plane and gather the refuse in the charging box into a rectangular parallelepiped with cross-sectional dimensions of 16 by 20 inches, perpendicular to the vertical. During normal operation, the cover was locked into position and, after gathering and initial compaction by the first two rams, these rams were locked into position. The high-pressure ram was then operated in the vertical direction, usually until it reached a predetermined indicated gage pressure, and the compaction was completed. (It is shown below that not all balers are three-stroke balers and not all balers operate using a pressure control principle.)

To remove the bale from the press, all three rams were first moved slightly out of position to remove the pressure from the bale. After removal of the cover plate from the top of the box, the third ram was again activated to eject the finished bale out of the charging box. All of the bales produced were initially rectangular parallelepipeds, with initial cross-section dimensions of 16 by 20 inches. The height of individual bales produced by this experimental baler varied, but was usually in the range of 14

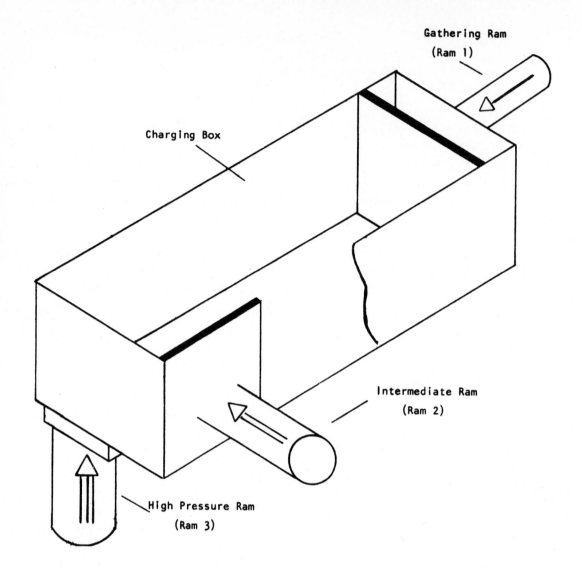

Figure 84. Compaction Press (Baler)

to 18 inches (commercial balers produce larger bales, 36 by 36 by 54 in. being a typical size).

During the American Public Works Association study, attempts were made to instrument the bales to obtain a better understanding of material flow and mechanisms of compaction internally, within the bale. The most promising results were obtained using penetration gages which showed that peak loads can be developed both at the bottom bale face (resting on the high-pressure ram) and at the top of the bale. It is not clear, however, whether the peaking occurs primarily as a result of individual material properties of the bale components or whether it is related to the compaction conditions.

The American Public Works Association study was performed using a variety of refuse types. Packer truck refuse from the City of Chicago was one source, with loads being received that typified two types of refuse: winter refuse and spring refuse. Samples of shredded refuse from Madison, Wisconsin were also studied. Densities of the incoming refuse samples are shown in Table 68, and Table 69 shows the effect of moisture on these densities. In addition, samples were synthesized to produce extremes in various properties such as moisture content and corrugated content.

Table 68

DENSITIES OF RESIDENTIAL REFUSE

Samples	Densities (lb/ft^3)			Comments
	Maximum	Minimum	Average	
204	15.9	3.8	9.0	Household refuse -- loose (contained appreciable number of samples of high water content)
27	10.1	4.8	7.2	Household refuse in paper sacks
34	11.0	8.2	9.6	Shredded household refuse (1-day delivery)

Table 69

EFFECT OF MOISTURE ON DENSITIES

Samples	Densities (lb/ft^3)			Comments
	Maximum	Minimum	Average	
106	15.9	7.1	10.5	Wet refuse
12	13.3	5.5	8.0	Damp spring cleaning refuse (containing clippings, leaves, and dirt)
86	11.0	3.8	7.6	Dry refuse

Tests were performed to determine the density attained at the baling, using the various feedstocks listed above, and to determine the springback with time after release from the baler. Samples were subjected to face pressures ranging from 500 to 3500 psi, 17 seconds being required to make the final high-pressure stroke. A few samples were subjected to 6000-psi face pressures.

Tests on unshredded municipal refuse showed that densities as high as 2916 pounds per cubic yard (108 lb/ft^3) could be obtained in the baler at a

3500-psi face pressure. Average densities at that pressure ran 2500 pounds per cubic yard (93 lb/ft³). Table 70 shows densities achieved in the press for various face pressures. A slight improvement in density was obtained by holding the bale at pressure, in the press, for a period of time.

Tables 71 and 72 show the percent of volume expansion of the bales after removal from the press and the final density of the bales after spring-back, respectively. Volume expansion after rapid baling is about 30 percent

Table 70

DENSITIES OF BALES OF RESIDENTIAL REFUSE
AS A FUNCTION OF INCOMING DENSITIES AND COMPACTION PRESSURE

Samples	Average Densities of Loose Refuse (1b/ft³)	Densities of Bales (lb/ft³)			Compaction Pressure (psi)
		Average	Maximum	Minimum	
31	6.7	93	108	78	3000-3500
29	10.8	93	102	78	
14	6.9	90	104	82	1850-2500
28	10.8	90	103	84	
3	6.8	82	86	77	1350-1500
5	10.6	86	95	83	
8	7.6	74	77	70	650-1000
13	10.0	74	90	66	

Table 71

VOLUME EXPANSION OF BALES AFTER COMPACTION

Samples	Average Loose Densities (lb/ft³)	Volume Reduction Ratios After 1-2 Min	Average Volume Increase After 1-2 Min (%)	Compaction Pressure (psi)
31	6.7	10.0:1	50	3000-3500
29	10.8	6.4:1	43	
14	6.9	7.9:1	66	1850-2500
28	10.8	6.1:1	56	
3	6.8	7.4:1	62	1350-1500
5	10.6	5.5:1	51	
8	7.6	6.6:1	49	650-1000
13	10.0	5.1:1	59	

Table 72

DENSITIES OF BALES AFTER VOLUME EXPANSION

Samples	Average Densities of Loose Refuse (lb/ft³)	Average Densities of Bales After 1-2 Min Volume Expansion (lb/ft³)	Compaction Pressure (psi)
31	6.7	61	3000-3500
29	10.8	65.5	
14	6.9	58.5	1850-2500
28	10.8	60.0	
3	6.8	50	1350-1500
5	10.6	54	
8	7.6	49.5	850-1000
13	10.0	48.0	

for the 3500-psi tests and 37 percent for the 1350- to 1500-psi test bales. Appreciable springback occurs, because air trapped in the bale during compaction does not have time to escape.

The American Public Works Association study did not show the effects on springback of municipal refuse produced by increasing holding time in the press. However, tests on synthetic refuse loads showed that holding the refuse at pressure for a period of time can reduce the springback considerably. This indication has been substantiated in commercial operations (Ref. 115), where holding times of only 10 seconds or so appreciably reduce springback. A generalization to be noted from the above tests is that the final density of the bale in the press was virtually unaffected by variations in the density of the incoming loose refuse. Stable bales were produced by compaction at high pressures in the absence of excessive moisture.

On the other hand, bales compacted at low pressures were quite fragile. Fragile bales were usually obtained after compaction at pressures between about 500 to 1000 psi. These bales occasionally fell apart immediately after removal from the baler. Some fell apart after being handled successively, several times. An improvement in stability was found for bales compacted at pressures between 1000 and 1500 psi; however, the stability of most bales was markedly improved after compaction at 1500 psi and up to 2000 psi. A further increase in pressure, up to 6000 psi, produced no apparent improvement in bale stability.

The overall stability of the bales compacted at about 1000 and 1500 psi usually improved appreciably if the force on the bales was held for several minutes. The bales compacted at about 1000 psi for 5 minutes seemed to be as stable as those compacted at 1500 to 1750 psi, without holding of pressure. Similarly, the bales compacted at about 1500 psi for 5 minutes appeared to have the properties of samples compacted at about 2000 psi without holding of the pressure.

The stability of bales of very high moisture content was always poor, irrespective of the compaction pressure applied. There were, however, indications that the stability of the bales containing an appreciable amount of moisture could be improved if the compaction pressure were lowered.

Tests were also conducted using plastic- and paper-sacked refuse from the Chicago area; these results are reported in Tables 73 and 74. The

Table 73

DENSITIES OF BALES OF PAPER- AND PLASTIC-SACKED REFUSE DURING COMPACTION

Densities of Uncompacted Refuse (lb/ft^3)			Densities of Bales (lb/ft^3)			Compaction Pressure (psi)
Maximum	Minimum	Average	Maximum	Minimum	Average	
10.1	4.8	7.4	106	90	94	3300-3500
9.7	6.3	7.7	93	81	85	1850-2000
Precompacted at 3500 psi		69.75	70	69	69.5	6000*

*Two bales only.

Table 74

VOLUME REDUCTION RATIOS AND BALE DENSITIES OF PAPER-SACKED RESIDENTIAL WASTES DURING AND AFTER COMPACTION

Volume Reduction Ratios (averages)		Average Volume Increase (%)	Densities of Bales (lb/ft^3, averages)		Compaction Pressure (psi)
During Compaction	After 1-2 Min Volume Expansion		During Compaction	After 1-2 Min Volume Expansion	
13:1	9:1	47	94	64	3300-3500
11:1	8:1	38	85	63	1850-2000

major difference between this sacked refuse and loose refuse is the more uniform moisture content in sacked refuse, due to protection from rain and snow. Average values of compacted density are quite similar to those for the loose refuse. Bale stability results as a function of pressure were similar to those for loose refuse for paper-sacked refuse. Stability of plastic-sacked refuse bales was poor because of poor adhesion of plastic bags to each other. Spraying with adhesives helped, in this respect.

Baling of the shredded refuse was investigated using various levels of added moisture. Table 75 shows the densities obtained in the press, and Table 76 shows the density after springback. In general, the shredded

Table 75

DENSITIES OF BALES OF SHREDDED REFUSE DURING COMPACTION

Samples	Average Densities of Loose Refuse (lb/ft³)	Densities of Bales (lb/ft³)			Compaction Pressure (psi)	Comments
		Maximum	Minimum	Average		
24	9.4	103	96	101	3400	Normal fast runs
1	8.3	--	--	99	3400	5-min pressure holding time
1	10.2	--	--	98	3400	Water added: 11% by wt
1	10.9	--	--	98	3400	Water added: 17% by wt
8	10.2	100	92	95	1850	Normal fast runs
1	9.9	--	--	98	1850	5-min pressure holding time
1	9.9	--	--	66	350	Normal fast runs

Table 76

VOLUME REDUCTION RATIOS AND BALE DENSITIES OF SHREDDED REFUSE DURING AND AFTER COMPACTION

Samples	Volume Reduction Ratios (averages)		Volume Expansion After 1-2 Min (%)	During Compaction	After 1-2 Min Volume Expansion	Compaction Pressure (psi)
	During Compaction	After 1-2 Min Volume Expansion				
24	10.7:1	7.4:1	44	101	68.6	3400
1	12.1:1	8.5:1	42	99	69.7	3400 (5')
1 (a)	9.9:1	5.7:1	73	98	56.0	3400 (H_2O)
1 (b)	9.2:1	5.1:1	80	98	55.0	3400 (H_2O)
8	9.4:1	6.7:1	45	95	67.3	1850
1	10.1:1	7.0:1	44	98	68.0	1850 (5')
1	6.8:1	3.5:1	93	66	34.0	350

waste allows about a 5- to 10-percent higher density both in the press and after springback. However, much more springback occurs when water is added. The added moisture probably fills the capillary passages that allow air to be released during compression of either shredded or unshredded refuse.

Stability of the shredded waste compacted at 1850 psi and 3400 psi was good. The bales produced from dry refuse had straight sides and sharp edges. Wet bales tended to bulge. The only bale what had poor stability was compacted at 350 psi.

LEACHATE AND GAS GENERATION

Leachate is produced both in the baling process and in the balefill. Some weight loss was observed during the compaction process. This loss ranged from 37-percent weight loss for loose refuse exposed to rain, 1- to 3-percent loss for moderately wet refuse, and no loss for dry refuse. An average 5-percent loss was observed in the tests of loose refuse.

In the American Public Works Association tests, the squeezings from the press were collected over a two-week period. Unfortunately, this was a period when relatively dry samples were being tested. The chemical tests were based on a 1.5-liter sample collected during this period and are reported in Table 77. Because time is a critical factor in microbiological

Table 77

CHEMICAL ANALYSIS OF LEACHINGS

Sludge/Liquid	Sample 1 (%)	Sample 2 (%)
Sludge		
Organic matter	38.9	76.7
Silica as SiO_2	30.8	6.9
Aluminum as Al_2O_3	15.2	5.8
Phosphates as P_2O_5	8.9	4.4
Calcium as CaO	2.0	1.3
Magnesium as MgO	1.4	0.0
Iron as Fe_2O_3	0.5	0.6
Sulphates as SO_3	Trace	Trace
Carbonates as CO_2	0.0	0.0
Liquid		
pH	7.8	4.5
Total dissolved solids	6090 ppm	6040 ppm
Organic matter	1720	4480
Bicarbonates	90	610
Suphates	250	200
Chlorides	21	218
Heavy metals (suspension)	Large % amount	Small % amount

analysis of samples, water was added to incoming loose refuse before compaction, in order to produce a 1.5-liter sample within a few hours of receipt of the refuse. Results of the microbiological analysis of the leachings from the compaction process are given in Table 78. No pathogenic organisms, with the exception of one virus, were detected in the leachings. This discovery is not surprising, because many pathogens cannot survive outside a living host organism for any length of time.

Table 78

MICROBIOLOGICAL ANALYSIS OF LEACHINGS

Category	Analysis
Bacteriology	
Aerobic plate count	8.3×10^6 organisms/ml
Anaerobic plate count	7.7×10^6 organisms/ml

Identified bacteria (to genus)	Bacillus sp
	Caliform group
	Proteus sp
	Streptococcus sp
	  (alpha hemolytic and nonhemolytic noted)
	Alcaligenes sp
	Micrococcus sp
	  (coagulese negative)
	Flavobacter sp
	Aerobacter sp

Yeasts	Yeast cells were observed.
Molds	Penicillium Streptomyces Paecilomyces Mucor
Parasitology	A number of free-living amoeba as well as ciliates were noted.
Water Bacteriology	
Standard plate count at 35°C	4.9×10^7 organisms/ml
MPN technique	9.2×10^8 organisms/100 ml
Fecal streptococci (membrane filter)	7.0×10^6 organisms/100 ml
Staphylococci (membrane filter)	5.7×10^7 organisms/100 ml
Virology	Virus isolated and identified as ECHO by antiserum neutralization
Chemistry	pH:7; D.O.:none

It is not surprising that the leachings contained an appreciable amount of some of the pollutants, especially since both the chemical and biological tests were carried out with concentrated samples designed primarily to allow detection of the type of pollutants that could be encountered in the refuse. It should also be remembered that the investigation was restricted to a few tests and that a more quantitative study would be required to assess the pollution potential of the leachings.

With respect to compaction, however, it must be emphasized that the chemical and microbiological constituents detected in the leachate were not introduced into the refuse by the compaction process. These constituents are part of the normal makeup of solid wastes disposed of in everyday operations. Because some of the pollutants are extracted from the wastes during compaction, the compacted wastes should, in fact, contain fewer contaminants than the wastes entering the compaction plant.

Because balefilling is a very new process, little data exist on leachate generation in balefills. One installation in San Diego receives only 8 inches of rainfall per year, and leachate is negligible. Another installation, in St. Paul, Minnesota, is equipped with a leachate collection system. No leachate has yet been observed from over 100,000 tons of baled waste. This situation can probably be explained by the very low permeability of highly compressed waste bales.

Tests performed on baled refuse by Northwestern University under contract to the American Public Works Association (Ref. 115) showed the following percolation rates for baled refuse:

Sample	Wet Density (lb/yd^3)	Coefficient of Permeability (ft/day)
1	965.2	42.6
2	1,323.0	13.6
3	1,409.4	10.0
4	1,917.0	2.0

Sample 1 is typical of a conventional landfill with good compaction. Sample 4 is typical of the more dense bales produced by a high-pressure baler. The American Public Works Association concluded that the 20:1 difference in permeability should result in an equivalently lower flow of water through the refuse for a given hydraulic gradient and will produce a lower quantitative rate of leachate. Tests were also conducted to measure gas production by taking compacted samples, immersing them in water baths of several different temperatures, and subsequently buffering them to higher pH values to encourage alkaline fermentation and gas production.

In even the conditions most favorable to gas production, the gas generation ceased after three days, as organic acids formed and lowered the pH. The low permeability of the waste prevented penetration of the alkaline buffer at a rate fast enough to counteract the internally generated organic acids. The American Public Works Association conclusion was that balefilling "does not require, with respect to water pollution and gas control, environmental control measures different from those employed in sanitary landfilling today." They guardedly suggest that balefilling may present a lesser degree of potential environmental control problems than the sanitary landfills now in use.

STABILITY IN HANDLING AND RAIL HAULING OF BALED REFUSE

A series of vibration and drop tests, along with rail hauling tests, was performed and reported in Reference 114. Bales made from loose refuse, paper-sacked refuse, and paper refuse were used; strapped and unstrapped bales were also used. The American Public Works Association concluded (Ref. 115) that strapping offered no real advantage in high-pressure bales, either with regard to bale stability or volume reduction. The unstrapped material held together as well as the strapped material, in the handling and transport tests, and nonshredded material was as stable as shredded material. Drop tests from a height of 10 feet and vibration table tests simulating 600-mile rail hauls were conducted.

Rail haul tests of 700 miles (from Chicago to Cleveland and back) were conducted on Penn-Central rail lines. The cars were instrumented to record impact. In one case, there was a near-collision impact. In that particular test run, there were more than 300 bales in the cars, and none returned broken, even though some of the impact was sufficiently severe to damage the cars. The test series pointed out, however, that the bales should be tightly loaded and braced in the cars.

BALEFILLS

Balefill operation is considerably simpler than sanitary landfill operation. Because compaction has already been accomplished in the baler, the major function at a balefill is to position the bales. This positioning is usually done by the front-end loader, which removes the bales from the transfer truck. Tight placement of the bales prevents springback.

An experimental balefill in Georgia (Ref. 116) contains bales that have been in place for six years. When placed, surveyors' stakes were driven into them, and these stakes have been periodically monitored to detect motion due to swelling or settling. To date, no shifting of any kind has been detected. Typical in-place density of baled refuse is 1600 pounds per cubic yard, compared to 800 to 1000 pounds per cubic yard in a well run sanitary landfill.

Because baling is a new process, state and Federal regulations governing operation of balefills, themselves, are not yet firm. The two balefills that were visited were operating under variances that permitted them to operate without cover. In both cases, the face of the balefill was open, but the operators did place cover material on top of the bale-stack, to reduce abrasion by vehicles driving on top of the stack.

Typically, the bales are stacked three deep, which is the usual lift of a fork lift truck. Once an area has been covered by a three-bale lift, more lifts are stacked on top, in three-bale depths, until the desired final level is reached.

ST. PAUL BALING FACILITY

American Solid Waste Systems, a subsidiary of the American Hoist and Derrick Company, has built and now operates a high-pressure solid waste baling facility in St. Paul, Minnesota. The Company owns the baling facility and also a balefill, which it operates. These facilities accept 95 percent of the municipal waste produced by the city of St. Paul and a large fraction of its commercial waste. The St. Paul baling facility started in January 1971. It reached a capacity of 500 tons per day in July 1972. The current operating rate is 550 to 600 tons per day of input.

Refuse is received at the baling facility from 6 a.m. to 8 p.m. The station is cleaned and shut down by 1 a.m. The baling facility requires six employees on the first shift and five on second shift. Solid waste that arrives early in the morning is usually received from commercial sources and is high in dry corrugated content. Some of this refuse is set aside for mixing with later loads of municipal solid waste that may have a high moisture content. A moisture content of 15 to 25 percent results in the best bales.

Operation of the baling facility is now quite routine. Refuse is dumped onto a flat unloading floor having an 800-ton accumulation capacity. A front-end loader pushes the refuse onto a horizontal slat conveyor, which transfers it to an inclined slat conveyor that lifts the refuse up to the receiving hopper of the baler. The receiving hopper is mounted on load cells, and an automatic system shuts off the feed conveyor when 3000 pounds of refuse is contained in the receiving hopper. (This amount is adjustable.) The baler then receives this load and uses a three-stage (x, y, and z components) squeezing process to produce a 3 by 3 by 4.5-foot bale, which is then fed out and automatically pushed onto the waiting transfer trailer. Cycle time for each bale was 120 seconds; newer models accomplish this task in 90 seconds. Figure 85 shows the baling sequence. This approach does not use strapping. The American Hoist and Derrick Company had to use the above layout because of a high water table at their baling facility site. They would have preferred eliminating the inclined conveyor and placing the baler and transfer trailer pad in a pit. The balers are derived from scrap baler designs. Hydraulic systems have 3500-psi pressures, and have face pressures at the ram reach of about 3000 psi. The American Hoist and Derrick Company has baled automobiles and white goods. They have even baled a Mercury station wagon into a 3 by 3 by 4.5-foot bale.

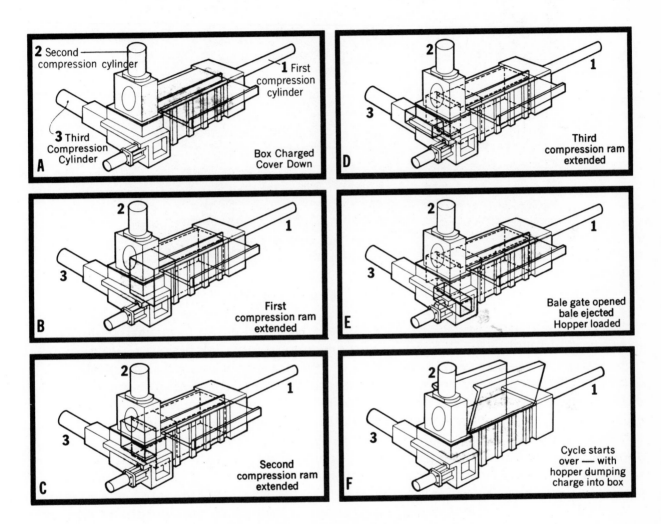

Figure 85. Baling Sequence

The baling plant is housed in a 120 by 240-foot building that has a 53-foot peak and is fully equipped with sprinklers (on a flood basis).

The baled refuse is trucked by a contract hauler to the balefill on flatbed semitrailers of 24-ton capacity. Four truck-trailer combinations and two drivers are used. Normally, a driver drives a full truck from the baler to the landfill, where he picks up an empty truck-trailer to bring back. On this basis, a driver makes the 23-mile round trip in one hour. The route includes use of some city streets, a two-lane state highway through residential areas, and about 40- to 50-percent interstate highway. The sides of the trailer are covered with camouflage netting; the top is open.

Figure 86. St. Paul Balefill

The balefill (Figure 86) is a 37-acre site intersected by a railroad spur. It is operated during two shifts, using one man during each shift. Portable lights allow two-shift operation. Equipment presently consists of an Allis Chalmers 840 front-end loader, converted to a fork lift, which can pile the bales three high. One 8.8-acre section of the balefill on one side of the railroad spur is now completed and contains 100,000 tons of refuse piled seven to ten bales deep. Some additional bales will be brought in to fill low spots. The American Hoist and Derrick Company then plans to build a slab-mounted, one-story commercial building on the site. The St. Paul balefill now contains about 140,000 tons of bales. The State of Minnesota has granted a variance allowing operation without cover. Bales are stacked three high, with no spacing between bales. Some soil cover is used on top of the bales, to permit trucks to drive over them without abrading the bales. A very small amount of loose trash was observed at the balefill, much less than is typical of standard landfills. The 28-acre portion of the balefill now in use has no leachate collection system, because it is located on clay soil. About a 40-foot depth of fill will be achieved, with State approval. The 8.8-acre portion did have a leachate collection system using perforated pipe emptying into a clay-lined catch basin. To date, no leachate has been observed in the catch basin.

Capital cost of the St. Paul baling facility was $1.3 million, excluding land. A tipping fee of $1.20 per cubic yard is currently in effect and will be raised to $1.40 per cubic yard on June 1. This charge covers the entire cost of the baling, transfer trucking, and balefill operation, according to the American Hoise and Derrick Company.

Appendix III, "Cost Analysis of American Hoist and Derrick Company Baling Operation," presents a detailed cost analysis for two hypothetical baling systems, assuming 1971 Ohio costs. In one case, a facility built from scratch to handle an annual tonnage of 256,256 tons is analyzed. The capital cost is computed at $1,075,000. The cost per ton, including amortization and interest, is predicted to be $1.64. This cost does not include transporting the bales or operating the balefill. Appendix IV also describes a facility, of the same capacity and location, that is installed in an unused incinerator building. The incinerator holding pit and crane would be used to feed the baler. Here the capital cost is $806,330, including the building alterations. The cost per ton is $1.48 at a 100-percent-capacity utilization rate.

At a 90-percent utilization rate, the total predicted disposal cost, including rail hauling 329 miles and the balefill operation, is $6.66 per ton for the plant built from scratch and $6.49 per ton for the baler facility built in the incinerator building.

SAN DIEGO BALING FACILITY AND BALEFILL OPERATION

The San Diego experimental balefill operation is financed by an Environmental Protection Agency grant. The baling facility was built on a section of a San Diego Public Works Department facility, and the balefill is located in a ravine less than 1 mile from the baling facility, adjacent to the Balboa Park golf course.

The baling operation consists of shredding and then baling. Shredding allows slightly denser bales, in laboratory tests, but the actual experience at San Diego shows about the same density (50 to 70 lb/ft^3) as shown at the St. Paul facility, where baling is done without shredding.

San Diego uses a Williams Model 475 shredder, driven by a 500-hp Allis Chalmers motor. The shredder is located in a pit and is fed by a slat conveyor. An open-air, concrete dump pad is used, and a Caterpillar Model 920 front-end loader pushes incoming waste onto the conveyor. Hammers are pulled for refacing after every 1000 to 1200 tons of throughput. Pulling the hammers requires four hours. The shredder has a 25-ton-per-hour nominal capacity, and no serious problems have been encountered, except once when the rotor was bent by a rolled nylon rug. Now all rolled rugs, tires, and large iron pieces are pulled before shredding.

The shredded waste is brought from the shredder up to the baler by a vertical bucket elevator. This elevator has been the major problem in the installation, because of the low density of the shredded waste (2 to 3 lb/ft^3). Its capacity has been increased by use of a compression roller, which increases the density to 8 pounds per cubic foot. In addition to low through-put due to low density, jamming was encountered because of refuse spilling onto the lower pulley.

The baler is an American Baler Company unit that produces bales a layer at a time (undirectional compression). The extruded bale is strapped together by wires manually inserted through separator blocks containing guide holes. This entire operation is a bottleneck, and experiments have been run on a variety of separator block materials. Throughput rates of 12 tons per hour were typical until high-density polyethylene was adopted. This change produced 21-ton-per-hour peak rates. The baler was inoper-ative when the facility was visited, dur to a variety of problems including excessive erosion of guide skids. (The bales are very abrasive.)

Design capacity of the baler is 175 tons per day. The highest capacity achieved to date is 110 tons per day, with 80 to 90 tons per day being typical.

The balefill capacity is 64,000 tons, but it is just beginning to be filled. There have been no vector problems, and a variance has been obtained to permit open-face operation. The tops of the bales are covered daily by 6 inches of soil.

Because of the low rainfall in San Diego (8-1/2 in./yr), leachate is not a problem, and no provisions were made for leachate collection or testing. The capital cost of the baling facility is just under $370,000:

Category	Amount
Equipment	$195,974
Building, construction, and engineering	164,177
Balefill site preparation	9,143
Total	$369,294

Total costs, when running at capacity, are calculated to be $4.74 per ton, including amortization.

Section 9

FACILITIES AND COMPONENTS

In this section, the nuts-and-bolts aspect of elements common to many of the processes are discussed. Particular emphasis is placed on shredders, because these components are common to virtually all resource recovery processes, except wet pulping and some slagging pyrolysis processes. Very brief discussions of hydrapulpers, air classifiers, and conveyors are also included. Discussion of specific components is preceded by a review of receiving facility requirements.

RECEIVING AREA DESIGN REQUIREMENTS

A common factor in all refuse processing systems is the need for an area where the refuse can be received from the incoming packer trucks or transfer vehicles. In this discussion, the point of reference will be a plant having the capacity to receive 1000 tons per day of refuse.

The receiving area for incoming refuse is a prime factor in determining the total land area required for a facility. Other factors include access roads and scales, the process equipment itself, intermediate process storage, product storage, and shipping facilities. In the process studies described above, area requirements were specified.

In some instances, data for a 1000-ton-per-day facility were furnished by the equipment manufacturer. In other instances, an extrapolation from a different capacity, estimated by an equipment manufacturer, was utilized. For some of the processes, where no estimates were available from the manufacturer, a best guess was presented.

In designing a receiving area, proper enclosures are required to protect raw refuse and the equipment from the weather, especially rain and snow. This requirement influences the effectiveness of shredding, classification, and process reactions (in the case of pyrolysis and composting processes). It also has an impact on environmental elements (e.g., surface water, odor, and cleanliness).

In this evaluation, it is assumed that a given commercial facility should have adequate provision to store at least one day's collection, or 1000 tons of municipal waste. Two options are considered for the storage of raw refuse:

- Pit storage

- Level floor receiving area storage

287

Pit storage has been the more common method used in waste disposal facilities, especially in municipal incinerators designed ten or more years ago. The obvious advantage of pit storage is the minimum floor space required to store raw refuse. Incoming refuse can be piled 25 feet high, or higher, in a pit with retaining walls. Sufficient barrier height above the floor should be provided in the receiving area of the storage pit, where trucks are backed up to unload. This barrier should be high enough to prevent a truck from tipping over backwards into the pit. At the same time, the barrier should be low enough to enable complete unloading of the refuse into the pit and not onto the receiving area floor. Sufficient overhead clearance should be provided to allow for various means of truck unloading.

Removal of the refuse from the pit can be accomplished by:

- Crane and grapple system.

- Inclined conveyor at the bottom of the pit, bring the material out to the desired machine for processing.

- Combination of horizontal conveyors on the pit floor, feeding refuse onto a bucket conveyor or elevator. An overhead crane and grapple system allows more of a pit to be used for storage of refuse, although ample head clearance is required to accommodate overhead equipment. Pit conveyors reduce the effective volume of a storage pit, because of the false bottom required to install the conveyor and provide access space. In addition, some pits are designed to have small front-end loaders operating on the pit floor to load horizontal conveyors. Area must be provided to permit maneuvering of the loaders.

While pit storage appears to offer the advantage of requiring less floor space than level floor storage, it has several major disadvantages:

- Pit corrosion.

- Lack of adequate accessibility to pit conveyors for repair and maintenance.

- Unusually high ceiling requirements, if an overhead crane is used.

- Refuse first loaded in is unloaded last from the pit.

- Need for major safety provisions for unloading trucks and for crane operators.

- Cost of pit.

Pit storage does concentrate the refuse in a smaller area, permitting easier sprinkling for fire protection. However, the crane operator in a traveling cab of an overhead crane is vulnerable to fires in the pit. Modern European practice in pit receiving areas is to use closed circuit television and remotely operated cranes. This, of course, adds cost to the installation.

A floor-level waste storage system speeds the flow of trucks through the facility. Raw refuse (1000 tons), stored at floor level, should be piled to no more than 6 to 8 feet, in order to enable a front-end loader to feed it into the conveyors or machines with relative ease. Assuming a density of 200 to 250 pounds per cubic yard for incoming packer truck refuse, it is estimated that the floor-level storage would require approximately 50,000 square feet of floor area just for the refuse. Added to this, 1-1/4 acres would be required for smooth flow of truck traffic into and out of the facility.

Based on the above estimate of storage space needed for 1000 tons of raw refuse, together with the space required for a given processing system, it appears that total land requirements for a 1000-ton-per-day plant would be from 3 to 6 acres. The more complex or the less reliable a given process, the higher its total land requirements are likely to be, because redundant operating lines will be needed. In addition, the nature of output products from a given process would also greatly influence the total acreage required.

SHREDDERS

INTRODUCTION

The term "shredder" is undoubtedly a misnomer in this subsection. Shredding means tearing or cutting. Many of the size reduction machines discussed in this subsection do not tear or cut. There is a variety of types of size reduction equipment, and they have been defined below under "Glossary." This subsection will describe those devices that reduce the size of individual particles in a dry process applied to residential solid waste.

Assessment Approach

The study of these size reduction machines has previously been approached on a basis of the method of reduction. Each type of machine was studied with regard to the degree to which the forces of tension, compression, and shear are utilized in the process. Another type of study devoted itself strictly to the construction principles of the machine, and a third type devoted itself to the nature of the material and the product resulting from the process. All three approaches are utilized in this assessment.

Horizontal and Vertical Shredders

Shredders generally are either vertical or horizontal machines. The horizontal-shaft machine has been used for at least 100 years. The newer, vertical-shaft machine has enjoyed some degree of success for certain aplications and is included as part of this study. Figure 87 shows an external view of the horizontal-shaft hammermill used in the City of St. Louis installation, and Figure 88 shows another horizontal-shaft machine with the access door open. Figures 89 and 90 show comparable views of vertical-shaft machines.

Figure 87. Hammermill Installed for Shredding of Municipal Solid Waste
(Source: Gruendler Crusher and Pulverizer Co.)

Figure 88.   Horizontal-Shaft Hammermill for Solid Waste
(Source: Williams Patent Crusher and
Pulverizer Co.)

Figure 89.   Double-Drum Vertical-Shaft Waste Shredder
(Source: Eidal International Corp.)

Figure 90.   Ring Hammer Rotor and Liners Inside Vertical-Shaft Waste
Shredder (Source: Eidal International Corp.)

The power requirements and heat buildup in these machines are difficult to predict because of the varied nature of the waste material being
processed and other aspects pertaining to the inherent design of the machine. For instance, a horizontal-shaft machine generally has a relatively
small amount of product within the reduction chamber at any time. The
balance of the volume is filled with air, which helps to cool the equipment
and to transport the low-density product particles. A vertical-shaft machine has very little air within the reduction chamber and is densely loaded
with product in various stages of size reduction. The friction is a significant factor, both between adjacent particles and between particles and
parts of the machine. Heat buildup is significant and can cause fires and
premature failure of the lower thrust bearing.

Problem Materials

It has been discovered that most size reduction machines encounter
some objects that are difficult to process. It is surprising to discover
that a machine that fragmentizes automobiles has trouble digesting the
vinyl seat covers from the automobile. In some cases, these seat covers

must be removed before processing the automobile hulk through the frag-mentizer. Likewise, a machine that has no problems size-reducing concrete blocks and large pieces of metal has difficulty with plastic bleach bottles.

In some cases, the shape of the object is incompatible with the design of the machine. A machine that accepts automobile engines may have sufficient clearance at the ends of the rotor to allow wire cable and fencing to wrap around the ends of the rotor, causing a shutdown. One type of machine that utilizes a screen to limit the product size has no trouble size-reducing 2 by 6 and 4 by 4-inch lumber, but has difficulty processing wood excelsior. Resiliant items, such as automobile tires, respond better to shearing and tearing machines than to impact-type mills.

In machines that have been designed to take the variety of materials found in municipal solid wastes, there are sometimes problems resulting from individual loads of homogeneous materials. For instance, a load of very wet grass clippings may cause a problem in the shredder. Also, in one instance, an entire truckload of deposit slips from a bank caused a problem in the size reduction machine, because the deposit slips were carried around the rotor on the tips of the hammers rather than being passed out of the mach-ine. However, when other residential waste material was mixed with the deposit slips, there was no problem in processing the mixture through the same machine.

In general, the problem in dealing with municipal solid wastes stems from the variety of materials found in that waste.

Hammers

A machine that is designed to contend with the friable blocks and the large pieces of metal found in such waste will generally have wide, massive hammers. Often these hammers have huge tips, which tend to defeat the purpose of the swing hammer; the inertial forces tend to cause the hammer to react much like a rigid hammer, and the tendency for the hammer to swing back out of the way after contact with stubborn materials is minimized.

This type of hammer generally is not designed for optimum efficiency in size reducing cloth and wood materials. A wide hammer, for example, must displace a section of wood the entire width of the hammer. This means that the wood must be sheared along two surfaces, one at each side of the hammer, in order to cut through. By comparison, a narrow knife or blade can cut through the same wooden object with a single line of shear. On the other hand, the metals and friable materials that might be encountered in the waste would rapidly dull or destroy a narrow knife or blade.

Future designs might involve a combination of wide hammers close to the rotor shaft, with blades and knives pivoted from the hammers at greater diameters from the shaft. Rigid hammers allow a machine to operate at lower speeds, but are more subject to damage than are swing hammers.

Recent interest in size reduction of waste materials has prompted activities that are certain to result in better machinery for the purpose within the next few years. In addition to the considerations of horizontal versus vertical shafts and wide hammers versus blades and knives, there are also considerations concerning such matters as the method of feeding material into the shredder, the tip velocity required to size-reduce the material, the method of controlling the output product size, and the serviceability of the equipment.

Tip Velocity

The tip velocity required for adequate size reduction depends upon the nature of the material being processed and the inherent machine design. With friable materials (such as rocks and glass that shatter upon impact), and tough materials (such as corn cobs), high velocities are required to do a satisfactory job in impact-type machines. On the other hand, ductile materials (such as sheet steel and many of the plastics and textiles found in municipal wastes) will respond to slower moving members in machines that grind, shear, or cut. Friable materials in similar machines will respond to crushing action at slow speeds.

Input Feed

The method of introducing materials into the size reduction device has been a problem for applications accepting municipal wastes. The fact that some items are objectionable in any given machine means that some sort of segregation generally precedes loading into the size reduction device.

Great strides have been made in this area during the past few years, and there are some manufacturers who are willing to state today that they can supply machinery that requires absolutely no presegregation. However, because explosives and other hazardous materials may be included in the raw waste material, some safety precautions should be used when designing a size reduction facility. Further, the fact that a given machine will grind up large leaf spring assemblies and die plates does not mean that this operation is acceptable, because it will generally take its toll in operating costs. Therefore, common sense must prevail in planning the feed method for size reduction equipment.

Normally, the waste material is elevated to the entrance of the size reduction equipment by means of a conveyor or by using a crane to drop the material into a hopper, which allows the material to enter the throat of the machine by gravity.   There are two common problems with this type of operation:

- The material tends to enter the machine in slugs, which is inefficient and causes undue stress on the equipment.   The output stream is generally not consistent in its volume flow rate, and this inconsistency may create problems.   Input conveyor speed can be controlled by sensing the shredder motor current or weight on the conveyor, which reduces the surge problem.

- The rotating tips within a typical hammermill machine tend to take on the nature of a solid cylinder, and the material being introduced into the machine has difficulty entering the cylinder.   For instance, it is common to see an automobile tire dropped into a hammermill and thrown out with considerable force from the throat of the machine, rather than entering into the tip circle where it should have been size-reduced.

One solution to this problem is to drop the material from sufficient height to attain the speed that will allow the object to enter the tip circle. Generally, the chute through which the material is dropped will also contain any items that are not accepted into the tip circle on previous attempts, and they are free to fall back into the tip circle until they are finally received and size-reduced.   However, this method requires considerable overhead clearance above the machinery, and it is common to encounter total structure heights of 60 feet, including the charging area, the machine itself, and the discharge area underneath.

Another solution to the feeding problem is the compression feed mechanism.   This mechanism consists of a set of driven rollers or tracks above the feed conveyor. Generally, this mechanism is adjusted to not only push the material into the tip circle of the machine at a predetermined rate, but also to compress the material and thereby provide for more efficient operation by the size reduction machines.   This equipment, however, is expensive and must be designed for each given application.

A third method of solving the feeding problem is to design the machine in such a way that the material enters the tip circle near the bottom of the machine on an angle at which the material will be pulled into the tip circle (similar to the action in an undershot water wheel).

Outfeed -- Size Control

After the material has been processed by the moving members, it will continue to reside inside the size reduction chamber until an adequate opportunity to leave the machinery occurs.   The control of the output particle size depends upon the use to which this product will be put.   If the waste is size-

295

reduced prior to land filling, it is possible that no size control facilities are required in the machinery, and the device can be called a coarse shredder. There are flail-type machines on the market that do a reasonably good job of size reducing large corrugated boxes, bottles, and cans in preparation for landfills and recycling activities. These devices have high hourly capacities, in relation to the horsepower required, because the output size can be very large and still be acceptable. Further, it is expected that little damage will occur to the internal parts of the machine, because objectionable items are not contained within the size reduction chamber but are free to discharge through the bottom without significant size reduction.

Recent emphasis on resource recovery has generated a need for machinery that will produce a product of a consistent, small size. It appears that multiple-stage shredding with intermediate separation processes is a good solution to the problems resulting from this demand. The first machine in the series might have a large enough throat to accept bulky furniture items and large boxes. The output size would be regulated by grate bars, by a perforated screen, or by no mechanism at all, if the hammers were spaced properly and the tip velocity were sufficient to result in a product with a size range between 4 and 8 inches. This shredded material would then be processed through a separation device, which utilizes air or some other means to separate the low-density items from the high-density items.

Each of the two output streams from the separation device is subjected to subsequent size reduction equipment, which can generally operate more efficiently because the input is consistent in its size and is easier to feed. Vibrating aprons may be utilized, in the secondary size reduction feed equipment, to level any surges or lumps in the flow of material. The shredding machinery can be designed for the specific purpose of cutting or tearing the low-density items, as compared to densifying the metals and pulverizing the friable materials in the high-density stream.

Serviceability

Among the problems that accompany the application of size reduction equipment to the variety of materials in municipal waste, serviceability of the machinery has become paramount. Whereas it might have been perfectly acceptable to design machinery for a rock crusher in such a way that a multitude of bolts required removal to provide access to the hammer replacement portion of the machinery, today that same section of the machine must be accessible within minutes, to provide for continuous operation. Rather than allow items that are not easily size-reduced to remain in the machinery long enough to cause severe damage, it is wise to include some provision for a trap to take this material out of the machine for secondary treatment or for discarding. The hammers must be available for replacement, build-up of the worn tip areas, or a complete replacement of the tip portion of the hammer. If the hammer shaft is long and must be removed to allow replacement of the hammers, it is important that sufficient clearance be provided at the side of the machine to allow for this operation. The use of hydraulic cylinders, as in Figure 88, is gaining popularity as a means of providing fast access to the service areas of the size reduction machinery without sacrificing strength and close tolerances in the critical dusty areas. Hydraulic hammer bolt pullers are also gaining acceptance.

Safety

Recent Federal regulations concerning noise and safety make it imperative that the machinery be situated in enclosures that comply with the regulations. Control of odors and blowing debris also justifies an enclosure. It is very difficult to design the machine itself to ensure compliance under all conditions. Local regulations should be considered for any proposed installation.

Advantages of Size Reduction

There are several distinct advantages to size reduction. First, size reduction facilitates handling of the waste material. It is very difficult to move municipal waste materials (as they are received from packer trucks) on conveyors and through surge bins or storage bins. Size reduction also reduces the net volume of the material. It is true that some objects, such as wooden boards and bricks, take up more volume after size reduction than before. However, the net result of shredding municipal waste produces the potential for volume reduction similar to that which might be achieved by a compaction operation.

It has been found that size reduction renders the waste material more acceptable for landfill use. Not only does it become more aesthetically acceptable in a community after it has been milled, but it has a practical value in that less compaction effort is required at the landfill site, which greatly reduces the operating problems and costs of the landfill.

In some cases, milled refuse needs no daily earth cover in landfills, although this point has not yet been definitely settled. Size reduction aids in the separation and resource recovery process. Likewise, size reduction permits suspension burning of the combustible portion of the solid waste. This aspect is becoming very important, in light of the impending energy crisis.

Disadvantages of Size Reduction

Size reduction is not without its disadvantages. The cost of operation is a significant factor. This cost can amount to as much as the total cost of landfilling the waste, prior to the present landfill regulations. In one vertical-shaft machine, the energy requirements are 24.1 kilowatt-hours per ton of throughput.

There is also a reliability problem with the present waste size reduction equipment. It is difficult to find a single facility that can operate daily without some type of shutdown involving the size reduction equipment. However, progress is being made in this area. The answers would come faster if the need for size reduction equipment would have justified greater research and development activities sooner. At the moment, a national landfill crisis, which would result in economic justification of the size reduction process, has not been reached.

Further, total resource recovery has not become an objective, and the markets have not adjusted themselves for this recycled material. Suspension burning is also in its infancy, as far as utilization of combustible solid wastes is concerned. It is hoped that answers will soon be found to the questions in these areas.

GLOSSARY

The following alphabetical listing of terms and definitions has general approval of the leading manufacturers of size reduction equipment for solid waste applications.

Basic Machine Types

Following are terms and definitions for basic machine types:

Term	Definition
Bale breaker	Basically a hammermill with special provision to control the rate and direction of the feed of a bale into the tip circle.
Cage disintegrator	High-speed impact machine in which friable material is size-reduced to 1/2 inch or less while escaping from the center of a rotating horizontal cage. (This machine is popular in the chemical industry.)
Chipper (knife hog)	Size reduction device having sharp blades or teeth, attached to a rotating shaft, that shave or chip off pieces of objects such as tree branches or brush.
Crusher	Size reduction machine that crushes friable material (i.e., coal and rock) to 1/2- to 10-inch size between two jaws or between a slowly rotating member and a fixed member.
Cutter	Size reduction machine in which wide, rotating blades shear thin sections of ductile material past a fixed, sharp edge.

Term	Definition
Disk mill	High-speed single or contrarotating disks between which pulpable material is processed. Input size is generally 2 inches, maximum; the output is fine grain or pulp.
Drum pulverizer	Grinding machine in the form of a large rotating drum, with a horizontal axis containing heavy, hinged arms that rotate inside. The drum has perforations, and the result is a rasp and sieve effect. This machine accepts a wide range of materials.
Flail	Basically a hammermill with excessively long hammers (or jointed hammers) or chains, in relation to the tip circle diameter (i.e., hammer length $\geq$ 1/3 tip circle diameter).
Fragmentizer	Hammermill or shredder used for size reducing automobiles, major appliances, and other large metal fabrications.
Grinder	A machine that reduces particle size by abrading material between a low-speed rotating member and a fixed member.
Hammermill	Broad category of size reduction equipment utilizing pivoted or fixed hammers with a tip velocity from 8000 to 14,000 feet per minute.
Hog	Basically a hammermill used specifically for size reducing wood and bark.
Hydropulper	Similar to a rasp mill, except this machine is designed for a wet process, generally utilizing water to aid in the size reduction process.
Impactor (impact mill)	A machine that grinds material by throwing it against heavy metal projections rigidly attached to a rapidly rotating shaft.
Log splitter	Mechanically or hydraulically driven wedge used to split logs.
Nuggetizer	Same as a fragmentizer.
Pulp lap shredder	Hammermill-type device specifically designed to shred pulp lap.
Pulper	Machine designed to reduce fibrous material to pulp (i.e., wood and paper).

Term	Definition
Pulverizer	High-speed (tip velocity > 14,000 ft/min) machine utilizing blades attached to a rotating shaft for size reduction to 1 inch or less.
Rasp mill (rasper)	Grinding machine in the form of a vertical drum containing heavy, hinged arms that rotate horizontally over a rasp-and-sieve floor.
Roller mill	A crusher that pulverizes material between a roller and a bull ring, or between two rollers, at a low rotating velocity.
Shear	Size reduction machine that cuts material between two large blades or between a blade and a stationary edge.
Shredder	Machine that reduces the size of solid objects primarily by cutting and tearing.
Size reduction chamber	That portion of a size reduction machine between the material inlet (throat) and the outlet, in which size reduction processing takes place.
Stump cutter	Portable machine that is positioned over a tree stump and chews the stump into chips using carbide-tipped teeth mounted on the periphery of a vertical cutting wheel.
Turnings crusher	Hammermill-type machine designed to size reduce metal turnings to facilitate handling and cutting oil removal.

Parts and Characteristics

Following are terms and definitions for parts and characteristics:

Term	Definition
Blades	Impact or shearing members attached (fixed or pivoted) to the disks in a shredder -- generally less than 1/2-inch wide.
Breaker bar	Bar attached to the frame, with projections that fit between the hammers to break up long items in a hammermill-type machine. (Heavy-duty-impact fixed hammers at the entrance of a vertical shaft machine.)
Breaker plate	Impact area against which material is propelled by action of the hammers in a hammermill-type machine.

Term	Definition
Cage bar	Same as grate bar.
Cage frame	Structure that supports cage or grate bars.
Center feed	Horizontal hammermill in which the material enters the tip circle at the top, dead center.
Chute	Passage that controls the maximum size and direction of flow of the material.
Compression feeder	Power-driven mechanism that compresses material as it force feeds it into the size reduction machine at a controlled rate.
Coupler (coupling)	Device that connects the drive shaft from the power source to the main shaft of the size reduction machine.
Disks	Plates that transmit the power in a size reduction machine from the main shaft to the cutting, tearing, or impact members.
Down running	Horizontal shaft hammermill in which material enters the tip circle as the hammer travels downward.
Drive	The pulleys, sheeves, belts, gears, and couplers required to transmit the power from the power supply to the main shaft of the size reduction machine.
Feeder	Mechanism that transports material into the machine at a controlled rate.
Frame	Basic structure of a size reduction machine, which encloses the moving parts and the feed inlet and product discharge areas.
Grate bars	Adjustable members between the rotor and the discharge area of a hammermill-type machine. Spacing of the bars regulates output size. Further reduction also results from impact of material against the bars.
Grinding plates	Wear resistant plates attached to the inside of the frame to provide a grinding surface.
Hammers	Impact members in a hammermill. These members may be straight or shaped, pivoted from one end or fixed, or the rings may be free to rotate about their centers.

Term	Definition
Hopper	Funnel-shaped receptacle for introducing material into the machine or into another receptacle.
Impact blocks	Plates against which the material is thrown by the hammers in a hammermill or impactor.
Infeed conveyor	Mechanism used to transport material to be processed into the size reduction machine.
Jaw	Movable member that applies the crushing force.
Knives	Impact or shearing members attached to the disk(s) in a shredder, hog, or chipper. The knives generally have sharp leading edges.
Liner	Replaceable wear plate fastened inside the impact area of the frame of a size reduction machine.
Metal catcher (metal trap)	Receptacle into which heavy metal and other difficult-to-size-reduce items are discharged ballistically in a hammermill-type machine.
Outfeed conveyor	Mechanical means of transporting size-reduced material from the machine outlet.
Over-running	Horizontal-shaft hammermill in which material enters the tip circle before hammers reach the topmost position.
Ring hammers	Hammers shaped like gears, stars, or rings that pivot about their centers, thus maximizing the wearing surface in contact with the material.
Rotor assembly	Rotating assembly consisting of a main shaft, disks, hammer bolts, and hammers in a hammermill-type machine -- with either a vertical or horizontal shaft.
Screen	Perforated formed plate, just outside of the tip circle, which regulates the size of the output product.
Shaft (hammer or hammerbolt)	Structural member near the periphery of the disk from which the hammers swing.
Shaft (main)	Mandrel, either horizontal or vertical, through which power is transmitted to the disks and thence to the hammers, blades, or knives in a size reduction machine.

Term	Definition
Suspension bar (hammer suspension bar)	Same as shaft (hammer or hammerbolt).
Swing hammer	Impact member that is attached to the disks in a hammermill, generally pivoted from a point near one end, and often made of manganese steel alloy.
Throat	Feed entrance to a size reduction machine that limits the maximum size of the material to be processed.
Tip circle	Path followed by the outermost tip of the rotating member in a size reduction machine.
Trap	Same as metal catcher.

HISTORY

A study of United States patents reveals that there have been very few changes in the design of size reduction equipment during the past 100 years. Hammermills and crushers have been used to reduce the size of coal, rocks, and grain products, as well as certain chemicals, since the early years in the development of the United States. The designs were generally conservative and were adequate for the homogeneous materials that were being processed. Ruggedness was the keynote rather than sophistication of the design.

There is not a great deal of engineering information available concerning the design of this machinery, other than certain basic rules concerning the vibration and inertial forces. The result is that it is difficult to design machinery for new types of materials, such as the industrial and residential solid waste application. The general attitude by the industry has been to attempt to apply traditional designs to the new applications. This attempt has not worked in many instances. There is also a great deal of skepticism among the size reduction equipment manufacturers regarding the market potential for the solid waste size reduction equipment. This attitude has restricted the research and development efforts, and the machines in use today at solid waste size reduction facilities are generally minor modifications of the traditional line of apparatus already produced.

Recent innovations have been attempted but have met with limited success. A machine that involved piercing, tearing, and impact was marketed by the Joy Manufacturing Company from 1967 through 1969. This horizontal-shaft machine employed a compression feeder that did a commendable job of crushing and uniformly feeding the input material. The waste material was simultaneously stretched by a piercing star wheel and was chopped by a rotating series of hammers, with the resultant products being discharged horizontally and automatically separated ballistically as to density and aerodynamic characteristics. Unfortunately, the equipment line was discarded by the Joy Manufacturing Company.

Another recent innovation is the use of a vertical-shaft rotating member, compared to the traditional horizontal shaft. By a strict interpretation of the definition, this machine should probably be called a grinder rather than a shredder. Some of the characteristics of this equipment were discussed above under "Horizontal and Vertical Machines". The Federal Government has sponsored a demonstration project in Madison, Wisconsin, involving the use of shredders to landfill milled refuse without additional cover. That project has been switched from a horizontal-shaft machine to a Heil-Tollemache vertical-shaft machine, with considerable success. Some of the problems involving this innovation pertain to the proper entrance of the waste material into the top of the rotor area. However, there are indications that this problem is being solved in some of the applications involving municipal solid waste today, and reliability appears to be similar to that expected from horizontal-shaft machines.

As the need for solid waste size reduction grows, the interest by the traditional manufacturers grows with it, and many of them are willing to quote on this type of application today, whereas five years ago they were not interested. However, there appears to be some discouragement in the industry, resulting from the long lead time involved in selling this equipment to municipalities compared to the rather short cycle between initial customer contacts and placement of an order by manufacturing firms and agricultural companies. Another problem that has developed stems from the lack of reliability of the size reduction equipment to meet the demands of the solid waste applications. The purchasers are demanding guaranteed results from the manufacturers. Some of the manufacturers are shying away from such contracts, and this may result in some slowing of progress.

European countries have been size-reducing solid waste materials for several years and have developed some machinery that may have advantages over domestic equipment. (Lanway of England and Svedala-Arbra of Sweden are two examples of European shredders. See Appendix IV, "Trip Reports Describing European Shredders.") On the other hand, it appears that domestic attempts to meet the challenge have resulted in some advantages over the European equipment. Undoubtedly the next few years will find modifications to domestic equipment to incorporate any apparent features from the foreign designs that would result in a better machine.

ASSESSMENT OF EQUIPMENT AVAILABLE

This subsection is devoted to answering questions relative to the availability and feasibility of shredding equipment for a material processing plant designed to handle 1000 tons per day, five days per week, on a two-shift-per-day basis. The source of information includes a comprehensive study made during the past five years by the General Electric Company, the in-depth study of various systems (as indicated in Sections 2 through 8 of this report), and a survey of leading manufacturers of equipment who might be considered candidates for the hybrid material separation system. These municipal waste shredder vendors are:

Vendor	Plant Location
American Pulverizer Company	St. Louis, Missouri
Beloit Corporation Jones Division	Dalton, Massachusetts
Eidal International Corporation	Albuquerque, New Mexico
Gruendler Crusher & Pulverizer Company	St. Louis, Missouri
Hammermills, Inc. Division of the Pettibone Mulliken Corporation	Cedar Rapids, Iowa
Heil Company	Milwaukee, Wisconsin
Jeffrey Manufacturing Company	Columbus, Ohio
Lanway Refuse Pulverisers Francis & John S. Lane, Ltd.	Staffs, England
Pennsylvania Crusher Division	Broomal, Pennsylvania (Bath Iron Works Corporation)
Rietz Manufacturing Company	Santa Rosa, California
Stedman Foundry & Machine Company, Inc.	Aurora, Indiana
Williams Patent Crusher & Pulverizer Company, Inc.	St. Louis, Missouri

Primary Shredders

The vendor candidates were asked to propose equipment that would have a high degree of reliability in meeting the daily requirements of the 1000-ton-per-day hybrid system without shutdown. This equipment would either include a great deal of overcapacity or a spare piece of equipment. The input material was stated as packer truck solid waste, the the output material from the shredder was to be 6 to 8 inches, nominal size. In addition to a description of the equipment proposed and the estimated cost, information concerning the vendors experience with similar applications was requested.

Replies were received from Eidal, Gruendler, Hammermills, Heil, Stedman, and Williams, indicating an interest in supplying this equipment. All vendors proposed horizontal-shaft, hammermill-type equipment except Eidal and Heil, who proposed their standard vertical-shaft equipment.

Diameters of the rotors proposed varied from a minimum of 40 inches to a maximum of 74 inches. Width of the feed opening varied from 75 to 104 inches. Hourly capacities of individual machines varied from 28 tons per hour to 100 tons per hour. There were proposals for combinations of equipment with possibilities of handling bulky wastes as well as normal residential refuse from packer trucks. Horsepower ratings ranged from 400 to 2000. Because the vendors were instructed to place maximum emphasis on reliability, it was observed that the horsepower ratings tended to be higher per ton of refuse than might normally be expected. It is also interesting to note that the horsepower ratings proposed for this equipment are considerably higher than ratings commonly used with European shredding devices.

Unit prices for this equipment, including motors and electrical controls but not including conveyors to feed material into the shredders or to discharge it from the shredders, ranged from $90,000 to $310,000. The lower number was proposed by a company with no previous experience shredding municipal waste in large hammermills. The total price for equipment that would provide a spare system for reliability and sufficient capacity for up to 25-percent bulky materials would cost from $270,000 to $375,000, not including installation costs. An additional $50,000 would provide a reversible feature on the shredder, which would provide greater dependability and lower operating costs. Very little information was obtained concerning operating costs, except that maintenance costs generally were reported at anywhere from 10 to 42 cents per ton, including parts and labor. Electrical costs were reported at 22 cents per ton for one machine powered by a 500-hp motor. (A study during the month of October 1972, at the Milford, Connecticut shredding facility, indicates the total electrical cost at that facility to be 83-1/2 cents per ton [Ref. 117].)

Another study (Ref. 118) places the overall cost of shredding between $1.40 and $3.60 per ton, breaking the cost down as follows:

| Category | Cost per Ton | |
	Low	High
Capital cost debt provision	$ .80	$2.00
Personnel	.40	1.00
Maintenance	.20	.60
	$1.40	$3.60

Routine maintenance usually is accomplished in less than four hours. This maintenance includes hammer or breaker bar faceplate changes. These changes must be made fairly often, due to the hard and abrasive materials in the solid waste. Major overhauls generally require 100 man-hours or more, or a weekend shutdown, and are performed approximately twice each year under normal operating conditions.

One supplier responding the the General Electric survey indicated that the maintenance cost runs 50 percent higher for a nonreversible shredder than for a reversible shredder.

All of these vendors, except Stedman, reported considerable experience in shredding municipal wastes. However, this experience is still very meager, compared to the experience they have with the size reduction of other types of material.

Secondary Shredders

For this application, the vendors were requested to propose high-reliability equipment to receive primarily combustible waste material from the low-density outlet of an air classifier with a nominal size of 6 to 8 inches. Vendors were requested to reduce this size to a nominal size of 1-1/4 inches. The average flow rate would be on the order of 30 tons per hour, but the equipment should have provisions for up to 50-percent surge capacity. Proposals were received from Eidal, Gruendler, Stedman, Williams, and Hammermill. All of these proposed horizontal-shaft hammermills, with the exception of Eidal, who proposed their vertical-shaft machine, and Stedman, who proposed a four-row cage mill. Rotor diameters varied from 40 to 60 inches, and feed opening widths varied from 60 to 100 inches. Shredder motor ratings ranged from 400 to 1250 hp.

The total cost for this equipment, including motors and electrical controls but excluding installation for a redundant, 1000-ton-per-day system ranged from $145,000 to $375,000. Again, an extra $50,000 would buy a reversible feature on the shredders. Operating costs were generally the same as for the primary shredders described above. One vendor stated a total average of $1.50 per ton, on a similar piece of equipment used for crushing coal, over a period of 11 years.

Gruendler and Williams reported considerable experience in similar applications. Stedman has also gained some experience, with their cage mill, at Puerto Rico and Altoona, Pennsylvania compost plants. These plants have been in operation from one to three years, and the vendor states that his units have run 8000 tons of materials with no repair or down time.

Tertiary Shredders

In this application, the vendor was asked to receive the output from the secondary shredder and process it through high equipment to result in a 0.015-inch nominal size, in preparation for the Garrett process. Only three vendors indicated an interest, and there were serious reservations included in the replies. The Gruendler Crusher and Pulverizer Company supplied samples of various materials that had been shredded to the point that allowed them to pass through a 1/8-inch perforated plate. The material was fluffy and fibrous. However, the wood particles appeared to be much larger than 0.015 inch diameter. Gruendler pointed out that they have considerable experience in producing equipment for the wood flour industry, and for newsprint and plastic film as well as fabrics such as cotton rags. However, they do not claim to be able to attain the 0.015-inch size.

The Williams Patent Crusher and Pulverizer Company supplied a sample of a product that appears to meet the requirements. They also demonstrated their equipment and indicated an interest in obtaining answers to cost and equipment questions if they can be assured that the material indicated by the sample is suitable. Their general observation was that it is <u>possible</u> to attain the desired size, but it may not be <u>feasible</u>.

At the time of this writing, the Reitz Manufacturing Company had been approached for information but had not had adequate time to reply. However, they have supplied the equipment for the Garrett process and should be able to respond to this inquiry.

Ferrous Material Densifier

The vendors were requested to process approximately 7 tons of ferrous material hourly, with an input size of 6 to 8 inches and the output size open to the vendors' recommendations. Replies were received from Eidal, Gruendler, Hammermills, and Williams. Spokesmen for the metal industry advised that the material at this stage would need to be open in configuration, or in a butterfly shape, compared to a balled configuration. Generally, the vertical-shaft machine has a greater tendency to produce the densified ferrous metal in a balled fashion and the hammermill-type machines tend to tear the material into the open, butterfly shape. The open shape is required for detinning operations on the tin can and container content of municipal waste. Foundries do not object to the material being in the balled condition. The information concerning the difference between output configurations between the two types of machines has not yet been confirmed.

Equipment sizes ranged from rotor diameters of 36 to 50 inches and feed opening widths of 30 to 84 inches. Shredder motor horsepower ranged from 150 to 700. The minimum total cost was approximately $40,000, and the highest cost was nearly $150,000. Gruendler indicated an output size of less than 2 inches, with a density of 70 pounds per cubic foot. Williams indicated that their price included their VIP density separator, which they feel is a very necessary accessory. The inclusion of this equipment did not push their price out of line. They specified an output size of 3 inches at a density of 20 to 30 pounds per cubic foot. The material would be in the butterfly configuration, acceptable for detinning. No operating costs were included with the proposals, but a sufficient degree of experience was indicated to establish confidence that the equipment is available for the given application.

Bulky Materials

Although this requirement was not specifically requested of the vendors, a general statement in the cover letter solicited comments concerning advisability of using primary shredders for bulky oversized waste. The equipment specified above under "Primary Shredders" describes equipment that will handle up to 20 percent bulky wastes, which should be sufficient.

SUMMARY

Design improvements are required for improved reliability of size reduction equipment as applied to solid waste materials. The market potential has not justified this effort in the past, but it may in the future, because of increased emphasis on resource recovery, suspension burning of combustibles as encouraged by the energy crisis, and by the fact that landfill sites will soon be at a premium.

It is also apparent that more operating experience is required for this type of equipment, with comprehensive record keeping needed to allow proper assessment of such items as the effect of multistage size reduction with interim classifications and segregations. These records could also help assess the operating costs of various processes, the value added to the waste by size reduction, and the best combination of equipment for a given application. To a better extent, the reliability of this equipment on large-scale installations and the required provisions for redundancy could also be determined.

The state-of-the-art of size reduction equipment must be appraised as being at the brink of adequacy for the need.

WET PULPERS

Another form of size reduction equipment common to the paper industry is the wet pulper. This equipment is usually a vertical-shaft machine similar in many respects to a scaled-up home garbage Disposall® unit. In operation, water is added to the solid waste to produce a slurry of about 90- to 95-percent water. The slurry is impacted at high speed, hardened, with rotating members usually running at about 5000 feet per minute. A perforated bottom plate, in conjunction with the rotating members, produces a rasp and sieve effect. In solid waste applications, provision is made to ballistically effect heavy objects.

One basic drawback to wet pulping is its inability to size-reduce non-pulpable materials such as rubber, plastics, and metals. Any process using wet pulping must provide a separate dry shredder to handle bulky wastes.

® Registered trademark of the General Electric Company

Wet pulping does, however, offer a uniform output that is easily transportable by slurry to other units in a process plant. Wet pulping also is more gentle to glass, with about 70 percent of the glass being broken into pieces large enough to color-sort.

Little operating data on wet pulpers are available, and the only known installation where solid waste is wet-pulped is at the A. M. Kinney/Black-Clawson operation in Franklin, Ohio. There, a 12-foot-diameter Hydrapulper is used. It is powered by a 300-hp motor and is rated at a capacity of 6.25 tons per hour. The Hydrapulper uses 1350 pounds of water per ton of waste. Cost figures were not available, but normal maintenance consists of replacing the rotor after every 100 hours of operation and the stator after every 1200 hours. Sources of wet pulpers include:

Vendor	Plant Location
Black-Clawson Company	Middletown, Ohio
French Oil Mill Company	Piqua, Ohio
SOMAT Corporation	Pomeroy, Pennsylvania
Conservomatic, Inc.	Oreland, Pennsylvania
Wascon, Inc.	Springfield, Ohio

AIR SEPARATORS

Air separation of the components of shredded solid waste is an art that is in its infancy (Ref. 119). A variety of air classifiers are in various stages of development. No units are yet known to have reached more than 10-ton-per-hour operating rates with what might be considered purity of the separated streams. The performance of these units was not investigated in detail, and quantifiable test data are known to be rare. Reference 119 is the only known published information. The better known types of air classifiers are described below.

VERTICAL AIR COLUMN CLASSIFIERS

The shredded waste stream is inserted, at the midpoint, into a vertical channel with an upward air flow. High-density objects tend to drop; low-density objects tend to rise and be conveyed pneumatically. This type of unit is being installed at the St. Louis facility by Rader Pneumatics, Inc.

ZIG-ZAG CLASSIFIERS

The zig-zag air classifier is similar to the vertical column unit, except the vertical column has a zig-zag shape with as many as 12 or more 45-degree turns in it. The resulting vortex action at each turn permits greater separating efficiency for each unit of vertical height. In addition, the physical agitation at each bend tends to break up agglomerations of waste.

Zig-zag classifiers have been built and operated by:

- Scientific Separators Company
- Stanford Research Institute
- Garrett Research and Development Corporation
- Combustion Power Corporation

A portable truck-mounted unit with 10-ton-per-hour capacity will soon be touring the country under the auspices of the National Center for Resource Recovery. This unit was designed and built for the National Center for Resource Recovery by the Combustion Power Corporation.

ROTATING DRUM AIR CLASSIFIER

The General Electric Company Industrial Heating Business Department in Shelbyville, Indiana is currently evaluating a promising air separator unit conceived at the General Electric Research and Development Center. This unit consists of an inclined horizontal drum, through which low-velocity air is drawn axially. Small axial vanes in the drum pick up the waste particles as it rotates and drops them across the air stream, which entrains the lower density particles. Initial test results are quite promising.

This approach, if successful, is believed to offer good separation at much lower air consumption rates than those of the vertical air column and zig-zag classifiers. Indications are that it will produce a better degree of purity of the separated streams as well.

CONVEYERS

A detailed study of conveyors was not performed; however, in visits to various facilities, it became obvious that conveying solid waste is no easy job, due to the variety of components involved. Figure 91, for instance, shows the input stream to a bulky waste shredder being handled by an inclined feed conveyor. In the background is another inclined conveyor, handling the output stream.

Slat-type conveyors are widely used in refuse processing and generally give good service. It is generally a good practice, at the top of inclined slat conveyors, to allow a short horizontal run before passing over the return pulley. This run allows the load to adjust and tends to lessen the carryover on the return side of the conveyor. Care must be given to the design of any directional change, whether vertical or horizontal. Also, when narrowing the flow path (e.g., when feeding a 4-foot shredder opening from a 6-foot conveyor), a mechanical squeezing device is usually needed.

Figure 91. Oversize Bulky Waste Size Reduction System
(Source: Williams Patent Crusher and Pulverizer Co.)

Bucket elevators, belt conveyors, and vibratory feed conveyors are also used. Vibratory feed conveyors serve a useful purpose in that the vibration tends to evenly distribute the material across the width of the conveyor, providing a more uniform feed to the shredders. Loading can be distributed more evenly along the length of the belts by utilizing two belts, with faster speed on the second belt.

Belt conveyors tend to concentrate the material in the center of the belt. When feeding a shredder by a belt conveyor, the center hammers of the shredder will wear at a faster rate than the outside hammers, causing earlier shutdown for maintenance. Uneven wear also defeats the wear-takeup mechanism on some hammermills, where clearance between hammers and grate bars is adjustable.

Vibratory conveyors can sometimes cause structural or acoustic interaction with building structures, as noted in Section 3 under "St. Louis Fuel Recovery Process." Vibratory conveyors also tend to be less effective with highly elastic material and with long, narrow pieces.

Another common problem observed in solid waste facilities is that of underdesign of the conveyors. Often conveyors are protected by shear pins to prevent damage to the conveyors by jams.   However, it is not an easy job to clear a jam on an underdesigned conveyor.

Another form of underdesign has been encountered in several conveyor applications, particularly those involving bucket elevators. As noted in this section under "Shredders," shredded solid waste can usually be compacted to a higher density than the unshredded material. However, as it leaves the shredder, it is usually of low density, often as low as 1 pound per cubic foot, and volumetric conveyors, such as bucket elevators that have been designed on the assumption of higher density values, cannot provide the desired throughput capacity.

APPENDIX I

Appendix I

TRIP REPORTS DESCRIBING MUNICH AND STUTTGART INCINERATORS

MUNICH NORTH PLANT

The incinerator at the Munich North Plant is an integral portion of a power station whose operational priorities are:

- Power production
- District heating
- Refuse incineration

In 1972, a total of 2.5×10^6 gigacalories were supplied to the district heating system.

The municipality that operates the utilities heats about 30 percent of Munich by natural gas, district heating, or electricity, with about 70 percent being heated privately. Stated another way, of all the heat generated in Munich, 50 percent is from natural gas, 31 percent is from coal, 10 percent is from trash, and 9 percent is from oil.

Munich is a city of 1,326,000 people and generates 430,000 tons of municipal refuse per year (from 600 to 660 pounds per person per year or 1.7 pounds per person per day).

The charges for heat from the central heating plant range from 18DM per gigacalorie to 23DM per gigacalorie, or from $1.60 per million Btu to $2 per million Btu. The cost of operating the incinerator was quoted as 25DM ($8.40) per metric ton. Tipping fees are 15DM ($5) per ton to private firms and 12DM ($4) per ton to the City. The difference of 10 to 15DM per ton is made up from the sale of electricity and district heating.

The Munich electrostatic precipitators are built by Lurgi and operate at about 99.5-percent efficiency.

The post-precipitator particulare loading is 60 to 70 milligrams per cubic meter (STP). Present regulations require 150 milligrams per cubic meter (STP). With regard to sulfur, the coal used at this plant runs up to 1-percent sulfur, while the trash is believed to sometimes run as high as 2-percent sulfur. Tests on the flue gas show 800 to 1200 milligrams of sulfur per standard cubic meter, while chlorine runs half of that at 400 to 600 milligrams per standard cubic meter.

This performance will not meet future requirements of 400 milligrams per cubic meter (STP) for these elements.

315

STUTTGART INCINERATOR/POWER PLANT

The boilers at the Stuttgart plant were built by EVT, a German affiliate of Combustion Engineering, Inc. The grates are Martin grates.

The Stuttgart facility has three boilers that derive heat from a combination of refuse and oil. The temperature of the flue gas just over the rtash grate is normally 1100°C to 1200°C (2000°F to 2150°F). The temperature falls through the radiant section, and after combining with the combustion gases from the oil-fired side of the boiler, the temperatures as it enters the super heater is from 700°C to 800°C (1290°F to 1470°F).

In the trash burning side of the boiler, the water tubes are studded and are covered with a coat of silicon carbide refractory, up to a height of 6 to 8 meters above the grate. This coating is applied to ensure that the flame from the fire cannot impinge upon the metal of the boiler tubes. This separation is necessary in order to eliminate the corrosion caused by the alternating oxidizing-reduction characteristic of the flame and atmosphere.

After the two gas streams combine, it is felt by F. Nowak (of the Stuttgart facility) that metal temperatures should be held below 700°C. In the beginning of this boiler's operation, they did encounter some corrosion on the hottest portion of the super heater. Nowak feels that corrosion was possibly caused by the chlorine reducing the iron oxides, but he feels that this chlorine corrosion process in inhibited by adequate oxygen.

Using oil firing, metal temperatures above 600°C are dangerous, due to vanadium. Therefore, their steam temperatures are limited to 525°C (977°F). A Bunker C oil with very little vanadium and a sulfur content between 0.8 and 1.0 percent is burned.

The fly ash precipitators are operated between 150°C and 320°C, to avoid hydrochloric dewpoint problems. The precipitators were provided by Rothemuhle and have an efficiency of 98.8 percent. According to Nowak, "dust from trash alone is better than dust from coal to collect," and "efficiency is better with oil than without oil."

This boiler has a heat release of 2×10^5 kilocalories per cubic meter per hour in the refuse furnace and 3.5×10^5 kilocalories per cubic meter per hour in the oil-fired furnace. Burning the trash in the high temperature of the oil fire, rather than in the double furnace arrangement, might push the sulfur into sulfate rather than sulfur dioxide.

The total sulfur dioxide in the combined flue gases is 300 milligrams per cubic meter (STP). Analysis shows that from 7.5 to 9 percent of the total sulfur is contained in the flue gas, while more than 90 percent of the sulfur remains in the bottom ash and fly ash. Experience indicates that fly ash has a tendency to absorb the sulfur dioxide and sulfur trioxide and that fly ash from refuse absorbs sulfur better than fly ash from coal.

A suitably designed new boiler could operate on a 50-50 split between heat energy from oil and heat energy from refuse. It is believed that in present existing boilers it is satisfactory to supply 33 percent of the heat from refuse and 67 percent from oil.

In the Stuttgart plant, it costs 30DM ($10) or more to burn 1 ton of refuse.

This plant is connected to a district heating network that requires about 100 gigacalories per hour. The energy output from this plant is distributed as follows:

Distribution	Percent
To electric railroad and tram cars	8
To city electric power system	23
As superheated steam, at 60 psig (via steam lines)	31
As hot water (via hot water pipes)	30
As auxiliary power pumps	8

The price of 33DM per gigacalories ($3/million Btu) includes the cost of pipelines.

The city of Stuttgart, with 650,000 people, produces 250,000 metric tons of refuse per year and is doubling this quantity every ten years. Twenty-five percent of all municipal refuse in West Germany is burned. Of the 25 percent burned, 75 percent is used for energy recovery and 50 percent of the incinerators are operated in conjunction with powerplants. The powerplants are operated by the city utility authority that operates the water, sewage, and gas systems.

The cities in Germany are pushing incinerator-fired district heating as a means of pollution control. The amount of sulfur and particulates from a well designed incinerator are less than those from a large number of individual stoves and furnaces.

Many towns are banding together to build incinerator district heating systems. This operation usually is desirable whenever a total population of more than 100,000 exists in a suitably small area.

Utilization of refuse energy is being carried out extensively in Europe. The optimum manner is to burn refuse in conjunction with fossil-fuel-fired furnaces.

APPENDIX II

Appendix II

HYBRID FACILITY DESCRIPTION AND CAPITAL COST ESTIMATE

This report was prepared by the Technology Assessment Team to establish an order-of-magnitude estimate for the capital cost of a building and process of a typical hybrid resource recovery plant located in the eastern United States. The report does not include any costs associated with the site or transportation of products, or any costs at the powerplant or other product user facilities. This plant is intended to be a commercial facility, not a demonstration plant. All costs are shown in 1972 dollars.

PLANT

The plant has been defined by the Technology Assessment Team as having the ability to receive and process 1000 tons per day of packer truck municipal refuse (not bulky refuse). The plant will operate two shifts per day, 5 days per week, and will produce:

- Fuel for a powerplant.

- Product containing a high percentage of ferrous metal.

- Product containing a high percentage of glass.

- Product containing a high percentage of noncombustibles for landfill.

Product quality is not clearly defined at this time.

SITE

Site costs are not included in this estimate. These costs are characteristic of the specific site and must be added after the site has been identified. Factors to be considered for any site are:

- Land

- Grading

- Roads

- Powerlines and substations

- Water and sewers

- Fencing

- Surface preparation

- Communications

BUILDING

The building has been defined as containing 70,000 square feet of floor space:

Area	Square Feet
Receiving	40,000
Processing	20,000
Office, utility, and personnel	10,000

The receiving area will hold a one-day inventory of incoming refuse (the 1000 tons will average 5.4 ft above the 40,000 ft^2). The assumption was made that the refuse will be diverted to another disposal site if a plant shutdown exceeds one day.

The processing area has been roughly estimated to be adequate to handle the defined processing steps. No layouts were attempted. The remaining 10,000 square feet will house offices, a locker room, personnel facilities, a utility room, and a maintenance shop.

General characteristics of the building will be:

- Conventional foundations (no piles)
- Rugged floor and bumping walls
- High ventilation requirements (four air changes per hour)
- Adequate lighting
- Low heat (50°F) in the receiving area
- Sprinklers
- Adequate building height (30-ft clearance)
- Large truck doors (15 ft wide by 25 ft high)
- Minimum column spacing (50 ft)
- Prefabricated metal construction

PROCESS

The process was defined by the Technology Assessment Team, and details are shown in Figure 92. This process was broken down into the individual pieces of processing equipment shown in Figure 93. The steps included in Figure 93 are:

Equipment	Description
Primary shredding (Steps 1-8)	Two lines operate; one does not operate.
	Each line averages 31.25 tons/hr (1000 tons/16-hr day).
	The product is a nominal 8-in. size.
Air separation (Steps 8-14)	Two lines operate; one does not operate.
	This separation is single-stage separation into dense and light phases, in proportions shown in Figure 92.
	Units of this size have not been demonstrated.
Secondary shredding and screening (Steps 15-19 and 22)	Two lines operate; one does not operate.
	Product has 1-in. maximum particle size.

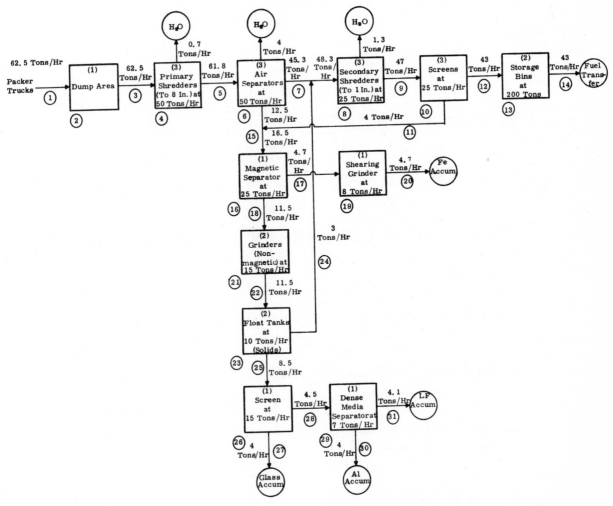

Figure 92. Block Diagram of Hybrid Solid Waste Management System

Equipment	Description
Secondary shredding and screening (Steps 15-19 and 22)(Cont'd)	Screens separate small-size particles, which are sent to the magnetic separator.
Fuel storage (Steps 20 and 21)	A single conveyor feeds two 200-ton Atlas storage silos (2/3 of one day's production).
	There are no processes included for unloading or transporting the fuel (i. e., fuel compaction and transport trucks).
Magnetic separator (Steps 22A, 23, 24, and 26)	Conveyors feed a single belt-type magnetic separator, which generator magnetic and nonmagnetic streams.
Magnetic metal shear and store (Steps 25 and 27-29)	The magnetic metal stream is run through a single mill to reduce the maximum size from 8 in. to 1 in. (Step 7). There is no additional processing.

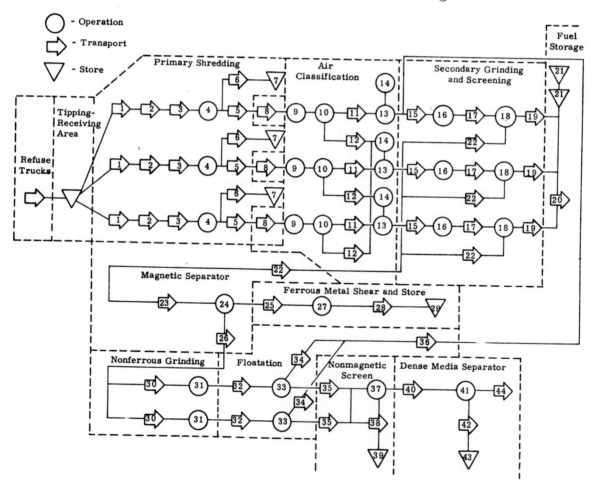

Figure 93. Preliminary Process Flow Diagram for Hybrid System

Equipment	Description
Magnetic metal shear and store (Steps 25 and 27-29)(Cont'd)	The product is piled on the floor or is discharged into trucks.
	The trucks are not included in the scope.
Nonmagnetic grinding (Steps 30 and 31)	One line operates; the other line does not.
	The girnders reduce the maximum size from 8 in. to 1 in. (?).
Flotation (Steps 32-34 and 36)	One line operates; the other line does not.
	This process produces two streams, based on a water density separation. The light stream (paper and wood) is transferred to the secondary grinding and screening process (Steps 15-19 and 22)
Nonmagnetic screening (Steps 37-39)	A single line operates.
	Small particles and water are separated in a screen process. Large particles move on to the dense media separators.
	The small particles (primarily glass) are pumped and separated (liquid cyclones) to a pile on the floor, to be loaded on trucks for sale. The liquid is returned to the flotation process.
Dense media separation for aluminum	A single line operates.
	Aluminum and other products are produced.
	Both products are stored on the floor for shipment.

At the present time, only shredding, grinding, and storage have demonstrated commercial-size processes on municipal waste material. Other processes have been demonstrated on pilot facilities, if demonstrated at all.

ESCALATION

Factors used to raise base costs from the source data to 1973 were:

- Building and site: 8 percent per year
- Equipment and processes: 6 percent per year
- Equipment shipping cost: 5 percent of FOB cost
- Sales tax: 7 percent of FOB cost

The base year has been established at 1972. A forecast escalator of 8 percent per year is recommended to estimate the costs for the construction of this plant in the future. Three years out would be a reasonable estimate at this time, or an increase of 26 percent over the base 1973 cost.

CONTINGENCIES

Contingencies reflect the state of engineering detail in defining a cost. A well defined project with an approved scope, including site layout, building layout, process flow diagrams, budget equipment quotations, and build-up cost estimates, would carry a 10 to 15 percent contingency for appropriation purposes.

Contingencies on this project are assumed to be from 10 to 50 percent, plus the 15 percent for a well defined project, depending on the degree of definition.

ARCHITECT-ENGINEER COSTS

Architect-engineer costs include functions related to translating a well-defined project scope into a finished facility. Normally, working with the client, the architect-engineer develops layouts of the site and buildings and firms up the process scope. This work is followed by detailed engineering design, equipment specification, bidding, selection, and purchase. Finally, a bid package is assembled with completed construction specifications, drawings, and vendor coordination necessary to allow competitive construction bidding, if desired. An average fee for this type of service is 15 percent.

SUMMARY

This estimate for $13,800,000 represents an order of magnitude estimate to be included in a request for funds (appropriation request) to construct a resource recovery building and process with a substantial assurance that this amount would cover the scope as defined. This estimate does not include site costs or any costs associated with transportation of the products, or additional processing of products, at user plants. The large contingency factors used on this estimate can be reduced by more detailed engineering definition of the building and processes, with corresponding changes in the base cost estimate:

Base cost	$ 9,650,000*
Contingencies	4,150,000
Total	$13,800,000

Table 79 lists estimated capital costs for the hybrid system; Table 80 describes equipment to be used in the system.

*Includes $240,000 sales tax on equipment.

Table 79

ESTIMATED CAPITAL COSTS
FOR HYBRID SYSTEM

Category	$ Thousands		
	Base	Contingency for State of Engineering Development	Total
Building			
Refuse receiving and process building (power plant facility not included)	$2,100	$ 530 (25%)	$ 2,630
Process Cost			
Primary shredding	1,760	790 (45%)	2,550
Air classification	2,020	1,010 (50%)	3,030
Secondary shredding and screen	700	320 (45%)	1,020
Fuel storage (no compaction or shipping)	760	340 (45%)	1,100
Magnetic metal separator	100	50 (45%)	150
Ferrous metal shearing**	100	50 (50%)	150
Nonferrous metal grinding	190	100 (50%)	290
Floatation	280	190 (65%)	470
Nonmagnetic screening**	100	60 (65%)	160
Aluminum separation**	290	180 (65%)	470
Total	$6,300	$3,080	$ 9,370
Total building and process	8,400	3,610	12,000
A/E cost (15%)	1,260	540	1,800
Total cost	$9,650	$4,150	$13,800

*Sales tax on equipment ($240,000) included in costs.
**No storage or shipment provided.

Table 80

EQUIPMENT LIST FOR HYBRID SYSTEM

Item	Material Flow Rate (tons/hr)		Equipment		Description
	Average	X1.5	Nominal Capital (tons/hr)	Quantity	
1	62.5	93.8	--	--	Packer trucks
2	62.5	93.8	--	1	Dump area
3	62.5	93.8	50	3	Input conveyors, slat type, 6 x 120 ft
4	62.5	93.8	50	3	Primary shredders (to 8 in.)
5	61.8	92.7	50	3	Exit conveyor, primary shredder
6	61.8	92.7	50	3	Air separators
7	48.3	72.5	25	3	Feed conveyor, secondary shredder
8	48.3	72.5	25	3	Secondary shredders (to 1 in.)
9	47	70.5	25	3	Exit conveyor, secondary shredder
10	47	70.5	25	3	Screens
11	4	6	6	1	Doors conveyor, screens to magnetic separator
12	43	65	65	2	Fuel conveyor to storage bins
13	43	65	200	2	Storage bins
14*	--	--	--	--	Exit conveyors, storage bin
15	16.5	24.8	25	1	Feed conveyor, magnetic separator
16	16.5	24.8	25	1	Magnetic separator
17	4.7	7.6	8	1	Feed conveyor, shearing grinder
18	11.5	17.3	15	2	Feed conveyor, grinder (nonmagnetic)
19	4.7	7.6	8	1	Shearing grinder
20	4.7	7.6	8	1	Exit conveyor, shearing grinder
21	11.5	17.3	15	2	Grinders (nonmagnetic)
22	11.5	17.3	15	2	Exit conveyors, grinders (nonmagnetic)
23	Solids 11.5	Solids 17.3	Solids 15	2	Float tanks
24	Solids 3	Solids 4.5	Solids 4.5	1	Exit conveyor, float tanks to secondary shredder
25	8.5	12.8	13	1	Exit conveyor, float tanks to screen
26	8.5	12.8	13	1	Screen (glass)
27	4	6	6	1	Exit conveyor, screen to glass accumulator
28	4.5	6.8	7	1	Feed conveyor -- dense media separator -- aluminum
29	4.5	6.8	7	1	Dense media separator -- aluminum
30	0.4	0.6	1	1	Exit conveyor, dense media separator to aluminum accumulator
31	4.1	6.2	7	1	Feed conveyor -- dense media separator -- copper
32	4.1	6.2	7	1	Dense media separator -- copper
33	0.1	0.2	0.2	1	Exit conveyor -- dense media separator to copper aluminum

APPENDIX III

Appendix III

COST ANALYSIS OF AMERICAN HOIST AND DERRICK COMPANY BALING OPERATION

This cost analysis was presented at the American Public Works Association Congress in Philadelphia on September 15, 1971. It was prepared by C.H. Scott of Control Systems, Inc. in Cincinnati, Ohio, who noted that:

- Section 1 unfolds total costs, operational and construction, of the facility installed on the purchaser's footings. Section 2 covers total costs, operation and installation, in an existing incinerator complex utilizing the holding pit and overhead crane for purposes of feeding the baler.

- The pages that follow are the facts applicable to a particular Ohio community and are supplied for illustration purposes only.

- Each Governmental subdivision presents its own particular problems, and the economic means by which these cost analysis systems can be applied can only come about as a result of an on-the-spot study.

The calculations and costs presented in the following schedules apply certain basic criteria averages known by Control Systems, Inc. from experience as well as other basic information applicable to the individual Governmental subdivision involved.

Schedule A sets forth the information pertaining to the American Press Model No. SWC-2528. Schedule B illustrates the manpower requirements with the hourly labor rates applicable in the labor market that was considered. This schedule gives consideration to hospital care, paid holidays, vacation reserve, and other employer costs.

Schedule C describes burden and overhead costs and allows for accounting and the application of data processing. Schedule D is a reserve account for replacement parts on the maintenance of basic hardware and the building.

Schedule E applies the heat, light, and power load, using the existing rates in the community considered. Schedule F is a contingency account and/or a reserve and covers purely those items of an unknown nature. Schedule G sets forth the facility costs and is based on a turnkey quotation. Schedule H provides the interest cost and/or debt service and is calculated at a 9-percent simple interest rate on a monthly declining balance.

Conclusions are prepared for a total quick-glance analysis and leave only a landfill cost and transportation charge to be added to arrive at a total per-ton cost of baled solid waste.

SCHEDULE A*

Operating hours (daily)		16
Annual operating hours based on a 5. 5-day week		4576
Bales/hr		40
Average weight/bale		2800 lb (1. 40 tons)
Annual bales		183, 040
Annual tons		256, 256
Annual yr^3 of fill required		243, 443
Daily tons average		896
95% utilization	(851. 20) 808. 64 yd^3	243, 443. 2 tons
90% utilization	(806. 40) 766. 08 yd^3	230, 630. 4 tons
85% utilization	(761. 40) 723. 52 yd^3	217, 817. 6 tons
80% utilization	(716. 80) 680. 96 yd^3	205, 004. 8 tons

*Application of American Press Model SWC-2528

Bale capacity = 40 bales/hr

Bale weight is controllable from 2200 lb through 3200 lb

Suggested bale weight of 2800 lb = 1. 4 tons

Bale configuration 3 × 3 × 4 = 1. 33 yd^3

Section 1

COMPLETELY NEW BALING FACILITY

SCHEDULE B*

Employees	Hospital Care	Holidays (8 Paid)	Vacation Reserve (3-Wk Base)	Wages
Gate keeper or weight master				
1st shift at $3.41	$ 636.00	$ 218.24	$ 409.20	$ 8,156.72
2nd shift at $3.61	636.00	231.04	433.20	8,635.12
Press operator				
1st shift at $4.36	636.00	279.04	523.20	10,429.12
2nd shift at $4.56	636.00	291.84	547.20	10,907.52
Tractor operator				
1st shift at $3.68	636.00	235.52	441.60	8,802.56
2nd shift at $3.88	636.00	248.32	465.60	9,280.96
Laborers (2)				
1st shift at $2.00 = $4.00	1,272.00	256.00	480.00	9,568.00
2nd shift at $2.20 = $4.40	1,272.00	281.60	528.00	10,524.80
Managers (2)				
1st shift at $5.25	636.00	336.00	630.00	12,558.00
2nd shift at $5.50	636.00	352.00	660.00	13,156.00
	$7,632.00	$2,729.60	$5,118.00	$102,018.80*
8 paid holidays		$2,729.60		
Vacation reserve			$5,118.00	
Based labor (see above)				$102,018.80
Social security				5,712.64
State unemployment				3,405.61
Federal unemployment				549.29
Workmen's compensation				2,021.40
Retirement contribution				9,887.26
Hospital care				7,632.00
Total labor and fringe costs -- annual				$139,074.60

*Per-ton cost   at 100% = 0.54271
         at 95% = 0.57130
         at 90% = 0.6030
         at 85% = 0.6385
         at 80% = 0.6784

This schedule sets forth the manpower requirements, with all considerations applied to each individual operation. Each person would work 44 hr/wk, be paid for holidays worked, and be paid time and a half for all hours over 40/wk. Hospital care and fringe benefits costs are scheduled.

SCHEDULE C

Fire insurance		$ 6,000.00	
PL and PD insurance		3,600.00	
Accounting and data processing		26,000.00	
Supplies		2,200.00	
	Annual	$38,400.00	
	Per-ton cost		
	at 100% =		0.1499
	at 95% =		0.1577
	at 90% =		0.1665
	at 85% =		0.1763
	at 80% =		0.1873

SCHEDULE D

Hardware replacement parts reserve		$12,000.00	
Reserve for building maintenance		6,000.00	
	Annual	$18,000.00	
	Per-ton cost		
	at 100% =		0.0702
	at 95% =		0.0739
	at 90% =		0.0780
	at 85% =		0.0826
	at 80% =		0.0878

SCHEDULE E

Heat, light, and power	Annual	$36,003.32	
Equipment requirement			
609-kva max			
430-kva operating			
	Per-ton cost		
	at 100% =		0.1405
	at 95% =		0.1479
	at 90% =		0.1561
	at 85% =		0.1653
	at 80% =		0.1756

SCHEDULE F

Contingencies	Annual	$25,523.12	
	Per-ton cost		
	at 100% =		0.0996
	at 95% =		0.1048
	at 90% =		0.1107
	at 85% =		0.1172
	at 80% =		0.1245

330

Baler Model SWC-2528			
Freight			
Installation			

Factory technician unloading,
       assembly, and startup
    Electrical
		$ 722,830.00	

Conveyors			
Freight and installation Metal skirting		108,800.00	
Building			

Special footer for baler
       weight 600,000 lb
    Footer for building
    10-in. reinforced concrete with drains
    Heating and lighting
		205,000.00	

Rubber-tired front-end loader		28,450.00	
Misc. items		9,920.00	
		$1,075,000.00	
	Per-ton cost		
	at 100% =		0.4195
	at 95% =		0.4415
	at 90% =		0.4661
	at 85% =		0.4935
	at 80% =		0.5243

SCHEDULE H

Interest cost		$ 559,117.20	
Annual		55,911.72	
Monthly payback		4,659.31	
	Per-ton cost		
	at 100% =		0.2181
	at 95% =		0.2296
	at 90% =		0.2424
	at 85% =		0.2566
	at 80% =		0.2727
	Schedule G	$ 8,958.33	
	Schedule H	4,659.31	
Total monthly payback		$13,617.64	

CONCLUSION

Schedule	100%	95%	90%	85%
A -- Tons	256, 256	243, 443.2	230, 630.4	217, 817.6
B -- Labor	0.54271	0.57130	0.6030	0.6385
C -- Overhead	0.1499	0.1577	0.1665	0.1763
D -- Maintenance -- parts reserve	0.0702	0.0739	0.0780	0.0826
E -- Heat, light, and power	0.1405	0.1479	1.561	0.1653
F -- Contingencies	0.0996	0.1048	0.1107	0.1172
G -- Facility and equipment	0.4195	0.4415	0.4661	0.4935
H -- Interest cost	0.2181	0.2296	0.2424	0.2566
Per ton	1.6405	1.7267	1.8228	1.9300

Daily operating hours = 16
Annual operating hours = 4576

ANALYSIS

At 90% utilization	$1, 8228
Known landfill cost FOB rail car at landfill siding (per ton)	1.75
Quoted rail rate for distance of 329 mi. -- per ton	3.09
Per-ton cost from collection trucks and/or transfer trailer -- totals	$ 6,6628

Section 2

EXISTING INCINERATOR CONVERSION

SCHEDULE B

Employees	Hospital Care	Holidays (8 Paid)	Vacation Reserve (3-Wk Base)	Wages
Equipment maintenance and managers				
1st shift at $5.50	$ 636.00	$ 352.00	$ 660.00	$ 13,156.00
2nd shift at $5.25	636.00	336.00	630.00	12,558.00
Weight master				
1st shift at $3.41	636.00	218.24	409.20	8,156.72
2nd shift at $3.61	636.00	231.04	433.20	8,635.12
Crane operator				
1st shift at $4.36	636.00	279.04	523.20	10,429.12
2nd shift at $4.56	636.00	291.84	547.20	10,907.52
Press operators and maintenance				
1st shift at $3.68	636.00	235.52	441.60	8,802.56
2nd shift at $3.88	636.00	248.32	465.60	9,280.96
Floor labor and maintenance				
1st shift at $2.00	636.00	128.00	240.00	4,784.00
2nd shift at $2.20	636.00	140.80	264.00	5,262.40
Bale loader and maintenance				
1st shift at $2.00	636.00	128.00	240.00	4,784.00
2nd shift at $2.20	636.00	140.80	264.00	5,262.40
	$7,632.00	$2,729.60	$5,118.00	$102,018.80

	Hospital Care	Holidays (8 Paid)	Vacation Reserve (3-Wk Base)	Wages
8 paid holidays		$2,729.60		
Vacation reserve			$5,118.00	
Based labor (see above)				$102,018.80
Social security				5,712.64
State unemployment				3,405.61
Federal unemployment				549.29
Workmen's compensation				2,021.40
Retirement contribution				9,887.26
Hospital care				7,632.00
Total labor and fringe costs -- annual				$139,074.60

*Per-ton cost   at 100% = 0.54271
                      at 95% = 0.57130
                      at 90% = 0.6030
                      at 85% = 0.6385
                      at 80% = 0.6784

This schedule sets forth the manpower requirements, with all considerations applied to each individual operation. Each person would work 44 hr/wk, be paid for holidays worked, and be paid time and a half for all hours over 40/wk. Hospital care and fringe benefits costs are scheduled.

SCHEDULE C

Fire insurance		$ 6,000.00	
PL and PD Insurance		3,600.00	
Accounting and data processing		26,000.00	
Supplies		2,200.00	
	Annual	$38,400.00	
	Per-ton cost		
	at 100% =		0.1499
	at 95% =		0.1577
	at 90% =		0.1665
	at 85% =		0.1763
	at 80% =		0.1873

SCHEDULE D

Hardware replacement parts reserve		$12,000.00	
Reserve for building maintenance		6,000.00	
	Annual	$18,000.00	
	Per-ton cost		
	at 100% =		0.0702
	at 95% =		0.0739
	at 90% =		0.0780
	at 85% =		0.0826
	at 80% =		0.0878

SCHEDULE E

Heat, light, and power	Annual	$36,003.32	
Equipment requirement			
609-kva max			
430-kva operating			
	Per-ton cost		
	at 100% =		0.1405
	at 95% =		0.1479
	at 90% =		0.1561
	at 85% =		0.1653
	at 80% =		0.1756

SCHEDULE F

Contingencies	Annual	$25,523.12	
	Per-ton cost		
	at 100% =		0.0996
	at 95% =		0.1048
	at 90% =		0.1107
	at 85% =		0.1172
	at 80% =		0.1245

SCHEDULE G

Baler Model SWC-2528		
Freight		
Installation	$722,830.00	
Factory technician unloading, assembly, and startup		
Electrical	33,500.00	
Conveyor	50,000.00	
Building alterations (est)	$806,330.00	
Monthly payback	$6,719.42	
Per-ton cost		
at 100% =		0.3147
at 95% =		0.3312
at 90% =		0.3496
at 85% =		0.3702
at 80% =		0.3933

SCHEDULE H

Interest cost	$419,291.60	
Annual	41,929.16	
Monthly payback	3,494.10	
Per-ton cost		
at 100% =		0.1636
at 95% =		0.1722
at 90% =		0.1818
at 85% =		0.1925
at 80% =		0.2045
Schedule G	$ 6,719.42	
Schedule H	3,494.10	
Total monthly payback	$10,213.50	

CONCLUSIONS

Schedule	100%	95%	90%	85%
A -- Tons	256, 256	243, 443.2	230, 630.4	217, 817.6
B -- Labor	0.54271	0.57130	0.6030	0.6385
C -- Overhead	0.1499	0.1577	1.1665	0.1763
D -- Maintenance -- parts reserve	0.0702	0.0739	0.0780	0.0826
E -- Heat, light, and power	0.1405	0.1479	1.561	0.1653
F -- Contingencies	0.0996	0.1048	0.1107	0.1172
G -- Facility and equipment	0.3147	0.3312	0.3496	0.3702
H -- Interest cost	0.1636	0.1722	0.1818	0.1925
Per ton	1.4812	1.5590	1.6457	1.7426

Daily operating hours = 16
Annual operating hours = 4576

ANALYSIS

At 90% utilization	$1, 6457
Known landfill cost FOB rail car at landfill siding (per ton)	1.75
Quoted rail rate for distance of 329 mi. (per ton)	3.09
Per-ton cost from collection trucks and or transfer trailer -- totals	$6, 4857

APPENDIX IV

Appendix IV

TRIP REPORTS DESCRIBING EUROPEAN SHREDDERS

LANWAY COMPANY

The design of this Lanway shredder in Brierley Hill, England, grew from the experience of the Lanway Company in the mineral hammermill business. The primary argument advanced for the efficiency of this design over that of ordinary hammermills and mills of the Eidal type is that, being designed from the start for the shredding of municipal trash, it requires less horsepower than those mills derived directly from mineral mills.

The mineral mill must have sufficient installed horsepower to pulverize everything that enters, without stalling; otherwise, the problem of removing large unshreddable objects would unduly increase downtime. The Lanway shredder has overcome this problem by providing sufficient room inside the shredder for most large, shred-resistant objects to pass through and be re-jected without stalling the shredder.

Large, realtively brittle items, such as car engine blocks, will be shat-tered by the impact of the special manganese steel hammerheads. These manganese steel hammerheads are replaceable and are replaced after about 3 inches have been worn away. The pattern is such that a toothed cutting edge is maintained even in the face of the wearing. The wearing of these manganese steel hammerheads is estimated, based on 2000 tons of refuse that has been shredded, to require replacement every 3000 to 15,000 tons, depending upon the fineness of the setting and of course on the amount of hard material in the rubbish. This replacement rate converts to running maintenance costs of from 30 to 60 cents per ton (in the U.S.), according to John Giglio, United States representative of the Lanway Company.

The Lanway people are somewhat disturbed that they have frightened possible customers by suggesting an underpowered system. As G. Pow, at the Lanway Company, said: "We don't believe we need more power, but we will, of course, put on any motor that a customer specifies."

An interesting feature of this shredder is the adjustment provided to com-pensate for the wear on the hammers and to allow for variations in clearance desired between the hammers and the grating structure. This adjustment is an external screw-jack type of arrangement.

The Lanway Company had arranged to have 2 tons of city refuse to run through the shredder, including a large divan, on the day the plant was visited. The demonstration did not include any objects tough enough to be rejected. Bulky objects, such as white goods, would be shredded by this system, although an electric motor, for example, might only be mangled and not completely shredded.

The Lanway Company does not have any shredders of this type running on an operational basis. They are fabricating shredders and conveying equipment for an installation at Margate. These shredders at Margate are interesting, because the shredded material will be carried across the river on a conveyor and will be landfilled on the other side of the river.

The Lanway Company has recently submitted a proposal to New York City's environmental agency for barge-mounted shredders that would chop fuel that would be burned in a Consolidated Edison powerplant. The shredders would be barge-mounted, to eliminate the need for finding and acquiring a site for a land-based system.

The Lanway Company has designed their installations as a total package, including the conveyors, becuase they build conveyors for other purposes (crushed rock). Because Lanway does have an American representative (J. Giglio, in Pawtucket, Rhode Island), they declined to discuss installation prices, indicating that all such requests should be addressed to their representative.

Whereas they want to do the on-site erection and checkout in England, they would contemplate sending only a field installation supervisor to the United States and the unit would be shipped in knocked-down form for on-site assembly. At this time, Lanway does not believe it necessary to house the equipment in a building.

SVEDALA ARBRA COMPANY

Svedala Arbra is a Swedish company that specializes in stone crushing and conveying equipment. Svedala has some working arrangement with Allis-Chalmers, in the United States, and uses Allis-Chalmers Hydro-Cone rock crushing units.

The Svedala Arbra shredders that were seen in operation are modified rock hammermills, modified to specifically design them to shred municipal solid waste.

The Svedala Arbra concept, like the Lanway concept, avoids shredding dense, very tough objects, although the items that it will reject, unshredded, range down to objects the size of tin cans. The Svedala Arbra shredder has a set of breaker bars. These bars are spaced so that the rotating hammers pass between the breaker bars. A large object is either broken and forced between the breaker bars into the region of the regular grate, or the object is driven over the top of the breaker bars and out past a set of spring-loaded doors. The action of this shredder on an object such as an old home freezer would probably be to tear it into major chunks and then reject the major chunks, in much the same way as the Lanway shredder operates.

The swing hammers in this mill are four-cornered; the steel hammers themselves have very tough teeth welded onto each of the two narrow faces at each end. As a corner wears down, the hard teeth wear less and maintain the toothed configuration. When worn out, the rods holding the hammers are pulled and the hammers are reversed. After all four corners are worn out, a new set of hammers is installed.

These shredders are supplied with overhead rails on which one or both of the top quarters of the shredder housing can be rolled back for maintenance. At the present time, the two shredders at this landfill operation are shredding about 2000 tons per week, and the swing hammers are changed to a new corner about once a week. Two to four man-hours are said to suffice for changing a set of hammers.

The shredders are at Kovic, a combined municipal and industrial dump operated by a private collection and disposal company (Sellbergs). The refuse brought to this dump is brought from the areas southeast of Stockholm.

The initial installation costs of this double shredder were about 2.5×10^6 Swedish croner, and a duplicate installation built today in Sweden was estimated at 4×10^6 Swedish croner (about \$850,000). The installation includes a steel building (Butler type) and two steel slat conveyors onto which the rubbish is dumped directly from the trucks (Sellbergs is switching from European to Heil trucks). The conveyors feed the two shredders directly, and a single conveyor carries the shredded rubbish out to dump into a 24-cubic-meter dumping carryoll (rubber-tried). This carryall moves the shredded material to the landfill. These is no packing in the landfill, other than that done incidentally by the carryall running over the top of it.

A number of interesting points can be made about this landfill. First, the top foot or so of the shredded rubbish proceeds to compost. Such a large number of earthworms (standard fishing-type earthworms) are seldom seen; the presence of these earthworms indicates nonseptic aerobic composting rather than anaerobic fermentation. There was almost no odor, and there was no evidence of rats.

The material was damp only to a depth of 1 to 1-1/2 feet, even though there had been fall rains and snows. There has been a great deal of interest in expediting the open composting procedure as a method of further volume reduction in the landfill.

S. Mattsson, of Svedala Arbra, believes that adjusting the shredder for a coarser cut may open the top layers for a better penetration of air and water. At the present time, the top layers are tight enough so that a heavy rain will run off rather than percolating into the landfill. This landfill demonstrates the exothermic reaction of the composting, melting the snow and even feeling warm 1 foot below the surface.

No attempt has been made to cover this landfill with dirt. In some areas, grass seed has been spread on the surface; it germinates and grows readily. In one area, a faint green indicated germination of the seed during winter, the seed being warmed by the composting heat.

On the same area, about 1/4 mile away, Sellbergs operates an industrial landfill into which everything, including liquid wastes, was being dumped. A large, wheeled (steel-lugged) landfill compactor was operating to compact the fill. Again, the shredded waste and grass technique is being used to cover the compacted industrial waste in the dump. This dump was somewhat isolated, with regard to leachate, by having concrete retaining walls built across the drainage ways by which water percolating through the landfill would be able to move into the neighboring streams, lakes, and inlets. The low collection basins in front of these concrete dams have suction pipes installed in them for intakes to circulating pumps, so if the basins begin to fill up, the water can be sprayed on top of the fill. The assumption is that in the long run, all of the water percolating into the fill can be reevaporated or transpired by the vegetation growing on the fill. It would appear that, in the long run, this concept might be aided by planting water loving trees in the vicinity of the seepage basins or on the face of the fill. Tests carried out at this landfill indicate that the volume of the fill, if composting proceeds well, will be reduced by at least 50 percent.

This volume reduction is not currently taking place, because the material is spread on too thickly and compacts too much to allow the penetration of sufficient oxygen and water for the composting to proceed in the lower portions of the landfill. The operator of this landfill is planning to enclose the fill on those sides exposed to public view by building a ridge of shredded trash that will be planted with grass on the visible side. One advantage noted is that the shredded trash does not create as much wind-borne paper as the unshredded material from the industrial waste area.

Ultimately, this area will become a ski slope. It may be many years before the material compacts to a final volume and temperature.

APPENDIX V

Appendix V

REFERENCES

1. K. McCartney, <u>Preliminary Assessment of Gasification Technology</u>, Connecticut Southern Gas Company, December 29, 1972.

2. W. E. Franklin, D. Bendersky, L. J. Shannon, and W. R. Park, <u>Resource Recovery from Mixed Municipal Solid Wastes -- A Survey Analysis, and Catalog of Existing and Emerging Systems -- Part I -- Assessment of Resource Recovery Technology</u>, prepared for the President's Council on Environmental Quality, Projects 3523-D and 3634-D, Midwest Research Institute, Kansas City, Missouri, August 10, 1972.

3. <u>Greater Bridgeport Regional Solid Waste Management Study</u>, Metcalf and Eddy, November 1972.

4. D. Bendersky, et al, <u>Quality of Products from Selected Resource Recovery Processes and Market Specifications</u>, prepared for the General Electric Company, Project No. MRI-3700-D, Midwest Research Institute, Kansas City, Missouri.

5. <u>Preliminary Report on Burning Refuse as a Supplementary Fuel in a Utility Boiler</u>, Northeast Utilities Service Company, July 1972.

6. D. L. Klumb (Union Electric Company), telephone conversation with W. A. Boothe, March 8, 1973.

7. G. W. Sutterfield, <u>City of St. Louis Facilities and Operating Experience -- Solid Waste Disposal Project</u>, paper presented at the Solid Waste Disposal Seminar, Union Electric Company, St. Louis, Missouri, October 1972.

8. G. W. Sutterfield (City of St. Louis Refuse Commissioner), interview with W. A. Boothe, May 25, 1972.

9. D. L. Klumb, <u>Union Electric Facilities and Operating Experience -- Solid Waste Disposal Project</u>, paper presented at the Solid Waste Disposal Seminar, Union Electric Company, St. Louis, Missouri, October 1972.

10. W. H. Doty, T. P. Nemik, G. V. Stolzer, and G. M. Thomson, <u>The Analysis of Refuse</u>, paper presented at the Solid Waste Disposal Seminar, Union Electric Company, St. Louis, Missouri, October 1972.

11. D. L. Klumb, <u>Report on St. Louis/Union Electric Solid Waste Disposal Experiment</u>, paper presented at the Incinerator Division Meeting, American Society of Mechanical Engineers, November 28, 1972.

12. D. L. Klumb (Union Electric Company), telephone conversation with W. A. Boothe, December 4, 1972.

13. E. J. Lasch, <u>Order of Magnitude Estimate for Capital Cost of a Typical</u> <u>1000 Ton/Day Refuse Fuel Plant (Basis City of St. Louis) in Connecticut</u> <u>1972,</u> December 7, 1972.

14. R. L. Ahlbrand (A. M. Kinney, Inc.) correspondence with W. A. Boothe, August 21, 1972.

15. C. K. Miller (A. M. Kinney, Inc.), questionnaire completed for Midwest Research Institute Project No. 3634-D, April 19, 1972.

16. R. A. Wilson, "'Clean' Town Cashes in on Daily Dirt," <u>Iron Age</u>, September 2, 1971.

17. <u>Black-Clawson Hydraspososal/Fibreclaim</u>, Publication No. YK-10M-372, Black-Clawson Company, Middletown, Ohio.

18. E. J. Lasch, <u>Notes on Conference on Glass Recovery Process</u>, based on conference with Dr. J. P. Cummings, Director of the Franklin Glass Recovery Project, March 1, 1973.

19. <u>The Kinney Thermal Recovery System</u>, A. M. Kinney, Inc., Cincinnati, Ohio, January 1972.

20. C. Spencer (Louisville Dryer Company), telephone conversation with R. Hasselbring, August 25, 1972.

21. D. A. Furlong (Combustion Power Corporation), questionnaire completed for Midwest Research Institute Project No. 3634-D, May 11, 1972.

22. F. Walton (Combustion Power Corporation), telephone conversation with W. A. Boothe, December 7, 1972.

23. E. J. Lasch, correspondence with W. A. Boothe and R. P. Shah, September 25, 1972.

24. R. M. Roberts and E. M. Wilson, <u>Systems Evaluation of Refuse as a Low</u> <u>Sulfur Fuel</u>, Vol. I, National Technical Information Service Report No. PB 209271 prepared for the Environmental Protection Agency, Envirogenics Company, November 1971.

25. W. R. Niessen and A. F. Alsobrook, <u>Municipal and Industrial Refuse:</u> <u>Compositions and Rates</u>, paper presented at the National Incinerator Conference, American Society of Mechanical Engineers, June 1972.

26. E. R. Kaiser, "Refuse Composition and Flue Gas Analyses from Municipal Incinerators," <u>Proceedings of the National Incinerator Conference,</u> American Society of Mechanical Engineers, New York, New York, 1964.

27. G. Stabenow, <u>Performance of the New Chicago Northwest Incinerator,</u> paper presented at the National Incinerator Conference, American Society of Mechanical Engineers, June 1972.

28. E. R. Kaiser, <u>Municipal Incinerator Refuse and Residue</u>, paper presented at the National Incinerator Conference, American Society of Mechanical Engineers, 1968.

29. W. R. Niessen and S. H. Chansky, <u>The Nature of Refuse</u>, paper presented at the National Incinerator Conference, American Society of Mechanical Engineers, 1970.

30. E. D. Stewart (Monsanto Company), correspondence with W. A. Boothe, General Electric Company, September 5, 1972.

31. W. A. Boothe, <u>Visit to Ecology, Inc.</u>, trip report, October 1972.

32. H. Hollander, et al, <u>Beneficiated Solid Waste Cubbettes as Salvage Fuel for Steam Generation,</u> paper presented at the National Incinerator Conference, American Society of Mechanical Engineers, 1972.

33. R. M. Roberts and E. M. Wilson, <u>Systems Evaluation of Refuse as a Low Sulfur Fuel: Part I,</u> Paper No. 71-WA/Inc-3, American Society of Mechanical Enginners.

34. L. K. Davis, correspondence with P. G. Shanley, December 14, 1972.

35. M. B. Owen, "Sludge Incineration," <u>Journal of the Sanitary Engineering Division of ASCE,</u> Paper No. 1172, February 1957.

36. E. R. Kaiser, telephone conversation with R. P. Shah, March 2, 1973.

37. G. J. Trezek, et al, "Mechanical Properties of Some Refuse Components, <u>Compost Science,</u> November-December 1972.

38. R. P. Shah, <u>Discussion with Rothemühle Engineers on Electrofilters,</u> Trip Report, March 12, 1973.

39. R. P. Shah, <u>Composition and Heat Value of Average Municipal Refuse,</u> memo, November 10, 1972.

40. I. G. Bowen and L. Brealey, <u>Incinerator Ash -- Criteria of Performance,</u> paper presented at the National Incinerator Conference, American Society of Mechanical Engineers, 1968.

41. J. H. Fernandes, <u>Incineration Air Pollution Control,</u> paper presented at the National Incinerator Conference, American Society of Mechanical Engineers, 1968.

42. S. Gilette, <u>Air Compliance Particulate Emission Data for Municipal Incinerators,</u> memo, November 16, 1972.

43. <u>Systems Evaluation of Refuse as a Low Sulfur Fuel,</u> Vol. II, National Technical Information Service Report No. PB 209272, prepared for the Environmental Protection Agency, Envirogenics Company, November 1971.

44. K. H. Thoemen, <u>Contribution to the Control of Corrosion Problems on Incinerators with Waterwall Steam Generators,</u> paper presented at the National Incinerator Conference, American Society of Mechanical Engineers, 1972.

45. R. W. Bryers and Z. Kerekes, <u>Corrosion and Fouling in Refuse-Fired Steam Generators</u>, paper presented at the Symposium on Solid Waste Disposal, Seventh National Meeting, American Institute of Chemical Engineers, August 1971.

46. P. D. Miller and H. H. Krause, <u>Corrosion of Carbon and Stainless Steels in Flue Gases from a Municipal Incinerator</u>, paper presented at the National Incinerator Conference, American Society of Mechanical Engineers, 1972.

47. H. H. Krause, D. A. Vaughan, and P. D. Miller, <u>Corrosion and Deposits from Combustion of Solid Waste</u>, Paper No. 72-WA/CD-2 presented at the 1972 Winter Annual Meeting, American Society of Mechanical Engineers, November 1972.

48. F. Nowak, "Corrosion Problems in Incinerators," <u>Combustion</u>, November 1968, pp. 32-40.

49. R. P. Shah, <u>Air Pollution Potential from Mixed Refuse Burning in a Utility Boiler -- Particulate Emission</u>, memo, November 17, 1972.

50. R. E. Sommerlad, "Discussion of Paper by P. D. Miller, et al" (Ref. 46), <u>National Incinerator Conference -- Discussions</u>, American Society of Mechanical Engineers, 1972.

51. G. M. Mallan and C. S. Finney, <u>New Techniques in the Pyrolysis of Solid Wastes</u>, Report No. GR&D 72-035, Garrett Research and Development Company, LaVerne, California (also paper presented at the 1973 National Meeting of the American Institute of Chemical Engineers, August 29, 1972).

52. G. Mallan (Garrett Research and Development Company), telephone conversation with W. A. Boothe, August 11, 1972.

53. <u>The Occidental Report</u>, July-August 1972.

54. W. A. Boothe, <u>Visit to Garrett Research and Development Company</u>, trip report, February 12, 1973.

55. B. Morey and J. P. Cummings, "Glass Recovery from Municipal Trash by Froth Flotation," <u>Proceedings of the Third Mineral Waste Utilization Symposium</u>, March 14-16, 1972.

56. <u>A Report on the Garrett Research and Development Co., Inc. Municipal Solid Waste Resource Recovery System</u>, Garrett Research and Development Company, LaVerne, California, January 1972.

57. G. M. Mallan (Garrett Research and Development Company), questionnaire completed for Midwest Research Institute Project No. 3634-D, April 21, 1972.

58. K. McCartney, <u>Preliminary Assessment of Gasification Technology</u>, Southern Connecticut Gas Company, December 29, 1972.

59. L. Shannon, <u>Pyrolysis Processes</u>, Midwest Research Institute, Kansas City, Missouri, February 28, 1973.

60. R.W. Borio, <u>Combustion and Handling Characteristics of Garrett's Pyrolytic Oil</u>, Project 900127, Combustion Engineering,Inc., Windsor, Connecticut.

61. W.E. Franklin and W.R. Park, <u>Economic Analysis of Pyrolysis System for Recovery Resource Value from Mixed Municipal Wastes,</u> prepared for the General Electric Company, Midwest Research Institute, Kansas City, Missouri, April 1972.

62. E.D. Stewart (Monsanto Company), correspondence with J.A. Mirabal, September 28, 1972.

63. <u>LANDGARD Solid Waste Disposal System</u>, Publication No. IGI-LRA 572/5M/LID, Monsanto Company, St. Louis, Missouri.

64. P.H. Kydd, <u>An Evaluation of Refuse Pyrolysis</u>, General Electric Company, Schenectady, New York (Class IV report).

65. <u>Monsanto Envirochem and You,</u> Publication No. IGI-LRA 572/5M/LID, Monsanto Company, St. Louis, Missouri.

66. W.A. Boothe, <u>Visit by Union Carbide</u>, memo, March 2, 1973.

67. R.S. Paul (Union Carbide Corporation),correspondence with W.A. Boothe, October 18, 1972.

68. R.S. Paul (Union Carbide Corporation),correspondence with W.A. Boothe, November 15, 1972.

69. <u>URDC Background Data,</u> Urban Research and Development Corporation, East Granby, Connecticut, April 30, 1971.

70. W.A. Boothe, Urban Research and Development Corporation Trip Report, East Granby, Connecticut, Obtober 6, 1972.

71. W.A. Boothe, <u>Trip Report, Environmental Protection Agency,</u> September 7, 1972.

72. <u>A Better Way</u>, American Gas Association, October 1971.

73. A.A. Carotti and E.R. Kaiser, 64th Annual Meeting PACA, Atlantic City, New Jersey, June 27-July 2, 1971.

74. E.R. Kaiser, <u>Evaluation of the Melt-Zit High-Temperature Incinerator</u>, U.S. Department of Health, Education and Welfare, Washington, D.C., 1969.

75. W.A. Boothe, <u>Trip Report, American Thermogen, Inc.,</u> February 3, 1973

76. R.L. Lewis (American Thermogen, Inc.) correspondence with W.A. Boothe (including enclosures), October 19, 1972.

77. W. A. Boothe, <u>Trip Report -- Environmental Protection Agency</u>, October 8, 1972.

78. A. Howard, "The Manufacture of Humus by the Indore Process," <u>Journal of the Royal Society of Arts</u>, Vol. 84, November 22, 1935, pp. 26-59.

79. <u>Municipal Refuse Disposal,</u> American Public Works Association, Public Works Administration Service, Chicago, Illinois, 1966.

80. <u>Composting Municipal Solid Wastes in the United States</u>, Publication No. SW-47r, U.S. Environmental Protection Agency, Cincinnati, Ohio, 1971.

81. A. S. Cheema, "India Develops Urban Compost-Sewage Use Plan," <u>Compost Science</u>, Vol. 8, No. 2, Autumn 1967-Winter 1968, pp. 13-15.

82. J. S. Coulson (Fairfield Engineering Company), correspondence with W. A. Boothe, August 31, 1972.

83. S. Varro (Ecology, Inc.), telephone conversation with W. A. Boothe, December 1, 1972.

84. D. T. Knuth (Environmental Engineering, Inc.), correspondence with H. W. Burr, June 19, 1970.

85. <u>Pesticide Analytical Manual</u>, Vol. 1, U.S. Food and Drug Administration Washington, D.C., January 1968, Sections 212 and 311.

86. H. W. Burr (Conservation International, Inc.), correspondence with W. A. Boothe, December 7, 1972, January 5, 1973, February 9, 1973, and March 7, 1973.

87. C. Tietjen and S. S. Hart, "Compost for Agricultural Land?," <u>Journal of the Sanitary Engineering Division, Proceedings of the American Society of Civil Engineers</u>, Vol. 95 (SA2), April 1969, pp. 269-287.

88. S. A. Hart, <u>Solid Waste Management/Composting: European Activity and American Potential</u>, Report No. SW-2c, Environmental Protection Agency, Cincinnati, Ohio.

89. R. E. Brooks, correspondence with W. A. Boothe, December 15, 1972.

90. W. A. Boothe, <u>Minutes of Connecticut Solid Waste Management Program</u>, Industrial Advisory Council, Reprocessing Committee, March 1, 1973.

91. E. J. Lasch, <u>Order of Magnitude Estimate for Typical 1000 Ton/Day Garrett Pyrolysis Plant</u>, memo, January 30, 1973.

92. E. J. Lasch, <u>Order of Magnitude Estimate for Typical 1000 Ton/Day Hybrid Resource Recovery Plant in Connecticut -- 1972</u>, December 3, 1972.

93. E. A. Glysson, et al, <u>The Problem of Solid Waste Disposal</u>, University of Michigan, Ann Arbor, Michigan, 1972

94. R.J.A. Black, et al, <u>The National Solid Wastes Survey: An Interim Report</u>, U.S. Department of Health, Education and Welfare, Washington, D.C., 1968.

95. <u>Refuse Collection and Operational Methods Sanitary Landfill</u>, Report No. TE 080007, Caterpillar Tractor Company.

96. <u>Sanitary Landfill Design and Operation</u>, Report No. SW65ts, Environmental Protection Agency, Cincinnati, Ohio, 1972.

97. <u>Solid Waste Management Plan -- Metropolitan Planning Commission</u>, Aerojet General Corporation and Black and Veatch, May 1971.

98. J.A. Fischer and D.L. Woodford, <u>Environmental Considerations of Sanitary Landfill Sites</u>, paper presented at the 22nd Annual Sanitary Engineering Conference, February 1972.

99. <u>Recommended Standards for Sanitary Landfill Operations</u>, Report No. TE 080014, Caterpillar Tractor Company.

100. J.M. Phillips, <u>General Guidelines for Landfilling Chemical Wastes</u>, General Electric Company, September 12, 1972.

101. G.M. Hughes, et al, <u>Hydrogeology of Solid Waste Disposal Sites in Northeastern Illinois</u>, Final Report, U.S. Government Printing Office, 1971.

102. <u>In-Site Investigation of Movements of Gases Produced from Decomposing Refuse</u>, Publication No. 35, California State Water Quality Control Board, Sacramento, California, 1967.

103. <u>Landfill Decomposition Gases</u>, Report No. Sw-72-1-1, Environmental Protection Agency, Cincinnati, Ohio, June 1972.

104. J.C. Burchinall and H.A. Wilson, <u>Sanitary Landfill Investigation</u>, Final Report No. SW-00040-01, 02, 03, Department of Health, Education and Welfare, Washington, D.C., 1966.

105. W.D. Bishop, et al "Gas Movement in Landfill Rubbish," <u>Public Works</u>, November 1965.

106. G.P. Callahan and R.H. Gurske, <u>The Design and Installation of a Gas Migration Control System for a Sanitary Landfill</u>, May 1971.

107. R.L. Steiner, et al, "Criteria for Sanitary Landfill Development," <u>Public Works</u>, March 1971.

108. <u>Demonstration Sanitary Landfill Project, Second Annual Report</u>, Kansas City, Kansas, MARC, 1972.

109. <u>Omaha-Council Bluffs Solid Waste Management Plan Status Report</u>, 1969.

110. <u>Closing Open Dumps</u>, Report No. Sw-61ts, Environmental Protection Agency, Cincinnati, Ohio, 1971.

111. A. E. Zanoni, Ground Water Pollution from Sanitary Landfills and Refuse Dump Grounds, Research Report No. 69, Department of Natural Resources, Madison, Wisconsin, 1971.

112. Guidelines for the Land Disposal of Solid Wastes, Processing and Disposal Division, Environmental Protection Agency, Cincinnati, Ohio, January 26, 1973.

113. Investigate the Potential Benefits of Rail Haul as an Integral Part of Waste Disposal Systems, Demonstration Grant No. GO6-EC-00073, U. S. Department of Health, Education and Welfare, Washington, D. C.

114. High Pressure Compaction and Baling of Solid Waste, Report No. Sw-32d, U S. Environmental Protection Agency, Cincinnati, Ohio, 1972.

115. K. Wolf, APWA Research Foundation and City of Chicago Study of High-Pressure Compaction and Baling of Solid Waste, paper presented at the American Public Works Association Congress, September 15, 1971.

116. W. A. Boothe, Trip Report -- American Hoist and Baling Plant, February 14, 1973.

117. E. F. Taylor, Jr. (Connecticut Power and Light Corporation), correspondence with W. A. Boothe, December 11, 1972.

118. NCRR Bulletin, Vol. III, No. 1, Winter 1973.

119. R. A. Boettcher, Air Classification of Solid Wastes, Report No. Sw-30c, Environmental Protection Agency, Cincinnati, Ohio, 1972.

INDEX

INDEX

349